Food Safety for the 21st Century

To Christopher, Renate and Lawrence for their encouragement to develop
this book and for their steadfast support of our careers and families

and

To all participants in the global food supply chain from farm to table whose
combined efforts are essential to provide a safe supply
of food for all consumers

Food Safety for the 21st Century

Managing HACCP and Food Safety
Throughout the Global Supply Chain

Carol A. Wallace

William H. Sperber

Sara E. Mortimore

A John Wiley & Sons, Ltd., Publication

This edition first published 2011
© 2011 Carol A. Wallace, William H. Sperber and Sara E. Mortimore

Blackwell Publishing was acquired by John Wiley & Sons in February 2007. Blackwell's publishing programme has been merged with Wiley's global Scientific, Technical, and Medical business to form Wiley-Blackwell.

Registered office
John Wiley & Sons Ltd, The Atrium, Southern Gate, Chichester, West Sussex, PO19 8SQ, United Kingdom

Editorial offices
9600 Garsington Road, Oxford, OX4 2DQ, United Kingdom
2121 State Avenue, Ames, Iowa 50014-8300, USA

For details of our global editorial offices, for customer services and for information about how to apply for permission to reuse the copyright material in this book please see our website at www.wiley.com/wiley-blackwell.

Library of Congress Cataloging-in-Publication Data

Food safety for the 21st century : managing HACCP and food safety throughout the global supply chain / Carol A. Wallace, William H. Sperber, Sara E. Mortimore.
    p. cm.
  Includes bibliographical references and index.
    ISBN 978-1-4051-8911-8 (hardcover : alk. paper) 1. Food–Safety measures. 2. Food industry and trade–Quality control. 3. Hazard Analysis and Critical Control Point (Food safety system) I. Sperber, William H. II. Mortimore, Sara. III. Title: Food safety for the twenty-first century.
  TX531.F663 2011
  363.19′26–dc22

2010031132

A catalogue record for this book is available from the British Library.

Set in 10/12 pt Times by Aptara® Inc., New Delhi, India
Printed and bound in Singapore by Markono Print Media Pte Ltd.

1 2011

Contents

Preface

The effective development and management of food safety programmes are essential to minimise the occurrence of foodborne illnesses and outbreaks. However, that responsibility is increasingly difficult to fulfil because of the growing human population and the rapidly growing global food trade. With our diverse professional experiences, we three authors have a combined experience of about 90 years in food research, management and education focused on food safety and quality practices. We have undertaken to write *Food Safety for the 21st Century* in an effort to assist all participants in the global food supply chain from farm to table to fulfil their individual responsibilities for food safety assurance. This book should be an excellent textbook in academic food safety courses and an excellent reference book for food safety researchers, managers and regulators worldwide. We wrote this book to be comprehensive and forward looking, with sufficient technical detail to support the complete range of food safety activities from hazard analyses and training programmes to regulation and policy development.

Future demands on the global food supply will challenge our ability to provide a sufficient supply of food that is reliably safe for consumption. The human population, projected to increase by 3 billion people by 2050, and the improving economic status in developing countries mean that we will need to double food production over the next 40 years, and all this in the context of climate change, the diminishing availability of fresh water, fossil fuels and arable land, and the emergence and spread of new foodborne pathogens. The emergence and mismanagement of the bovine spongiform encephalopathy (BSE) epidemic 25 years ago vividly demonstrate the necessity of improving food safety management practices from farm to table and throughout the global supply chain.

The Hazard Analysis and Critical Control Point (HACCP) system of food safety management began as a voluntary food industry effort nearly 40 years ago. Assisted by the Codex recommended code of practice for good hygienic practices and HACCP, first published in 1992, global food corporations have implemented HACCP wherever possible in their particular parts of the supply chain. Yet, our industry efforts to maximise the benefits of effective food safety management programmes have been hampered by fragmented governmental regulatory responsibilities and practices in many countries, especially by the promulgation of counterproductive food safety regulations in some countries.

Some governmental regulatory bodies seem incapable of doing more to develop and promulgate effective food safety regulations that will support the global food industry in its mission to ensure the safety of the food supply. Therefore, achieving effective food safety assurance in the global supply chain will likely require intergovernmental harmonisation of food safety regulations and practices and creation of a global food protection organisation.

The challenges facing all of us in our quest to maintain and improve food safety practices may seem daunting, but they are not insurmountable. There are large reservoirs of available food safety talent in the industry, academia, public health organisations and regulatory bodies. We need to generate the collective political will to collaborate and provide competent management and effective food safety management practices, effective educational programmes, and practical regulations. Working together, we will meet our challenges.

Bon appétit!

Carol A. Wallace
William H. Sperber
Sara E. Mortimore

Acknowledgements

We are indebted to the following people for input into this book:

Jose Chipollini, MoArk LLC, USA
Erica Sheward, Global Food Standards, UK

Disclaimer

The material in this book is presented after the exercise of care in its compilation, preparation and issue. However, it is provided without any liability whatsoever in its application and use.

The contents reflect the personal views of the authors and are not intended to represent those of the University of Central Lancashire, Cargill Incorporated, or Land O'Lakes, Inc.

How to use this book

Food Safety for the 21st Century is split into three main parts:

Part 1: Food Safety Challenges in the 21st Century
Part 2: Foodborne Hazards and Their Control
Part 3: Systematic Food Safety Management in Practice

In addition, there are two appendices providing HACCP and food safety case studies of example links in the global food supply chain and a resource section to help the reader find information and help in applying food safety.

This book is intended to be a compendium of up-to-date thinking and best practice approaches to the development, implementation and maintenance of world-class food safety programmes. Whilst some readers may wish to read the book from cover to cover, we anticipate that many readers will dip into the specific sections, chapters and appendices at different parts of their food safety journey. The book is written both for those who are developing food safety management systems for the first time and for those who need to update, refresh and strengthen their existing systems. The following paragraphs provide an outline of the content of each part and ideas of how they may be used.

Part 1, *Food Safety Challenges in the 21st Century*, sets the scene by providing a discussion of the key considerations for food safety in our modern world. Starting with a considerations of where we have come from and how contemporary food safety programmes have evolved (Chapter 1), this part continues by considering lessons learned from food safety successes and failures (Chapter 2) and looks at challenges in the global food supply chain (Chapter 3). This part finishes with consideration of the future of food safety and HACCP in our changing world (Chapter 4), allowing us to look forward and predict some of the actions that need to be taken to continually improve and strengthen our food safety programmes and approaches in the global supply chain.

This part will provide the reader with a detailed understanding of the context within which food safety management must operate. It will outline the key food safety considerations for individuals, businesses and organisations involved in the global food supply chains of the 21st century.

Part 2, *Foodborne Hazards and Their Control*, consists of three chapters which together form a database of information enabling the reader to recognise food safety hazards and design safe products and processes. This will be useful at the product development stage to provide an understanding of some of the key hazards and control mechanisms available to the food business, and will also be invaluable to HACCP team members who need to understand the likely hazards in their operations.

Part 3, *Systematic Food Safety Management in Practice*, outlines how to develop, implement and maintain world-class food safety programmes based on safe product/process design, prerequisite programmes and HACCP. The seven chapters of this part provide a detailed understanding of current thinking on food safety management, drawing on the experiences and learnings of the past 40 years to offer best practice approaches for developing or strengthening an effective food safety programme.

The authors

Carol A. Wallace, School of Public Health and Clinical Sciences, University of Central Lancashire, Preston, PR1 2HE, UK

Carol A. Wallace is Principal Lecturer, Food Safety Management at the University of Central Lancashire, UK, where she is a course leader for postgraduate food safety and HACCP programmes and leads research themes in food safety effectiveness. Having entered the food industry as a microbiology graduate, she very soon became involved in the early days of HACCP and food safety management systems in the UK and went on to gain 20 years of practical experience in the UK and international food industry prior to joining academia 5 years ago. She gained a PhD for her study of factors impacting HACCP effectiveness and continues to work closely with international food companies and organisations for the ongoing improvement of food safety standards.

William H. Sperber, Cargill Inc., 5814 Oakview Circle, Minnetonka, MN 55345, USA

William H. Sperber studied biological and chemical sciences at the University of Wisconsin, Madison, culminating in a PhD degree in microbiology. This 'friendly microbiologist' has worked in research and management positions with major global food companies for more than 40 years, the majority with the Pillsbury Company, where the HACCP system of food safety management originated. In his current retirement career, Bill is Cargill's Global Ambassador for Food Protection, in which capacity he proposed and continues to promote the development of a global food protection organisation.

Sara E. Mortimore, Land O'Lakes Inc., MS 2350, St. Paul, MN 55164, USA

Sara E. Mortimore has many years of food manufacturing experience in food safety and quality management. She is currently Vice President of Quality Assurance and Regulatory Affairs, at Land O'Lakes, Inc., one of the US premier farmer-owned cooperatives operating in both the food and agricultural sectors, including dairy products, eggs, animal feed, seed and crop protection. Previously, she worked in various international roles covering quality, food safety and global sourcing for Pillsbury and General Mills. During this time, she gained a deep cultural understanding of the attitudes and behaviours of people towards food safety in manufacturing around the globe. She graduated in the UK in food science and went on to gain a master's degree in training and development. As a result of this, she has developed a major interest in the development of integrated food safety management using the HACCP approach and in adequate training interventions that are aligned to the needs of workers and the abilities of trainers.

Glossary of terms and acronyms

Aerobe A microorganism that can grow in the presence of oxygen. Obligate aerobes, for example moulds, cannot grow in the absence of oxygen.

Allergen A compound capable of inducing a repeatable immune-mediated hypersensitivity response in sensitive individuals.

Anaerobe A microorganism that can grow in the absence of oxygen. Obligate anaerobes, for example *Clostridium* spp., cannot grow in the presence of oxygen.

Audit A systematic, independent and documented process for obtaining audit evidence and evaluating it objectively to determine the extent to which the audit criteria are fulfilled (ISO, 2002).

Audit criteria A set of policies, procedures or requirements. Audit criteria are used as a reference against which the actual situation is compared (ISO, 2002).

Audit evidence Records, statements of fact or other information, which are relevant to the audit criteria and verifiable (ISO, 2002).

Audit findings Results of the evaluation of the collected audit evidence against audit criteria (ISO, 2002).

Auditee Organisation being audited (ISO, 2002).

Auditor The person with the competence to conduct an audit (ISO, 2002).

BRC British Retail Consortium, based in London, UK, and one of the GFSI benchmarked food safety certification scheme standard owners.

CFR Code of Federal Regulations, a repository of US regulations.

CFSA Canadian Food Safety Agency.

COA Certificate of Analysis that would accompany a product or raw material and indicate compliance to specification.

Codex Codex Alimentarius Commission (CAC), a United Nations organisation that supports FAO and WHO by developing food standards, guidelines and codes of practice.

Control measure An action or activity that can be used to prevent, eliminate or reduce a hazard to an acceptable level.

Corrective action Any action to be taken when the results of monitoring at the CCP indicate a loss of control (Codex 2009b).

Critical control point (CCP) A step at which control can be applied and is essential to prevent or eliminate a food safety hazard or reduce it to an acceptable level (Codex, 2009b).

Critical limit A criterion that separates acceptability from unacceptability (Codex, 2009b).

Crohn's disease A chronic inflammatory bowel disease of humans, thought to be caused by *Mycobacterium paratuberculosis.*

D-value The process time required to reduce a microbial population by 90%, or one $\log_{10}$ unit.

Dutch HACCP Code An auditable standard based on the principles of HACCP, prerequisite programmes and management procedures.

Emerging pathogen Typically, an uncommon pathogen that becomes more prevalent because of changes in the host, the environment or in food production and consumption practices.

Enterotoxin A toxic molecule produced by a microorganism that causes gastrointestinal illness symptoms such as vomiting and diarrhoea.

Essential management practices (for food safety) Management practices and procedures that support effective application of safe product/process design, prerequisite programmes and HACCP systems, and ensure their ongoing capability to protect the consumer.

Extremophile A microorganism that can survive and grow under very extreme conditions such as high temperature or pressure, and extreme acidity.

Extrinsic A factor or process that is applied externally to a food, such as heating or modified atmosphere packaging.

Facultative A microorganism that can grow in the presence or absence of oxygen, a class that includes most foodborne microbes.

FAO The Food and Agriculture Organization of the United Nations, primarily responsible for food security.

Food defence Having a system in place to prevent, protect, respond to and recover from the intentional introduction of contaminants into our nation's food supply, designed specifically to cause negative public health, psychological and/or economic consequences (Yoe *et al.*, 2008).

Food protection All measures and programmes in place to protect the safety of the food supply.

Food security The state existing when all people at all times have access to sufficient, safe, nutritious food to maintain a healthy and active life (WHO, 2010).

Gantt chart A diagrammatic representation of a project plan, including actions and timetable.

GFSI The Global Food Safety Initiative, organised through CIES, the Consumer Goods Forum.

GIFSL Global Initiative for Food Systems Leadership run by the University of Minnesota.

GMPs Good Manufacturing Practices.

Guillain–Barré syndrome A syndrome involving neurological complications that are often induced as a sequel to microbial infections, often attributed to *Campylobacter*.

HACCP Hazard Analysis and Critical Control Point, a preventive system of food safety management based on product design, hazard analysis and process control.

HACCP plan A document prepared in accordance with the principles of HACCP to ensure control of hazards that are significant for food safety in the segment of the food chain under consideration (Codex, 2009b).

HACCP team A specific group of individuals with multidisciplinary expertise and experience who work together to apply the HACCP principles.

Halophile A microorganism that can grow at very high sodium chloride concentrations, for example *Halobacterium* spp.

Hazard A biological, chemical or physical agent in, or condition of, food with the potential to cause an adverse health effect.

Hazard analysis The process of collecting and evaluating information on hazards and conditions leading to their presence to decide which are significant for food safety and therefore should be addressed in the HACCP plan.

Hydrophilic The tendency of a polar compound to be soluble in water.

ICD Industry Council for Development.

IFST The UK Institute of Food Science and Technology.

ILSI The International Life Sciences Institute.

Immunocompromised A condition in which the host's immunity to infection is diminished by factors such as age (very young or very old), illness or chemotherapy.

Infection An illness or condition caused by the growth of a microorganism in a host.

Infectious dose The number of microorganisms required to cause an infection.

Intoxication An illness or condition caused by the ingestion of a toxin.

Intrinsic A property that is an inherent characteristic of a food, such as pH or water activity.

ISO International Organization for Standardization.

Johne's disease A chronic disease of cattle characterised by diarrhoea and emaciation, caused by *Mycobacterium paratuberculosis*.

Lipophilic The tendency of a non-polar compound to be soluble in fats or oils.

Mesophile A microorganism that grows optimally at intermediate temperatures, for example 20–45°C.

Monitoring The act of conducting a planned sequence of observations or measurements of control parameters to assess whether a CCP is under control (Codex, 2009b).

NACMCF The US National Advisory Committee on Microbiological Criteria for Foods.

OIE World Organization for Animal Health.

Operational limit A value that is more stringent than a specific critical limit that is used in process management by providing a buffer zone for safety.

Operational PRP A PRP identified by the hazard analysis as essential in order to control the likelihood of introducing food safety hazards to and/or the contamination or proliferation of food safety hazards in the product(s) or in the processing environment (ISO, 2005a).

Opportunistic pathogen A relatively harmless microorganism that can more easily cause an infection in an immunocompromised person, or if it is accidentally inserted into a sterile host site.

Osmophile A microorganism, particularly a yeast, that can grow under conditions of high osmotic pressure, typically created by concentrated sugar solutions.

Osmotolerant A microorganism that can survive high osmotic pressure.

PAS Publicly Available Specification.

PMO The US Pasteurized Milk Ordinance.

Prion A misshapen cellular protein that causes the agglomeration of normal-shaped prion proteins which in turn can cause transmissible spongiform encephalopathies, fatal brain diseases, such as BSE ('mad cow disease').

Process flow diagram A diagrammatic representation of the process, identifying all processing activities, which is used as the basis for hazard analysis.

PRP Prerequisite programmes such as good agricultural, manufacturing and hygienic practices that create the foundation for a HACCP system.

Psychrotroph A microorganism that grows optimally at low temperatures, for example 0–20°C.

Sanitary operating practices A term describing certain hygienic practices that form part of prerequisite programmes.

Significant hazard Hazards that are of such a nature that their elimination or reduction to an acceptable level is essential to the production of safe foods (ILSI, 1999).

SQA Supplier quality assurance. The programmes used to manage suppliers of raw materials, packaging and contract manufacturing.

SQF Safe Quality Food, one of the GFSI benchmarked food safety certification schemes, originated in Australia but now based in the United States.

Thermophile A microorganism that grows optimally at high temperatures, for example 45–70°C.

Toxic dose The amount of toxin required to cause a food intoxication.

Toxin A chemical or microbial metabolite that can cause toxic effects when ingested.

Validate To investigate and prove the effectiveness of a control measure, such as the critical limits, at a critical control point.

Validation Obtaining evidence that the elements of the HACCP plan are effective (Codex, 2009b).

Verification The application of methods, procedures, tests and other evaluations, in addition to monitoring, to determine compliance with the HACCP plan (Codex, 2009b).

Verify To confirm the continuing effectiveness of a control measure through process or records observations or analytical testing.

WHO The United Nations World Health Organization, primarily responsible for public health.

World-class food safety programme A programme based on the principles of safe product/ process design, prerequisite programmes and HACCP that is supported by essential management practices, thus controlling the operational, environmental and process conditions necessary for consumer health protection through the consistent production of safe food.

WTO The United Nations World Trade Organization, where Codex guidelines and codes have the force of law among signatory members.

Xerotroph A microorganism, typically a mould, that can grow under very dry conditions.

Z-value The change in temperature ($^\circ$C) required to change the D-value by 90% or one $\log_{10}$ unit.

Zoonotic A pathogenic organism that can infect humans and animals.

Part One
Food Safety Challenges in the 21st Century

1 Origin and evolution of the modern system of food safety management: HACCP and prerequisite programmes

1.1 HISTORICAL PERSPECTIVES

Food safety management practices have been evolving continually in the food industries of developed nations, particularly since the end of World War II (WWII) in 1945. Nevertheless, despite more than 60 years of progress in the assurance of food safety, failures sometimes occur. The intent of this introduction is to summarise the principal events in the origin and evolution of modern food safety practices so that readers can better understand how to improve practices and to provide even greater food safety assurance in the future.

The beginning of WWII coincided with the end of the Great Depression that had hindered economic progress throughout the entire world during the 1930s. Western nations mobilised their economic resources during the early 1940s to manufacture the weapons of war. Upon the war's end, the energised economic and manufacturing bases were converted to build infrastructure and produce consumer goods, rather than war materials. Several of the principal innovations that impacted food safety were the development and widespread use of mechanical refrigeration and the construction of national transportation systems, such as the interstate highway system in the United States.

Before the widespread use of mechanical refrigeration, many perishable foodstuffs were stored in iceboxes that required frequent replenishment of the ice supply. Iceboxes could not provide uniform or steady cold temperatures. As a result, perishable foods often became unfit for consumption; consumers were forced to shop frequently for perishable goods. Mechanical refrigeration units were equipped to provide relatively uniform and steady cold temperatures, about 4–7°C, thereby substantially reducing the amount of food spoilage and potential food safety incidents. The application of mechanical refrigeration was quickly extended to most homes and commercial establishments and to road and rail vehicles for the transportation of refrigerated or frozen foods and food ingredients.

The ability to use refrigerated transportation was greatly facilitated by the construction of modern rail and highway systems. Eventually, the production of refrigerated ocean liners and airplanes facilitated the shipment of perishable foodstuffs across the oceans. These developments mean that the system of local food production and consumption that was widely used several generations ago has now been largely replaced by a massive global food supply chain in which foods and food ingredients are shipped amongst most nations of the world.

Food Safety for the 21st Century, First Edition By Carol A. Wallace, William H. Sperber and Sara E. Mortimore
© 2011 Carol A. Wallace, William H. Sperber and Sara E. Mortimore

Mechanical refrigeration and lengthened supply chains have enabled the concentration of food production operations into relatively few large facilities that can ship food products to very large geographical areas. This phenomenon has occasionally been responsible for very large foodborne illness outbreaks that would have been less likely when food production occurred in multiple smaller facilities, each of which supplied smaller geographical areas.

A trend towards more convenient foods accompanied the developments described above. In products such as dried cake mixes, for example, dried eggs and dried milk were added at the point of manufacture so that the consumer would not need to use shell eggs or fresh milk during the preparation of the cake batter. The use of dried ingredients in the place of fresh raw materials was quickly applied to the production of many manufactured foods. This practice brought with it an unanticipated problem – an increase both in the incidence of *Salmonella* contamination and in the number of outbreaks and cases of human salmonellosis.

The reasons for these increases proved to be analogous to the reasons for larger outbreaks of foodborne illnesses being associated with large, centralised food production facilities. In home kitchens, the use of *Salmonella*-contaminated fresh milk or shell eggs in family-sized food portions could, at most, be responsible for a few cases of salmonellosis. However, when *Salmonella*-contaminated dried eggs or dried milk were used in food manufacturing facilities in the production of massive quantities of food, many cases of salmonellosis could result.

The increased levels of pathogen-contaminated foods and foodborne illnesses caused great concern in the rapidly evolving and growing global food industry of the 1950s and 1960s. Government regulators and consumers demanded safer foods. These demands were followed by intensified efforts to manage food production in order to reduce the food safety risks. Early efforts to assure food safety attempted to use quality-control procedures that had been implemented with the modernisation of the food industry after WWII.

Manufacturers of many types of products, including foods and many household appliances, used similar procedures in their efforts to control quality. These procedures typically included the collection of a predetermined number of samples from a production shift, followed by the testing or analysis of the samples in a laboratory. Statistically based sampling plans were used to determine the acceptability of each production lot. If the number of defective samples exceeded the specification for a particular product, the entire production lot would be rejected. If the number of defective samples did not exceed the specified limit, the production lot would be accepted. The management of quality control was based upon product specifications, lot acceptance criteria and finished-product testing.

Despite the applications of contemporary quality-control procedures, foodborne illnesses caused by the new food ingredients and products continued to occur. It was discovered that food safety incidents, including foodborne illness outbreaks, were sometimes caused even when the implicated production lot of food was determined to be in compliance with all of its specifications. Repeated incidents revealed a fundamental flaw in quality-control procedures that prevented the detection and prevention of such incidents. That fundamental flaw was the inability of quality-control procedures to detect defects that occurred at low incidences.

Upon extensive investigations of production lots of food that were implicated in foodborne illnesses, it was determined that the foods were typically contaminated with a particular pathogen at a very low incidence. In many cases, the defect rate was about 0.1%, i.e. about 1 unit in 1000 analytical units was found to be contaminated. Of course, when many millions of analytical units are produced during a single shift, it is easy to understand how numerous illnesses could be caused by a lot of food that was contaminated at the seemingly trivial rate of 0.1%.

Subsequent statistical analyses revealed that 3000 analytical units would need to be tested and found to be negative in order to provide assurance at the 95% confidence limit that a particular

Table 1.1 Probability of rejecting a lot containing a known proportion of defective units.

Number of samples tested	Percentage of defective units in lot		
	0.1	**0.5**	**1.0**
300	0.26	0.78	0.95
500	0.39	0.92	0.99
1000	0.63	0.99	–
2000	0.86	–	–
3000	0.95	–	–
5000	0.99	–	–

Source: Adapted from ICMSF (2002).

lot of food was free of a particular pathogen or similar foodborne hazard (ICMSF, 2002; Table 1.1). Testing thousands of samples from each production lot of food was obviously impractical.

Additional factors were found to contribute to the inability of product testing to detect food safety defects. These included the uneven, or non-random, distribution of microorganism in food materials, the variability between different testing procedures and the competence of the laboratory personnel. In those days, it was not uncommon for plant production personnel to be promoted without training into laboratory positions.

For the reasons described above, reliance on product specifications and finished-product testing were clearly inadequate to assure food safety.

1.2 ORIGIN AND EVOLUTION OF HACCP

During this same time period, the 1960s, several entities were collaborating on the production of foods for US military personnel and for the manned space programmes. These were The Pillsbury Company, the US Army Laboratories at Natick, MA, and the National Aeronautics and Space Administration (NASA). In an effort to guarantee that astronauts would not become seriously ill during a space mission, NASA had enacted very strict specifications upon the foods that it used. All parties soon realised that a food safety guarantee could not be provided without 100% destructive testing of a given lot of food (Ross-Nazzal, 2007). Several engineers recognised that the failure modes and effects analysis (FMEA) used by the military to test the reliability of electrical components could be adapted to assess hazards and control measures in food production. The early seeds of the Hazard Analysis and Critical Control Points (HACCP) of food safety were planted. One of the astronaut foods developed at this time, Space Food Sticks, was briefly produced as a consumer product (Figure 1.1). Its development included elements of both the FMEA and HACCP systems. The Sticks were designed to be non-crumbling so that they could not contaminate and impair vital instruments in the space capsules. Additionally, they were produced under controlled conditions that provided a high degree of food safety assurance, both for astronauts and, later, for consumers.

Two coincidental events in 1971 hastened the development of HACCP and its use in the food industry. Americans learned of the first event when a national radio broadcaster intoned, 'Good morning, America, there's glass in your baby food.' Farina produced by The Pillsbury Company had been contaminated with shattered glass in its production facility (The New York Times, 1971). Pillsbury's Director of Research, Dr Howard Bauman, who led Pillsbury's production

Fig. 1.1   Space Food Sticks – designed for astronauts, later marketed to the public.

of space foods for NASA, decided to apply this new system of food safety management to all of Pillsbury's consumer food production. In the following month, Dr Bauman delivered a presentation at the second coincidental event, the 1971 National Conference on Food Protection, sponsored by the American Public Health Association (APHA, 1972). His remarks, and those of his fellow panel members, were limited to descriptions of critical control points and good manufacturing practices. The term 'HACCP' had not yet entered the professional lexicon but this was to become one of the key events in the global spread and acceptance of the HACCP system (Table 1.2).

Table 1.2   Events that fostered HACCP development and evolution through the 20th century.

Year	Event
1923	US Pasteurized Milk Ordinance first published
1960s	Pillsbury, NASA, US Army collaborations
1969	Current good manufacturing practices first published
1971	Pillsbury cereal recall
	National Conference on Food Protection
	Multiple canned foods recalls, *C. botulinum* contamination
1972	Pillsbury trains FDA inspectors to apply HACCP to canned foods
	Pillsbury begins application of HACCP to its consumer products
1973	Canned foods regulations first published
1975	Pillsbury internal HACCP system complete
1985	National Research Council recommends HACCP
1988	National Advisory Committee on Microbiological Criteria for Foods (NACMCF) formed
	ICMSF Book 4 on HACCP published
1992	NACMCF and Codex adopt seven HACCP principles
1994	HACCP: A practical approach published
1997	NACMCF and Codex HACCP documents harmonised

During the early 1970s, the US canning industry experienced a rapid succession of 12 or more incidents of contamination of canned foods by *Clostridium botulinum*. All were accompanied by product recalls and disposals, including one that cost approximately US$100 million (Howard, 1971). While few illnesses and one death were associated with these incidents, the US Food and Drug Administration (FDA) recognised that better controls needed to be developed and required for the production of canned foods. Having participated in the 1971 National Conference for Food Protection, the FDA, intrigued by the concept of critical control points, contracted with The Pillsbury Company to conduct a training programme for its personnel responsible for the safety of canned foods.

Pillsbury presented a training programme for 10 FDA inspectors in September 1972. Lasting 3 weeks, the programme was almost evenly split between classroom activities and in-plant orientation and inspections at four canning companies. The accompanying instructional materials seemed to represent the first substantial use of the term, HACCP (The Pillsbury Company, 1973). The newly trained inspectors returned to Washington, DC, and published the canned foods regulations in 1973 (CFR, 2002). Based, to a significant extent, upon time and temperature controls, the canned foods regulations bear striking resemblance to the Pasteurized Milk Ordinance (PMO) first published in 1923 (FDA, 1997). It seems to the authors that the concepts of food safety based on prevention by adequate controls had long been present, perhaps subconsciously, in the minds of food processors and regulators. It is somewhat daunting to consider that our modern system of food safety management is so young.

Upon completion of the FDA training programme, Pillsbury began, in earnest, to apply the HACCP system to the production of its consumer products – a goal that was achieved in 1975. Increasing awareness of Pillsbury's new system of food safety management and the obvious effectiveness of the canned food regulations in curtailing further incidents of *C. botulinum* contamination led to a steady adoption of HACCP by other US food processors. A fertile environment for food safety enhancement existed in the United States at this time because of these regulations and because of the 1969 promulgation by the FDA of current good manufacturing practices (CFR, 1969).

The adoption of HACCP beyond the US food industry received a major impetus by the 1985 publication of a National Research Council report: 'An evaluation of the role of microbiological criteria for foods and food ingredients' (NRC, 1985). Completely masked by its title, the report included several highly influential recommendations that propelled HACCP forward. The first of these recommended that food regulatory agencies should use proactive procedures to audit food safety compliance by records verification rather than the customary procedures of plant inspections and product testing.

The HACCP system fitted perfectly the description of a 'proactive procedure'. Unfortunately, 25 years later, there has been no significant adoption of this recommendation by US food regulatory agencies. However, this situation may change in the near future, given the frequent occurrence of unexpected outbreaks of foodborne illness. The report further recommended that the responsible agencies form an ad hoc Commission on Microbiological Criteria for Foods. Sponsored by four US federal government departments – Agriculture, Health & Human Services, Commerce and Defence – this commission emerged in 1988 as the National Advisory Committee on Microbiological Criteria for Foods (NACMCF). One of its first charges was to develop a report to guide industry and regulators on the structure and implementation of the HACCP system. At about the same time, the Codex Alimentarius Commission Committee on Food Hygiene (Codex) began working on a similar report.

Following an abortive NACMCF HACCP report in 1989, both NACMCF and Codex published definitive HACCP reports in 1992 and 1993, respectively (NACMCF, 1992, Codex, 1993).

Because the United States served as the permanent chair of the Codex CFH, there was some overlap of personnel between NACMCF and Codex CFH. Accordingly, the two reports were quite similar. They were almost completely harmonised and republished in 1997 (Codex, 1997; NACMCF, 1998).

As originally developed by Pillsbury in the 1970s, HACCP was based upon three principles:

1. Conduct a hazard analysis.
2. Determine critical control points.
3. Establish monitoring procedures.

Several food safety failures with this system after 1972 led to the gradual development and use of additional principles to facilitate better management practices. The 1992 and 1997 reports cited above describe the seven current HACCP principles:

1. Conduct a hazard analysis.
2. Determine the critical control points.
3. Establish critical limit(s).
4. Establish a system to monitor control of the CCP (critical control point).
5. Establish the corrective action to be taken when monitoring indicates that a particular CCP is not under control.
6. Establish procedures for verification to confirm that the HACCP system is working effectively.
7. Establish documentation concerning all procedures and records appropriate to these principles and their application.

The global spread of HACCP as the preeminent system of food safety management was greatly facilitated by the Codex report of 1997. Jointly chartered by the Food and Agriculture Organization and the World Health Organization of the United Nations (UN), the Codex Alimentarius Commission's reports have the effect of law between UN trading partners who are signatories to the World Trade Organization. Thus, the humble beginnings of HACCP as a voluntary programme within the US food industry in 1972 evolved into an effective global system. Prominent international publications also facilitated the understanding and acceptance of the HACCP system of food safety (ICMSF, 1988; Mortimore and Wallace, 1994). There is now a global understanding and implementation of a food safety management system that is the same in almost every country. This is a remarkable achievement that can serve as a model for international cooperation and improvement in additional areas such as animal, plant, human and environmental health – areas that interface with our efforts to assure food safety.

Despite this promising history, HACCP has sometimes been misused as it was incorporated into regulations. Three prominent examples illustrate this unfortunate situation in the United States (Sperber, 2005a).

The first of these was a final rule published by USDA – Pathogen reduction; Hazard Analysis and Critical Control Point (HACCP) systems (CFR, 1996). Commonly known as the 'megareg', this very lengthy document required no CCPs to enhance the safety of raw meat and poultry products. Rather, it required conformance to a number of statistical sampling plans that permitted the presence of salmonellae and certain levels of indicator microorganism. The *Salmonella* performance standards best exemplify this point (Table 1.3). The performance standards were developed from baseline surveys that were conducted in the early 1990s. In the case of ground

Table 1.3 *Salmonella* performance standards in USDA 'megareg' (CFR, 1996).

Species	Performance standard[a]	n[b]	c[c]
Broilers	20.0	51	12
Cows and bulls	2.7	58	2
Steers and heifers	1.0	82	1
Market hogs	8.7	55	6
Ground beef	7.5	53	5
Ground chicken	44.5	53	26
Ground turkey	49.9	53	29

[a]Per cent positive for *Salmonella*.
[b]Number of daily samples tested.
[c]Maximum acceptable number of positive samples.

beef, for example the performance standard was determined to be 7.5% *Salmonella* positives. To monitor compliance with this standard, a single 325-g sample (tested as (5 × 65)-g subsamples) of ground beef is analysed for the presence of salmonellae each day for 53 consecutive production days. If five or fewer samples are found to be positive for the presence of salmonellae during this period, the production facility is judged to be in compliance with its HACCP plan and no regulatory action is taken. If more than five samples are found to be positive, a second 53-day round of sampling is initiated. If a plant fails three consecutive rounds of such surveillance, regulatory action is considered. One or more years could pass before enforcement action was initiated. Clearly, such standards, sampling procedures and delayed or nonexistent enforcement actions are unrelated to HACCP. As most readers already know, HACCP is a real-time food safety management programme in which immediate corrective actions are taken when deviations occur at a CCP. Regrettably, the 'megareg' also institutionalised a major misuse of resources, as a great deal of money and labour is necessary to conduct such a programme. While statistically based sampling plans that monitor the effectiveness of sanitation programmes (which is a better characterisation of the 'megareg') are meritorious, they are more practically conducted with the use of far smaller samples and less expensive analytical methods for indicator microorganism and tests, such as the aerobic plate count. Moreover, the results of such a sanitation-monitoring programme would be closely linked in time to the in-plant cleaning and sanitation procedures.

Similar criticisms can be made of the FDA HACCP rules for the production of seafood (CFR, 1997) and juice (CFR, 2001). No CCPs were identified and required for the production of raw molluscan shellfish, the seafood category most identified with human illnesses. Unlike the PMO developed in 1923 for dairy products, no mandatory pasteurisation was required for juice products. Furthermore, exemptions were granted to small producers and retail operations, permitting the replacement of several recommended control measures to enhance juice safety with the weekly testing of a 20-ml sample of juice for the presence of generic *Escherichia coli*.

These three regulations bear no resemblance to the HACCP principles promulgated by NACMCF and Codex. Their promulgation as 'HACCP' regulations served to create confusion, misuse resources and undermine the well-deserved and excellent reputation of legitimate HACCP applications.

Despite these several regulatory missteps, numerous effective HACCP rules and regulations have been promulgated by regulators worldwide. Some of these will be highlighted throughout this book. As one example, the USDA (creator of the megareg) issued an effective rule to enhance control of *Listeria monocytogenes* in refrigerated ready-to-eat meat and poultry products. This rule recommends science-based alternatives that can be put into place as CCPs, e.g. the use of post-lethality surface heat treatments or combinations of food preservatives to inhibit listerial growth (CFR, 2003a).

1.3   THE NECESSITY OF PREREQUISITE PROGRAMMES

The global adoption of HACCP did not proceed smoothly without the recognition of the need for additional measures to enhance food safety protection (World Health Organization, 2007b). As a preface to some of our discussion in later chapters, it was learned that HACCP could not operate successfully in a vacuum. Even with HACCP plans in place, food safety failures sometimes occurred because of inadequate cleaning and sanitation procedures, for example. To be successful, HACCP must be supported by a number of prerequisite programmes (PRP) (Sperber *et al.*, 1998). We learned that food safety could not be assured by HACCP alone. Rather, food safety can be much more effectively assured by the combined implementation of HACCP and PRPs. Originally formed to develop and implement HACCP plans, HACCP teams evolved into Food Safety Teams that must consider and manage both HACCP and PRP responsibilities and activities. PRPs are discussed in detail in Chapter 10.

It was also learned that HACCP did not usually work from 'farm to table', as many had hoped (Sperber, 2005b). The types of CCPs that are available in the food processing industry, where HACCP originated, are usually not available at the 'farm' and 'table' ends of the farm-to-table spectrum. Rather than thinking about farm-to-table HACCP, we should be thinking about farm-to-table food safety. A hazard analysis can be conducted at every step of the farm-to-table supply chain. When no CCPs are available to control a significant hazard at the 'farm' end, for example pathogen colonisation of live animals, PRPs could be put into place to reduce the pathogen burden in the following links of the chain.

1.4   THE FUTURE OF HACCP

The evolution of HACCP in developed countries from 1972 to 1997, a period of 25 years, seems quite rapid in the flow of global political events. However, we are optimistic that the continued globalisation of HACCP throughout the developing countries will proceed much more quickly. A major reason for the more rapid implementation of HACCP in developing countries is the rapidly increasing globalisation of food trade. Global trading partners benefit by the uniform application of the most effective food safety procedures. In particular, aided by the inherent authority of the Codex HACCP document, global food corporations have been largely responsible for the globalisation of HACCP.

The HACCP system was expanded from three principles in 1972 to seven principles in 1992. It is reasonable to anticipate that additional principles will be developed and added in the future, as will be discussed in Chapter 4. Looking into the future, it is also quite likely that the HACCP system will continue to evolve. There is already an emerging recognition that even the broad matter of food safety cannot be managed in isolation from other health systems. Rather, food safety systems of the future will likely interface more directly with animal, human and environmental health and food security programmes. At the end of the 20th century, HACCP systems were positioned as the 'crown jewels' of a food safety programme, supported by PRPs. This particular arrangement should persist for a very long time, but it will likely become integrated into a much larger network that includes public health, animal health, food security and agricultural sustainability.

The reader should remain aware that almost all of the progress in the development of HACCP as an effective food safety management programme and its global acceptance and use has been accomplished by the voluntary efforts of global food companies, beginning with The Pillsbury Company in the 1970s and continuing today with the efforts of many dozens of responsible and progressive food companies. Except for the 1997 Codex document that gave guidance for the use of HACCP and hygienic practices, there has been very little contribution to this effort by federal and intergovernmental public health and food agencies. We will propose in Chapter 4 bold recommendations for the future effective involvement of federal and intergovernmental organisations in food safety matters. Such involvement will be essential in order to maintain food safety in the rapidly changing global food supply chain.

2 Lessons learned from food safety successes and failures

2.1 INTRODUCTION

If HACCP works so well then why do we still have so many cases of foodborne illness? This chapter will attempt to answer this question, drawing upon our own experiences together with observations made by others in the industry. Someone once said, 'A sign of madness is to do the same thing over and over and expect to get a different result.' It feels a little like this in the food industry. Only by learning from other people's mistakes, by understanding root cause and by doing things differently will we really start to see improved food safety.

2.2 BENEFITS OF USING HACCP – LESSONS LEARNED FROM A SUCCESSFUL IMPLEMENTATION

There are real benefits when HACCP is effectively designed, implemented and maintained. This is not just from HACCP alone; the benefits really come through having a well-designed, broad-reaching food safety programme that has HACCP at the core. Key benefits include the following:

- *Public health protection.* This has to be the number one priority for anyone in the food industry. All consumers have a right to safe food that will nourish and sustain them. Any business will want to ensure public trust and confidence in its products. The WHO recognises HACCP, when properly used, as the most effective way to ensure food safety and to protect public health (WHO, 2007).
- *Brand protection.* Comes second on this list not just because it is important to senior managers and business owners but also because brand protection is essential for the continuation of the business or product line. Some brands and companies are never able to fully recover from an adverse food safety event. Later, we will examine some of the recent cases of foodborne illness and the reasons why they occurred. Many of them involved global brands and many of them are or were household names.
- *Cost benefits.* Many publications list cost as a barrier to implementation, but in reality HACCP and a strong food safety programme can actually save a company a significant cost. Any reader who is sceptical should examine whether his or her business truly understands the costs and implications of failure. Figure 2.1 illustrates how the cost of prevention

Food Safety for the 21st Century, First Edition By Carol A. Wallace, William H. Sperber and Sara E. Mortimore
© 2011 Carol A. Wallace, William H. Sperber and Sara E. Mortimore

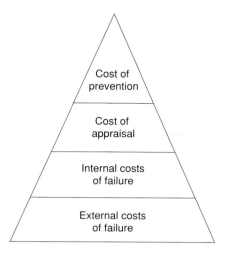

Fig. 2.1   The cost of quality.

is typically and significantly less than the costs associated with failure. The cost of prevention includes having an appropriate number of technically qualified staff, programmes such as HACCP, effective GMPs (good manufacturing practices), hygienic operating environment, training and research. Costs of appraisal will include multiple inspection and testing activities such as self-audit (both internal and external third party), supplier audit, environmental microbiological monitoring programmes, product testing and process control monitoring. Internal costs of failure are often underestimated but can be startling. This will include costs associated with holding product, testing product, destruction or downgrade of product that did not meet specification, consumer complaints resolution, claims and staff time involved in these activities. External cost of failure obviously depends on the size of the company, but recall costs if the product has national distribution can quickly rise to tens of millions of dollars. Brand equity damage is another loss, but of course the most important cost of all is the cost to human life in the event of severe food safety failure.

The cost of implementation really depends on how the company was being run prior to implementing HACCP (McAloon, 2001).

- *Increased confidence through reduced reliance on ingredient and end-product QC testing.* Safety cannot be instilled by testing, and the likelihood of finding a food safety hazard is much lower than many food business operators might realise. As discussed in Chapter 1, reliance on end-product testing for food safety assurance is not only impractical, given the number of samples needed, but it is also ineffective at low levels of contamination. Use of HACCP will move a company away from this retrospective quality-control approach to a way of working that is much more proactive. Companies not only increase confidence in food safety assurance through knowledge that controls are in place throughout the process, but also reduce the costs of ineffective end-product testing, and unnecessary waste generated by batch rejection.
- *Real-time monitoring.* This allows for *preventive action* to be taken, thereby reducing losses later in the process. In identifying controls that require monitoring throughout the production process, it is possible to take corrective action before finished products are made thereby

preventing waste. Once a product is made, it is usually too late to correct – other than by reworking. Later, in Chapter 12, we will see how identifying 'operational limits' in addition to 'critical limits' can be an aid to taking preventive action, i.e. having the ability to adjust the process before it goes out of control.

- *Meets regulatory obligations and customer expectations.* Whether a formal regulatory requirement or not, implementation of HACCP enables a company to meet regulatory obligations and customer expectations for safe food. Effective food safety management will *keep company executives out of jail.* Senior management has ultimate responsibility for food safety and there have been a number of well publicised failures where the management team were held accountable. In terms of regulation, countries vary in their requirements. Where HACCP is not regulated but is voluntary, the fact that it is recognised by WHO as being the best way to ensure food safety is a good reason to use it. HACCP is also recognised in global trade agreements through the World Trade Organization. Some countries are starting to require third-party certification as a means of demonstrating that the HACCP programme has been assessed.
- *Global food safety system.* As established through Codex, HACCP was one of the first truly global food safety systems. This gave a common language and expectation between customers, suppliers and regulatory enforcement authorities around the world and has become the basis for more recent global standards developments such as ISO 22000 (2005) and the Global Food Safety Initiative (2007).
- *Focused use of resources.* Use of HACCP enables a focused use of resources. HACCP is risk based and, therefore, helps a company shift mentality from a 'one size fits all' approach to a 'what is the risk and how do we reduce it through introduction of focused control measures' mindset. When we come to look at barriers and misconceptions, it will be clear that the understanding of HACCP, and what it can do for a company that is committed to food safety, is often lacking.
- *Unanticipated benefits.* Such benefits of using HACCP for food safety include a number of elements. There is usually a significant improvement in product quality as the preventive approach becomes the new culture of the organisation. The workforce can become highly motivated and empowered through greater involvement and team engagement. This has to be orchestrated as it rarely happens by accident. More discussion on this will take place in Chapter 9 when we review preparation activities. The same economic benefits can be seen for quality improvement as were seen for food safety. There will also be increased regulatory compliance – any food that is adulterated (leading to a food safety issue or not) will likely be withdrawn from the marketplace before it becomes a regulatory matter. Finally, there is a real competitive advantage that comes through having strong food safety programmes, with multiple functions, not just Quality Assurance personnel who can really talk confidently about their company's food safety programmes.

2.3 MISCONCEPTIONS OR 'FAILURE TO UNDERSTAND HACCP'

Controversial issues exist 50 years after the concept was first conceived. HACCP is a tool that was designed to *enhance*, not hinder, food safety management programmes. It was revolutionary in moving companies away from reliance on end-product testing to a systematic preventive approach. Whilst it is frustrating to see that many companies have yet to truly grasp the concept and how many misconceptions remain, this also means that the opportunity to implement a really good HACCP programme remains available to many in the industry. Others have also documented this (Motarjemi and Käferstein, 1999). The following are a number of often-heard misconceptions:

- *HACCP has been 'done' already.* This view is most commonly held by large food manufacturers in the developed countries and even by some regulatory agencies. For example, when a HACCP-based programme for small businesses was launched in the United Kingdom by the Food Standards Agency, it was reported by an official that larger manufacturers had 'done' HACCP and were not of concern. It is interesting to consider some of the recent outbreaks of foodborne illness that originated in large companies in developed countries (see Section 2.5.1; Table 2.1). These case studies serve as a constant reminder of never being complacent and taking food safety for granted. Food safety management is a continuum; it is never 'done'.

- *'Having a HACCP plan = HACCP'.* A HACCP plan is a document. The document must capture knowledge and activities that are implemented, maintained and supported by strong prerequisite programmes (PRPs), including having a hygienic operating environment and an educated and trained workforce. HACCP plans need to be owned by the business, which generally means that they should not have been written by an external consultant. They must be based on a hazard analysis that considers all aspects of the product and the manufacturing process at all times. A consultant will not normally have that depth of knowledge. The HACCP plan needs to be current, which means being regularly reviewed and always updated when anything changes either in the product formulation, the process or the operating environment. HACCP is not a paper exercise.

- *'HACCP costs too much'.* How can any business afford *not* to use HACCP? (See earlier discussion on cost benefits.) This issue needs more discussion in the public domain. Smaller businesses may need help to understand their true cost base. Public health professionals need to support education of consumers as well as their own teams to ensure both a well-informed customer base and enforcement authority. If evidence is needed that HACCP can save money, McAloon (2001) describes savings of $150 000 per year at Cargill through originally unintended cost savings that resulted from just one CCP improvement in process control. Costs associated with the perceived need to add human resource are a genuine concern, however. Often, large companies find that they already have sufficient personnel in place to develop HACCP programmes and can usually reallocate people to more productive activities related to HACCP development. The cost of implementing HACCP is often confused with the cost of needing to upgrade PRPs in order to comply with sanitary design requirements, and this is often a real challenge in both large and small companies (see Chapter 10).

- *'HACCP is complicated and requires a huge amount of documentation'.* Whenever this is stated, it is often a result of poor understanding of what is needed. Where HACCP is highly focused on food safety critical controls, the paperwork is often minimal. In many instances paper work can actually be reduced on a day-to-day basis when compared with previous systems. As a new approach, it can feel different – new terms will be introduced and care should be taken to avoid jargon when starting to implement it as this can overly concern employees. It is really important to invest the time in educating and training everyone in the business; particularly where increased documentation is a concern. The mystique surrounding HACCP should be removed and the concept should be presented in simple terms to ensure

Table 2.1 Examples of Major Food Incidents.

Year	Country	Food	Contamination	Known/suspected cause	Effect	Cost	Reference
1989	UK	Hazelnut yogurt	Botulinum toxin	Formulation change to reduced sugar version. Thermal process insufficient for new formula hazelnut puree **'Design'**	27 ill 1 death	$ millions across entire UK yogurt market	Shapton (1989)
1990	Worldwide	Bottled water	Benzene	Filter not checked in 18 months **'Preventive maintenance'**	Worldwide recall; 160 million bottles destroyed	$79 million	Reuter (1990)
1993	Germany	Potato chips	Salmonella (90 serotypes isolated)	Contaminated spice mix applied post cook step **'Supplier control'**	1000 cases – mainly children	Unknown	Lehmacher et al. (1995)
1994	USA	Ice cream	Salmonella Enteritidis	Ice cream mix ingredients were transported in a truck previously used to transport raw liquid eggs **'Cross contamination'**	Over 200 000 ill	Unreported	Hennessy et al. 1996
1998	USA	Toasted breakfast cereal	Salmonella agona	Cross contamination from ingredients or environment. Not proven. Plant cited for poor GMPs **'Cross contamination'**	Over 400 cases	Unknown	Breuer (1999)
2000	Japan	Milk products including yogurt	Staphylococcus aureus toxin	Lack of temperature control in raw milk during a power outage. Inadequate communication during a crisis **'Lack of knowledge regarding risk'**	About 10 000 ill	Unknown	Wrigley et al. (2006)

2006	UK	Chocolate	Salmonella montevideo	Leaking waste water pipe dripping into production area where chocolate crumb was manufactured. Regulators criticised the manufacturer for their risk assessment, inadequacy of their HACCP plan **'Cross Contamination' 'Lack of knowledge regarding risk'**	About 60 cases	Over $40 million Company was prosecuted by regulators and pleaded guilty	FPA Food Safety Update, July 2007
2006	USA	Carrot juice	Botulinum toxin	Inadequate pH control to prevent. Product was refrigerated but temperature abuse suspected **'Design'**	4 cases in 2 states (3; 1)	Unknown	Kaye (2006)
2006	Worldwide	Spices	Sudan red	Economic adulteration when spices were artificially coloured to look fresher and fetch a higher price	National recalls across many countries – mainly affecting Europe	Over £100 million estimated	Davies et al. (2006)
2008	Canada	Cooked sliced deli meats	Listeria monocytogenes	Inadequate sanitation of slicing machines. Insufficient verification testing **'Sanitary design'**	20 deaths	Over $20 million	CBC News 17 September 2008
2008	USA	Fresh jalapeno peppers	Salmonella St. Paul	Not determined – may have occurred on the farm, during processing or distribution	1442 ill; 286 hospitalised; may have contributed to 2 deaths	Unknown	US Centers for Disease Control and Prevention (2008)

(Continued)

Table 2.1 (Continued)

Year	Country	Food	Contamination	Known/suspected cause	Effect	Cost	Reference
2008	Australia	Frozen meal	Undeclared fish allergen	Company inadvertently packed a tuna meal in a package that did not declare allergen **'Mislabelling'**	National recall. No known ill effects	Unknown	FZANZ website
2008	China	Dried milk powder	Melamine	Economic adulteration	Estimated 54 000 children ill; 13 000 hospitalised and 4 died	$ Many millions	Congressional Research Service (2008)
2009	USA	Peanut butter	Salmonella typhimurium	Leaking roof Unsanitary process conditions Inadequate segregation between raw/roasted peanuts **'Cross contamination'**	700 ill 9 deaths	$100 millions Company filed for bankruptcy	FDA (2009)
2009	USA	Alfalfa sprouts	Salmonella St. Paul		228 cases in 13 states	Unknown	US Centers for Disease Control and Prevention (2009)

that it is seen as a practical straightforward approach. Records do need to be maintained both as evidence that the company has done the right thing and for peace of mind that products have been made safely and according to specifications. Records are required for government and customer inspection, as evidence that the process was under control. In these ways, records can be presented in a positive light rather than being seen as a burden.

- *'HACCP requires too many resources'*. This can be a real concern for small companies, but many large companies also operate lean organisations. Certainly, when starting to implement a programme, the resources required will be more than are needed for running an established programme. There will be a need for training of the HACCP team and team leader, and for the wider workforce. Access to additional technical expertise might be needed if it does not exist in-house, and temporary administrative help will probably be a benefit. However, the benefits once implemented (e.g. reduced waste, fewer consumer complaints, improved quality and reduced testing), should offset many of the resource concerns.

- *'HACCP by itself will control food safety'*. HACCP is at the core of the food safety programme but is part of many other, often interrelated, activities. When HACCP was first being discussed more widely, some companies mistakenly believed that it replaced the need for solid PRPs (Wallace and Williams, 2001). WHO (1998) defines PRPs as 'Practices and conditions needed prior to and during the implementation of HACCP and which are essential for food safety.' Most food safety failures are not failures of the HACCP system. Rather, they are usually caused by a lack of management commitment to provide adequate resources or a failure to properly manage PRPs.

- *'HACCP is a one-time activity'*. HACCP is anything but a task that is done only once; however, after the major effort of conducting the HACCP study, many companies breathe a huge sigh of relief, mistakenly believing that the job is done. This is far from true. HACCP needs to be alive within the organisation, constantly at the forefront of people's minds regarding any change to the product (including its ingredients), the process or the environment. External prompts such as emerging information on hazards, other industry food safety issues and availability of new technology need to be considered. Companies that fail often have not recently reviewed and updated their HACCP plans.

- *'HACCP is not suitable for small companies'*. The consumer has a right to safe food whoever produces it. All food should be safe. It is perfectly possible to undertake a HACCP study and to implement a programme in a small business. However, there are often resource constraints, not least the human resource in the form of technical knowledge. If the technical knowledge is available in a small business, then HACCP can be implemented. Some enterprising small companies have called upon their suppliers, customers or trade and professional organisations to help them (Route, 2001). Others have used consultants; however, it is important to ensure that they have the required competency, i.e. knowledge in both HACCP and the industry sector. Generic HACCP plans can also be a useful starting point (WHO, 1999), but need to be customised for the businesses' products and operating environment.

- *'Zero risk is possible'*. This is an interesting and ideal statement. Although zero risk is unattainable, HACCP and PRPs provide the most effective means of delivering safe food to the consumer (Mayes and Mortimore, 2001). Generally, too little attention is given to the fact that use of HACCP can only minimise risk by reducing the likelihood that a food safety hazard will occur (Cormier *et al.*, 2007). With microbiological hazards, we have fairly common agreement and we are able to estimate risk in quantitative terms, e.g. log reduction during processing or estimation of microbial growth under certain conditions. With chemical hazards, this becomes more difficult as technology advances (DeVries, 2006). For example, when we are unable to detect the presence of an adulterant, we can say that the food is acceptable. However, technological advances might mean that smaller levels of the adulterant are now detectable than previously and, whilst this does not necessarily mean that the food is any less safe than before, there is now a regulatory and consumer concern that the food is adulterated with chemicals that should not be there.

- *'Farm-to-table HACCP is not possible'*. The reported successes of HACCP systems in food processing facilities led to interest in applying it to all segments of the food supply chain, 'from farm to table.' As discussed in Chapter 1, whilst use of hazard analysis can be made and the process of conducting a HACCP study will provide a structured approach, the definitive control measures that are available to food processors in the centre of the supply chain are not often available at the 'farm' and 'table' ends of the supply chain, as well as at intermediate points. We cannot emphasise strongly enough that food safety is best assured by the simultaneous use of HACCP and PRPs. The latter must be used to the maximum possible extent in segments of the supply chain in which definitive HACCP controls cannot easily be used.

2.4   BARRIERS TO EFFECTIVE HACCP USE

Barriers to use of HACCP fall into two main categories:

- Those related to misconceptions, many of which have already been outlined.
- Those that are genuinely a hindrance to moving forward.

We need to articulate what the real and perceived barriers are in order to find solutions. If we are unable to do this, then the food supply chain remains at risk.

- *Lack of knowledge.* This is a real barrier and includes not just technical knowledge within the team but also broader managerial food safety ignorance leading to lack of support.
- *Lack of human resources.* For small- and medium-size businesses, this is frequently stated as a barrier. Even when support was provided through universities (Taylor, 2002), the perception of HACCP as not being crucial to their business was voiced by a sample of small companies in the United Kingdom.
- *Local language materials not being readily available.* This is a genuine problem in developing countries such as those in Asia. Books are being translated, but for many people a book will be insufficient for learning. There is a need for local language trainers who have experience and competence in HACCP.
- *Insufficient competent third-party consultants.* A consultant who has limited actual experience of working in a HACCP team can be detrimental to the HACCP process: firstly, if they do not have a good understanding themselves, such that weaknesses are built into the HACCP system, or secondly, if they are unable to transfer ownership to the plant team.
- *Insufficient expertise.* There is insufficient expertise for biological, chemical and physical hazard analysis (HACCP principle 1) within food companies particularly at manufacturing sites. This is compounded by insufficient guidance on the application of HACCP principle 1 (Wallace, 2009).
- *Misleading HACCP publications.* Unfortunately, whilst there are many excellent publications on HACCP, there are also quite a few that contain misleading information on HACCP. Some of these may be a result of knowledge having developed since the pieces were written but there

are a number that indicate poor knowledge and understanding by the authors. This makes it very difficult for those new to the concept to differentiate between what is a reputable source and what might be misleading when looking for good sources of information.

- *Lack of equipment and poor infrastructure.* This includes lack of control devices such as sifters, sieves, magnets, metal detectors, etc. or lack of monitoring equipment such as thermometers, pH meters or chart recorders. An incorrect plant layout is also a challenge (Panisello and Quantick, 2001). Plants must be designed to allow appropriate flow through the process with prevention of cross contamination as a key objective. If this is not the case, then HACCP implementation is a much more challenging task due to the need to control and upgrade to a hygienic layout through risk assessment.
- *Validation and verification difficulties.* Generally, verification is not too difficult once properly understood. However, validation needs to confirm that the elements of the HACCP plan will be effective in controlling the hazards of concern. It includes a review of scientific literature as well as possible microbiological challenge studies and in-plant process validation. Small companies certainly need help with this and may not be able to do much more than the literature review and some basic process confirmation.

2.5 REASONS FOR FAILURE

2.5.1 Lessons learned from major food safety events

With increasing expectations from consumers, the increasing ability to detect hazards and sources of foodborne illness, together with advancing food safety knowledge, the bar is raised in terms of food safety management. We should be seeing a fall in the incidence of foodborne illness and yet this is far from being the case in many countries.

What conclusions can we draw from reviewing some of the recent well publicised cases of failure? The first one is that companies all over the world have issues, and secondly, that food safety failure occurs in large well-known companies as well as smaller enterprises (Table 2.1).

In summarising the examples given (which is a very small sample of many), a number of conclusions can be drawn:

- That there is a worldwide need to improve our food safety programmes. No single country is exempt.
- That all hazard categories – biological, chemical and physical – are involved.
- That the various reasons for failure indicate the need for robust PRPs for assurance of:
  ○ cross-contamination prevention,
  ○ improved supplier control,
  ○ hygienic design of equipment,
  ○ adequate cleaning and sanitation practices and
  ○ preventive maintenance programmes.
- That knowledge of intrinsic product safety is essential when modifying the design.
- That economic and deliberate adulteration can be a food safety issue.
- That there are high direct and indirect costs.
- **That we have not learned from past mistakes.**

Most of these examples could have been prevented through safe product design and by avoidance of cross contamination. Knowledge of the HACCP principles and how to use them effectively is a valuable tool in meeting food safety obligations. However, there are many companies who, despite having learned the basic theory, fail to make the conceptual leap in terms of using it to develop and maintain a food safety culture.

2.5.2 Commonly observed mistakes in the *implementation* of HACCP and management of food safety programmes

In a company that has a good hygienic operating environment together with a positive culture aimed at doing things right, most of the requirements for HACCP will already be present; they simply need to be identified and brought into the HACCP framework. But even under these circumstances, there is possibility of having developed a less than adequate HACCP programme in the absence of the right advice and guidance. The following are some observations to consider and learn from:

- *Overcomplicated and, therefore, difficult-to-maintain programme.* Time needs to be spent in planning what the system will look like and thinking ahead about the ease of updating to keep the system up to date. Many companies find that a modular approach works best because it divides the plant and/or process up into manageable units. It is essential to check that there is a proper link between each unit and that all process activities are included.
- *Inaccurate process flow diagrams.* Many companies like to simplify process flow diagrams (PFDs), but in doing so the hazard analysis will be incomplete. Our recommendation is to create a simple outline PFD that can be shared externally along with a very detailed PFD that can be used internally for the hazard analysis. Another reason why they are inaccurate is that there may have been changes since the PFD was originally drawn up and, hence, it is outdated. The PFD must be confirmed as being correct and complete, by walking through it in the plant. This must be done *before* the hazard analysis begins. Consider the process 24/7 and also personnel traffic patterns, air and drain flow. If there are high-hygiene zones, they can be marked on the diagrams.
- *Lack of understanding of the products' intrinsic safety factors.* It is important to understand what is making your product safe. Is it low water activity (a_w), pH, preservatives, a heat kill step or something else? This knowledge is needed in order to make informed decisions in the event of cross contamination or requested formula or process changes.
- *HACCP principles 1 and 2: hazard analysis and determination of CCPs.* Too many CCPs through misunderstanding the relationship between HACCP and PRPs is a frequent error (Wallace and Williams, 2001). This is the most difficult area of HACCP and one where technical knowledge is needed. Some experts (Gaze, 2009) recommend early consideration of whether the identified hazard is controlled by a PRP. If so, then it is usually not a CCP. Another mistake, but very common, is the identification of hazards in general terms (e.g. 'biological' or 'pathogens') instead of specific terms that identify the specific hazard and its manifestation. Is it 'presence', 'cross contamination' or 'growth' of microbiological hazards that is the concern? If possible, identify the likely microorganism such as *Salmonella, Listeria monocytogenes* or *Staphylococcus aureus*. By making a specific identification of a hazard, it is much easier to determine the appropriate control measures for its prevention. Identification of hazard significance through risk evaluation (likelihood and severity) is a challenge where technical expertise is lacking.
- *HACCP principle 3: establishing critical limits.* Literature is limited in this area but experience indicates that some companies will write in the regulatory limit and many will use

their actual operating specification range at this point. This indicates a lack of understanding that the critical limit is exactly what it says – the limit that is *critical* for food safety (i.e. the edge of the cliff) – and that this limit needs to be based on scientific data (validated). Establishing what the margin of error is between the operational and critical limit is also common sense.

- *HACCP principles 4 and 5: monitoring and corrective action procedures.* Lack of clear instructions and properly trained monitoring personnel can have catastrophic effects. The CCP monitor is in the front line and must be well informed as to his or her responsibilities. There are some practical activities that can help:
  - ○ Ensure the monitoring frequency is appropriate.
  - ○ Keep the documentation simple and easy to use.
  - ○ Train designated CCP monitors well – be sure to verify their understanding and their ongoing behavioural competency on a periodic basis.
  - ○ Involve the CCP monitors in the design of any forms.
  - ○ Use verification activities to follow up with them with regard to performance.
  - ○ Be very clear on requirements for corrective action and required training. Ideally, 'inform QA Manager' will only be specified once other actions are complete. This should not be the main or only action required.
- *HACCP principle 6: verification.* HACCP principle 6 includes both validation and verification activities. Validation is usually the greater challenge for many. Lack of suitable references or other evidence such as challenge studies, to show that the HACCP plan will be effective against the hazards identified, is a common failure. Verification is seen as being more straightforward – many of the activities will already be familiar and in place. However, recent research (Wallace, 2009) indicates a wide range of weaknesses in HACCP verification systems at production sites of a multinational manufacturer. Although all these sites had a range of verification procedures in place, including internal and external audit, the audits conducted had failed to pick up the weaknesses identified during the research project, and this underlines the need for agreed standard audit approaches and effective training of HACCP auditors. It is, therefore, recommended both that food companies question the competency and experience of external HACCP auditors before their engagement, and that standard setters establish effective qualifications, training and experience standards for HACCP auditors. With recent developments in HACCP-based standards and auditor competency requirements (e.g. ISO22000 and GFSI), this may be addressed in the near future. Problems can also arise from misunderstanding that the requirements for HACCP also apply to PRPs, which also need validation and verification. The guidance here is to ensure sound training and education and seek reputable advice if possible.
- *Lack of management support.* Real management commitment is a key success factor in any food safety programme. This has to be more than a vocal assurance of support. There needs to be a number of other signs of alignment:
  - ○ Signed food safety or quality policy.
  - ○ Willingness to hold people accountable in the event of failure.
  - ○ Provision of resources for food safety activities – seeing it as a priority.
  - ○ Frequent and visible confirmation of commitment through staff briefings.
  - ○ Attendance at food safety related training.
  - ○ Proactive requests for status updates.
  - ○ Participation in review of performance indicators such as audits and consumer complaint data.

- *Lack of employee commitment.* This is just as important as management commitment. Employees can sometimes have a cynical, 'seen it all before' attitude and be reluctant to embrace new work practices. Good communication, an open and honest approach, sharing examples of failure and making it relevant can help. Real management commitment is, of course, an essential starting point.
- *Lack of motivation once the HACCP plan is complete.* Combine this with factors such as staff turnover, illness, absenteeism and competition for resources once new projects come along and this can be a real challenge. The vision of a proactive and sustainable programme takes a lot of effort to bring it to life. Again, education and genuine commitment are key, as is making it a team effort and everyone's responsibility.

2.6 DIFFICULTIES WITH APPLYING HACCP THROUGH THE ENTIRE FOOD SUPPLY CHAIN

HACCP is a tool used to systematically identify significant hazards and preventive control measures that eliminate or reduce them to an acceptable level. A fundamental problem with doing this from 'farm to fork', or in the case of animal feed, from 'field to trough', is that each link in the chain has no control over the next link and only limited control over the previous link.

On considering the supply chain (Figure 2.2), it is clear that even in this simplified diagram there are many stakeholders. In reality, there will be hundreds of them per product – adding up the farmers, the raw materials, the distributors and the numerous transport and storage stages. Communication from one to the next is often poor, with many opportunities for failure. In some countries, traceability is also weak and very few products use only local ingredients. However, each element does have a degree of control over the one that comes before it through supplier quality assurance (SQA) programme implementation, including having agreed specifications, conducting on site audits and inspections, and Certificates of Analysis (COAs). These are PRP elements, but HACCP can be used to focus efforts towards ensuring that vendors back up the chain have knowledge and control over the elements of their programmes that are critical to food safety. Again, it is HACCP and PRPs that are being used to assure food safety – not just

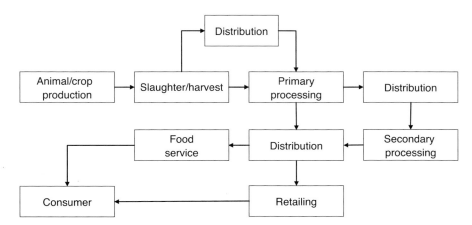

Fig. 2.2 Supply chain model. (Adapted from Sperber (2005).)

HACCP alone. There are liability issues with SQA programmes and, for that reason, it is not recommended that any formal approval be given of a vendor HACCP plan, but using the HACCP tool to audit their programme is a valuable exercise. Problems have occasionally arisen where a customer does this but then tells the vendor where the CCPs should be. Both should refer to the science and get independent expert advice if needed. Another option (if the vendor and customer disagree) may be to highlight the control as a CP (control point), or as part of a PRP.

A greater challenge comes in considering whether control can be applied *forward* through the chain. Arguably not, other than with clear communication of requirements using specifications and labelling. Some companies do go so far as to 'audit' who they do business with. In our experience, this can be a competitive advantage through ensuring that your distributors and customers have the capability to handle your product correctly.

It is no surprise that the HACCP approach to food safety management originated in the centre of the food supply chain with the food processors. Their success in applying definitive control measures such as pasteurisation, sterilisation, acidification and water activity reduction to processed foods led to the global acceptance and use of the approach. This in turn led to a move to apply HACCP from 'farm to table.' At the farm end, hazard analysis is extremely useful in identifying hazards that need to be controlled. Consider the situations with fresh produce and raw meat and poultry products, which are responsible for a significant proportion of current foodborne illness outbreaks when they are contaminated with *Salmonella* spp., *Escherichia coli* O157:H7 and *Campylobacter* spp. It is well established that cooking raw meat and poultry to a minimum centre temperature of 70°C will kill the above pathogens. Whilst raw meat and poultry producers cannot eliminate pathogens at the 'farm', a number of animal husbandry practices such as providing clean drinking water and using vaccination, competitive exclusion and hide cleaning before slaughter can reduce, but not eliminate, pathogens. At this stage, it is the PRPs that are used to great effect.

At the 'table' end of the supply chain is the consumer. Consumers often have limited knowledge and there is much evidence that poor practices abound in home kitchens (Griffiths, 1994). However, consumers can cook raw meat and poultry just as the processors can. Given the many millions of servings per day of cooked meat and poultry, it is not surprising that some consumers unintentionally or deliberately undercook meat or cross contaminate from raw meats to other foods, such as salad ingredients, in the kitchen. In the matter of fresh produce that can be consumed raw, definitive control measures to eliminate pathogens are not yet available even to food processors. Whilst good agricultural and sanitation practices can reduce the numbers of pathogens on fresh produce, they cannot assure elimination of all pathogens. Cross contamination, poor temperature control (cooling and storage) combined with their need to make choices (such as wanting meat rare) make this a challenge. As manufacturers, we have to validate and clearly communicate preparation instructions.

As will be discussed later, consumers and all stakeholders in the supply chain will need more realistic views on the presence of pathogens in foods that are traditionally consumed fresh or raw.

HACCP can be extremely useful in understanding where the significant hazards and control measures are across the entire supply chain. It is systematic, provides an excellent framework for discussion and we would highly recommend its use in this way despite the limitations. A key lesson learned over the years is that, as food manufacturers, we have the responsibility to do everything possible to assure the safety of the food we produce. Knowledge of how our raw materials are produced, likely hazards and controls at our suppliers, together with understanding the effect of likely abuse in distribution and sale, is vitally important as we design our products and set process parameters.

Despite the difficulties, HACCP can be a useful tool in helping to assure food safety from farm to table. However, HACCP alone is not enough, PRPs must be in place throughout, plus the willingness to share the responsibility and have good science-based dialogue between all stakeholders.

2.7   ROLES AND RESPONSIBILITIES – LESSONS LEARNED

Everyone has a role both as individuals and as organisations. WHO defined this as the concept of a 'shared responsibility' (Figure 2.3). As we have just seen, the global supply chain is highly complex and will only become more so over time. The range of hazards is broad and the food safety challenges that the industry has to address are enormous. Ensuring food safety in today's world is a formidable challenge and we have to work collaboratively. Whilst everyone has a role, the responsibilities need to be defined (Mortimore and Motarjemi, 2002).

2.7.1   Industry

We have a responsibility to do everything possible to prevent unsafe food being sold to the consumer. We have a long way to go with regard to proper implementation of tools such as HACCP. We need to do a better job with training and educating our workforce and we have a responsibility to communicate issues quickly and accurately in the event of a problem.

We have to carefully consider the science and do the right thing in designing products (ensure intrinsic safety), in operations (follow defined procedures) and in knowing when and how to take corrective action when needed. We need to be open to new and different approaches to food safety management. A key lesson learned is that there are best practices in industry in all

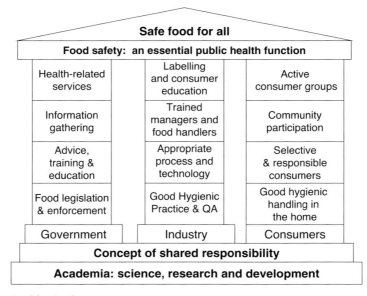

Fig. 2.3   Temple of food safety.

parts of the world. We have a responsibility to share our knowledge and be open to finding new solutions to old problems. Food safety should not be a competitive advantage.

2.7.2 Government

Government's primary responsibility is for *setting policy and regulations* and taking care to be transparent in the process. This role includes establishing initiatives for the regular evaluation of existing regulations through gap analysis, assessment of impact and review of likely effectiveness across the supply chain. Governments need to ensure that focus is maintained with regard to food safety policy as opposed to food policy. Enforcement of regulations is a key task, but arguably this can be done in a supportive as opposed to policing manner. Government is also responsible for considering consumer opinion (perception) in the risk analysis process and for risk communication.

Governments/public health agencies have a *hazard guidance role and responsibility*. They should be unified in establishing guidance on potential hazards by food category. They conduct risk assessments as significant new hazards emerge, for example *L. monocytogenes* in soft cheeses and deli meats in the 1980s. They are also responsible for monitoring environmental contaminants.

Consumer information and education is essential and government often plays a role here too, for example, in the home, in the school education system, in the workplace for food handlers, for health professionals and the media. Increasingly important is the need for the various national governments to liaise and agree key messages given the rapid developments in global communications.

Crisis management and communication in the event of major outbreaks or incidents (Table 2.1) is essential to ensure information flow and to maintain control as much as possible. Consumers as well as manufacturers and distributors need information in order to make informed decisions and to conduct investigations. Governments also have a responsibility to ensure that resources are available for rapid investigation, to share key learning and to give advice on preventive actions.

2.7.3 Retailers/food service establishments

This area is improving a lot, but we need a coordinated approach to stating required expectations, e.g. supplier questionnaires, third-party audits, specification format, COAs and testing. It is very inefficient to be operating to virtually the same standards but in many different formats.

Retailers and food service companies have a responsibility to have frequent dialogue with manufacturers in the event of a food safety market recall. A shared effort based on mutual understanding would be utopia.

Retailers and food service providers have the opportunity to talk directly to consumers. They have the ability to educate them and many do this well.

2.7.4 Trade and professional associations

Trade organisations can be extremely helpful in providing information across their membership. Whilst often largely financed by the larger companies, the forum enables both large and small companies to share knowledge, recognising that just one incident in a category can have a catastrophic effect across the whole sector. Professional associations enable exchange of knowledge across industry, academia and government – all having a common goal of safer food.

2.7.5 Consumers

Their primary responsibility is to recognise their role in food safety. Consumers must handle food in a hygienic manner and follow manufacturers' instructions for preparation and storage. Consumers and customers should quickly report any deficiencies to manufacturers. Consumers also have a responsibility to be open to new technologies that are valuable for food safety such as irradiation.

2.7.6 Academia

Whilst their primary responsibility may be to educate professionals to a higher level of knowledge, personnel working in academia also challenge the status quo by implementing research programmes that provide evidence to enable improvements in food safety and HACCP programme effectiveness. Academia also has a role in providing a neutral discussion forum between educators, industry and government.

2.7.7 The media

The media has a responsibility to report accurately and be fact based. They are usually the voice to the consumer in the event of a food safety issue. Overcommunication and scares will only confuse or make consumers numb to the information provided. Many are already weary of the amount of food safety information relating to what is safe to eat. Industry and the media must collaborate and build a more trusting relationship.

2.7.8 Advocacy and pressure groups

Like the media, pressure groups have a responsibility to report accurately and be fact based. They have an important role in promoting change where change may be needed and often tackle difficult topics. However, they need to be the voice of reason and avoid sensationalism.

We started this section by saying that everyone has a role. In summing up, we can say that everyone has a responsibility to *do things right* (as defined by documented HACCP programmes and PRPs in the case of industry) and to *do the right thing* (from the consumer's perspective).

2.8 CONCLUSIONS

Upon reading this chapter, there should be a few key conclusions drawn.

That we can always do better. HACCP has been around for so many years that many think it has been 'done'. This is far from the truth – so much so that perhaps we need to start over with a new name in order to galvanise the industry into action. HACCP is not a book or a document. It is an incredibly valuable tool (in the right hands) and we need to get back to basics in some ways. Get back to the science, the systematic approach, and develop some *real* and sustainable food safety programmes with HACCP at the core.

That we can learn from others' experiences and should be open to doing so. We must never think that we are better than anyone else, otherwise we will lose the opportunity to learn.

That people matter – enormously! Every single person from the CEO and chairman of the board to the very lowest operator. Each has a key role and responsibility and what each does matters. *Everyone* has to be committed to food safety and anyone can come up with a new idea for improvement or can identify a potential hazard.

$$\text{Food safety} = \text{HACCP} + \text{PRPs} + \text{People}$$
Doing Things Right
and
Doing the Right Thing
... Always

That we cannot take short cuts. Decisions must be based on a hazard analysis and a careful evaluation of risk. If we do accept that a different way of working is safe, then it should be capable of becoming the new norm and not just a one off time saver.

Hazard analysis and risk evaluation, risk communication and risk management need to be our modus operandi – everyday – 24/7.

That we need to understand intrinsic product safety. What is making our product safe? What would make it unsafe? What are the consequences if it becomes contaminated? Are we concerned about the presence of microorganisms, cross contamination or growth? How can we be more specific during the hazard analysis unless we really know our products?

That we cannot rely on others to fix our problems. Regulators, third-party auditors and customers are only in the plant for a short while and will not see everything. We have to take ownership for food safety and work together with others to find opportunities to improve.

That many misconceptions remain. Many misconceptions remain about HACCP and for many there are barriers (sometimes real, other times perceived) to its implementation. But these can be overcome by understanding what it takes to be successful.

That many food safety incidents were preventable. Many of these incidents were preventable if only the companies involved had the right knowledge and resources and were committed to food safety and to doing the right thing. It is astonishing to see examples where other companies had similar issues yet no action appeared to have been taken. We do not learn from others' experiences.

That there are real benefits. However, these will not be widely realised until we do a better job of implementing HACCP and PRPs. Many companies have yet to do this fully, which means that there is an opportunity to reduce foodborne illness by use of improved understanding.

3 Food safety challenges in the global supply chain

3.1 INTRODUCTION

Global trading in food is not new. Sir Walter Raleigh took potatoes from America to England in about 1590, and the Eastern spice trade has long been established. The reasons for us to continue trading with the world are expanding, but the basic driving forces of availability and innovation remain. Global sourcing today extends far beyond food, which probably lags behind industries such as apparel, appliances, consumer electronics and many others. Consumer-led demand for year-round produce, ethnic foods, innovative and organic foods, combined with industry's desire for improved productivity through low-cost sourcing, has led to increased momentum to move food around the world. Combined with the continuing growth in population and improved economies in some of the developing countries, resulting food safety challenges start to emerge.

Where does our 'global food' come from? The single ingredient and commodity foods mostly still come from the parts of the world that have traditionally grown them – spices from the orient, fruits from the tropics, grains from the great plains of North America, Asia, etc. Historically, there have been few problems with global food commodity trading. Increased demand, however, may lead to an intensification of commercial agricultural practices, and we may start to see crop contamination issues such as has been seen in the United States in recent times (e.g. *E. coli* O157:H7 in spinach crops). What has certainly changed is the amount of products that are being sourced and the diversity of companies and products that now use them. As an example, in the first 3 months of 2007, US imports of fresh fruits from China grew 279% to US$7.4 million, fresh vegetable imports grew 66% to US$32 million and fruit and vegetable juice imports grew 98% to US$109 million. Whilst these percentage increases sound high, the 'share' of the US food supply that comes from China is tiny, reportedly less than 1% (Fred Gale, Senior Economist at USDA's Economic Research Institute). The volume of exports and imports of 9 principal food commodities traded among 17 nations in all regions of the world are presented in Table 3.1. The total amount of exports substantially exceeds the total imports because these larger nations also export to more than 100 smaller nations not included in this tabulation. The total volumes of each of those 9 food commodities traded by the 17 nations are presented in Table 3.2.

Within this framework, there are pockets of intense trade. For example, China produces over half the world's pork, and a third of the world's horticultural output.

Food Safety for the 21st Century, First Edition By Carol A. Wallace, William H. Sperber and Sara E. Mortimore
© 2011 Carol A. Wallace, William H. Sperber and Sara E. Mortimore

Table 3.1   Food commodity trade among major nations, 2008 (in thousand metric tonnes).

Region/country	Imports	Exports	Deficit/surplus
Europe			
European Union (27)	61 503	30 957	−30 546
Russia	14 652	12 953	−1 699
Subtotal	76 155	43 910	−32 245
North America			
Canada	8 552	23 270	+14 718
United States	23 542	138 140	+114 598
Mexico	34 902	8 154	−26 748
Subtotal	66 996	169 564	+102 568
South America			
Brazil	8 608	64 209	+55 601
Argentina	3 406	45 517	+42 111
Venezuela	3 786	27	−3 759
Subtotal	15 800	109 753	+93 953
Asia–Pacific			
Japan	33 738	604	−33 134
China	52 599	49 237	−3 362
India	7 079	11 587	+4 508
Australia	19 727	10 709	−8 998
South Korea	16 470	645	−15 825
Indonesia	7 642	17 996	+10 354
Subtotal	137 255	90 798	−46 457
Africa			
South Africa	3 336	4 056	+720
Kenya	1 654	233	−1 421
Algeria	9 846	21	−9 825
Subtotal	14 836	4 310	−10 526
Total	311 042	418 335	+107 293

Source: Global Trade Information Services (2009).

Table 3.2   Volume of food commodities traded among major nations, 2008 (in thousand metric tonnes).

Commodity	Imports	Exports
Wheat	42 534	94 821
Maize	52 768	83 981
Rice	6 992	9 499
Soya beans	65 821	72 491
Vegetable oils	34 801	32 925
Poultry	3 907	8 902
Red meat	8 683	10 686
Fish and seafood	49 023	70 663
Fruits and vegetables	46 513	34 367
Total	311 042	418 335

Source: GTIS (2009).

It is a similar story for many other countries. In Europe, and the UK in particular, global sourcing was enabled by the trading between Commonwealth nations. Due to the advent of refrigeration, it was not just dry spices that were traded. The first cargo of frozen meat was reportedly shipped from Buenos Aires to France in 1877. In 1901, the first shipment of chilled bananas arrived in the UK. By 1910, the UK was importing 600 000 tons of frozen meat (James and James, 2006). Global sourcing of food has rapidly expanded since then. In this chapter, we will examine some of the challenges posed by the developing global food supply chain and then discuss strategies and tactics to promote food safety assurance.

3.2 INCREASED COMPLEXITY OF THE GLOBAL SUPPLY CHAIN

The drivers of change in the global food supply chain include economic, environmental and social factors. Each of these factors presents certain challenges, which are discussed below.

3.2.1 Economic factors

Land and labour

Both are lower cost in less developed countries though this differential with developed countries will likely diminish. Cost reduction has driven many established western companies to outsource from developing countries, resulting in a shift from trade being related primarily to commodity items, where climate played a key role in year-round sourcing, to the advent of finished, often highly processed, product manufacturing in developing countries which now have the emerging technological base combined with a workforce to undertake the task at a lower cost. The total value of processed food trade with the United States has nearly tripled in the past 16 years, with the amount of US imports exceeding the amount of exports (Table 3.3). Lower labour rates as a driving force mean that higher cost savings are realised in those processes which are highly manual, such as the hand peeling of shrimp. This is demonstrated by the reports that shrimp are now being exported from Scotland to Asia specifically for peeling and exported back to Europe. Figure 3.1 shows the staggering differences in labour costs from just a few years ago.

Construction and land costs are also still relatively low in developing countries, enabling companies to build manufacturing facilities for less than a tenth of the cost in developed countries.

Table 3.3 The US processed food trade (in billion dollars).

Year	Exports	Imports
1992	22.8	21.7
1996	30.1	27.8
2000	29.8	36.8
2004	30.3	47.2
2008	53.9	58.6

Source: ERS (2009).

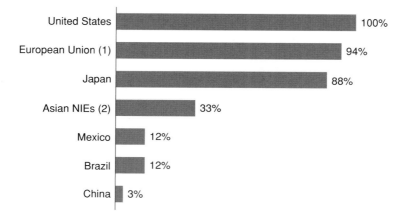

Fig. 3.1 Labour cost comparison. Indexed against United States at 100% (US$21.11). (1) European Union is the 15-member countries in 2002 prior to expansion in 2004. (2) Asian NIEs are the newly industrialised economies of Hong Kong, South Korea, Singapore and Taiwan. (Monthly Labor Review, August 2005, p. 32.)

Emerging economies

Large countries like China and India continue to have an impact on the globalisation of the supply chain. With rising incomes in emerging economies, however, combined with the global expansion of food service and retail companies, there appears to be a converging pattern in food consumption. Typically, this means that food spending increases through the purchase of more calories usually found in higher priced goods (Frazao *et al.*, 2008). Low-income countries typically have a diet high in starchy vegetables and low in animal protein, since meat and dairy products are considered a luxury due to their expense. The increase in income combined with the fact that multinational retail and food service chains have grown rapidly resulted in a convergence in consumption patterns. This change is happening at a much faster pace than in previous centuries. For example, Latin America national retail sales in food, as a percentage of food consumption, before the 1980s were 15–30%. By 2001, it had grown to 50–70%. In just 20 years, the growth was equivalent to what had previously taken 50 years in the United States. Asian trends in food consumption patterns are similar. Economic growth is accompanied by increased urbanisation – about one million people are moving every week from rural to urban areas where the need for labour is greatest and intensive construction of housing and manufacturing facilities and civic infrastructure are occurring.

3.2.2 Environmental factors

Expansion of pathogen range

About 50 years ago, there were only four major recognised foodborne pathogens: *Staphylo-coccus aureus*, *Salmonella* spp., *Clostridium botulinum*, and *Clostridium perfringens.* Today, there are nearly 30 recognised foodborne pathogens, including bacteria, viruses, protozoans and prions (Sperber, 2006). We can expect more. Epidemiologists have identified about 1400 human pathogens. About 800 are zoonotic, capable of infecting humans and animals, and 200 of these are considered to be emerging or re-emerging (Woolhouse and Gowtage-Sequeria, 2005). The H1N1 and H5N1 influenza strains are examples of re-emerging pathogens. Many factors are involved in spreading pathogens and extending their traditional ranges (Osterholm, 2006). These include the following:

- Modern transportation systems permit rapid worldwide movement of massive quantities of crops, animals and people, thereby enabling the spread of plant, animal, human and zoonotic diseases and invasive species of plants and animals.
- The well-known ability of microorganisms to mutate and adapt to changing environments is accelerated by their introduction into new environments.
- The changing climate and weather patterns are expanding the geographical range of pathogens and their vectors from tropical to temperate regions as the planet warms.
- Poverty, war and famine place enormous stresses on human populations, increasing their susceptibility to infectious diseases.
- A lack of political will at the governmental and intergovernmental levels to take effective measures to better protect the public health and the environment.

Representatives of the many diseases whose spread is widened by these factors are influenza, malaria, tuberculosis, Rift Valley fever and West Nile virus.

Decreasing arable land

Changes in developing countries had been so gradual that the loss of crop land through increased urbanisation did not seem to have a major impact on agriculture. Labour-intensive farming was acceptable when work was needed for the large numbers of people in rural areas of those countries. However, today, the use of more efficient western-style agricultural technologies will be necessary to offset the loss in land caused by urbanisation. Increased agricultural production efficiencies will displace even more farm workers, accelerating the urbanisation trend. Loss of arable land will also occur because of desertification and the increasing scarcity of water for crop irrigation. Advances in biotechnology to develop drought-resistant crops may help to ameliorate this issue. Sub-Saharan Africa and South America are the principal regions that have a surplus of arable land. Combined with their normally adequate rainfall, these regions will be important to maintain an adequate global food supply.

Climate change

Climate has always been important to agriculture and with change comes winners and losers. In 2007, the United States harvested record maize crops due to increased planting and favourable weather. In the same year, Australia suffered a drought and saw a drop in output. Predictions are being made by scientists regarding the effect of global warming on climate. Regions currently well-suited for agriculture may become drier or may become deserts. Cooler regions may benefit from longer growing seasons, should global warming continue. Rising sea levels will reduce crop land in coastal regions, a current example being flooding such as that seen in Bangladesh. These changes will require greater knowledge, resources and agility in the management of agricultural resources. Many traditional agricultural practices may have to be altered or abandoned. An example of a climate-induced crop change occurred in Panama. This country had a great climate for pineapple production, but did not grow it until they were taught by a global trader who saw the opportunity. We will certainly witness many changes of greater magnitude in the coming years.

Table 3.4 Global blue water withdrawals in the 20th century.

Water used for	Volume (km³/year)		Per cent	
	1900	**1995**	**1900**	**1995**
Irrigation	530	2600	88	69
Industrial	40	800	7	21
Municipal	30	400	5	10
Total	600	3800	100	100

Source: Derived from Falkenmark and Rockström (2005).

Water availability

The agricultural activities that support our civilisation are wholly dependent on an adequate supply of water, received either by rainfall or by irrigation. The amount of fresh water (blue water) contained in rivers, lakes and groundwater is about 0.8% of the total water on the planet. While most crops are fed by rainwater, irrigation is increasingly used to permit the production of crops in semiarid regions to feed the growing population. During the past century, the human population has increased fourfold, while the use of blue water for human activities has increased ninefold (Falkenmark and Rockström, 2005). More than two-thirds of the blue water is used for irrigation (Table 3.4). Major aquifers (groundwater) are being depleted in agricultural areas. Irrigation water is being already used to grow rice and maize in semiarid and desert regions, accelerating the groundwater depletion rate. It will be increasingly difficult to produce an adequate food supply as the human population increases and the long-accumulated natural stores of blue water are depleted.

Limited fossil fuels

The general consensus among petroleum geologists is that sometime in the past several years the world reached 'peak oil' production, that is the point at which half the estimated global oil reserves had been extracted and used. We are faced with shrinking reserves of all fossil energy sources, including oil, coal and natural gas. Modern agriculture is highly dependent on fossil fuels for cultivating the land, fertilizing and harvesting crops, transportation and storage. Shrinking fossil fuel reserves and the need for developing countries to modernise their agricultural capabilities will drive an upward spiral of increasing energy costs. This is occurring as much of the developing world is attempting to modernise production and provide more food and more food choices. These trends are certain to increase the cost of food. It is imperative that effective fuel conservation measures be enacted and that practical alternative and renewable energy sources be developed for widespread use. Currently, 7% of the US energy consumption is produced from renewable energy sources (Table 3.5).

Alternative energy sources

Much of the renewable energy used worldwide is produced from established sources and technologies – hydropower, geothermal and the burning of municipal waste. Continued development may enable these sources to provide a larger share of the global energy need. Wind and solar technologies hold promise for electricity generation, but the cost and the lack of distribution systems have limited them to less than 1% of all energy produced.

Table 3.5 Energy consumption in the United States, 2008.[a]

Energy source	Contribution (%)
Petroleum	37
Natural gas	24
Coal and coal coke	23
Nuclear	9
Biomass/waste	3.71
Hydropower	2.38
Wind	0.49
Geothermal	0.35
Solar	0.07
Total	100.00

Source: DOE (2009).
[a]Total consumption = 99.3 Btu.

Biofuels are receiving a lot of attention and governmental support because of their ability to reduce the use of gasoline and diesel oil. The production of ethanol from sugar crops appears to be practical when sucrose is the substrate. However, when dextrose derived from maize starch is used as the substrate, the productivity of ethanol production compared to sucrose is reduced by 95%. That is, the production of ethanol from cornstarch is at best a break-even proposition, with no net energy gain based on the petroleum inputs. As shown in 2008, the diversion of food crops such as maize to ethanol production will decrease the global food and feed supply while driving up prices. In 2007, 12% of the total world maize output was used for ethanol. This compares to 60% for animal food in the same year (FAO, 2002). Ten years earlier, virtually no maize was converted to ethanol.

The production of biodiesel from food crops could create a similar dilemma. Therefore, a great deal of effort is being expended to develop technologies in which bioethanol can be produced from cellulose and biodiesel can be produced from non-food plants or algal oils. Methane, the same molecule as fossil natural gas, is readily produced by the anaerobic digestion of manure and waste vegetation. In the very long term, hydrogen produced by the electrolysis of water or by microbial fermentation may become a practical fuel supply.

3.2.3 Social factors

Human overpopulation

World population is expected to reach 7 billion in 2010 and surpass 9 billion by 2050. Many developing and developed countries will not be able to feed their own people. This will drive the need for increased global trade in food – out of real necessity as opposed to being related primarily to lower labour costs. Today, Brazil, sub-Saharan Africa and some former Soviet bloc countries have adequate rainfall and large amounts of land available for cultivation. Brazil, for example, has 19% of the world's arable land, of which only 4% is currently in use (FAO, 2002).

On the basis of the current agricultural practices, these largely untapped agricultural regions may be able to provide food security to the expected 9 billion people. However, neither currently has the infrastructure to be able to move the food from the rural areas to ports. And what happens to the human population and its food supply after 2050? At this time, there has been limited political or social will to tackle the very difficult subject in terms of establishing human

Table 3.6 Comparative efficiencies of modes of transportation of food.

Transportation mode	Relative efficiency
Ocean shipping	100
Railroad	63
Highway trucking	19
Airplane	1
Automobile	0.03

Source: Derived from Bruhn (2009).

population control measures. This will greatly complicate our efforts to sustain and improve food safety and food security.

Increase in the numbers of immunocompromised people

With the ageing population, there will be an increase in the percentage of the population that fall into this category, posing an even greater emphasis on food safety (Section 5.2.2).

Year-round sourcing

Consumers in developed countries have come to expect a year-round supply of inexpensive fresh produce, as opposed to a seasonal supply. Environmental concerns are changing attitudes a little; for example, the European move to calculating the carbon footprint of products is raising consumer awareness and could drive a reduction in 'food miles'. However, this may be offset by the current trend for corporate social responsibility programmes. Reduction in global produce sourcing could negatively impact the lives of farmers in less developed countries who have come to rely on the income gained from western retail customers. Moreover, the long-distance transportation of foods by ocean ships or railroads is substantially more efficient than the local transportation by small vehicles (Table 3.6).

Improved living standards

Developing countries' improved standards of living have already been mentioned. But it does mean an increase in the demand for more processed foods – sometimes through imports from the West, increasingly through establishing local production capability, which can be a challenge as local raw materials have to be sourced, and the food safety managed back up through the supply chain.

All these economic, environmental and social factors mean increased pressure on the global food supply chain and therefore an increased need for a common approach to effective food safety management. The changing world will also require some new thinking in food safety management – what worked in the past may not be effective in the future.

3.3 FOOD SAFETY ISSUES IN GLOBAL TRADE

Let's consider recent issues. Microbial pathogens, presence of heavy metals, undeclared allergens, foreign material contamination and economic adulteration require focused food safety management strategies to prevent product failure. Many will be aware that, at the consumer

Table 3.7 FDA food import refusals by industry, 1998–2004.

Food	Number of violations	Violations (%)
Vegetables	14 496	20.6
Seafood	14 144	20.1
Fruits	8 233	11.7
Non-chocolate candy	5 137	7.3
Bakery products	3 800	5.4
Spices, flavours, salt	2 674	3.8
Soda, water	2 604	3.7
Multifood sauces	2 604	3.7
Cheese	2 604	3.7
Chocolate, cocoa	1 618	2.3
Pasta	1 548	2.2
Other	10 907	15.5
Totals	70 369	100.0

Source: Buzby et al. (2008).

level, the list includes non-permitted additives, avian influenza, and the use of biotechnology all of which present regulatory and genuine consumer concerns. Whilst also being a challenge to manage, these are not always the real food safety issues.

The rejection of imported foods because of food safety violations is an important economic consideration. A large number of food categories are affected by such regulatory actions (Table 3.7). Of over 70 000 violations reported in the United States from 1998 to 2004, 33% were for misbranding or lack of appropriate labelling and 65% were for adulteration, pesticide contamination, lack of packaging integrity or unregistered thermal processes for canned foods. Data indicate that the most common problems are the presence of foreign material (filth), whilst specific pathogens tend to be found in products which have a history, for example *Listeria monocytogenes* in cheese and cheese products.

The fact that 'filth' is one of the most commonly found violations is not surprising. Many food producers, particularly in developing countries, do not yet fully comply with the basic food manufacturing practices of operating in a hygienic environment and with equipment designed for sanitation. Temperature-controlled storage, pest control, allergen management, cross-contamination control and sanitation programmes are often inadequate. Neither do they understand the requirement or expectations of the supply chain once it leaves their shores. That combined with poor understanding of the Hazard Analysis and Critical Control Point (HACCP) system, microbiology and allergen control leads to a high likelihood of failure. The media in countries such as China report a high number of foodborne illness outbreaks domestically, especially in institutions such as schools and hospitals, which is evidence of this knowledge gap. So, there is a real desire not only for improvement in export but also for improvement in domestic food supply.

In many companies there is often a willingness to learn and often an appreciation that their customers are prepared to help. The lack of knowledge however is significant. As an example, one potential supplier when learning about Electric Insect Killers enthusiastically installed them not only inside the building but also the outside. Another company when informed that it was a good practice to leave an 18-inch gap between storage racking and the walls of a warehouse for cleaning and inspection purposes responded with a comment that they had very tiny people in their country and would not need such a wide gap. It would be hard to disagree with such logic.

Many Good Manufacturing Practice (GMP) standards are based on western manufacturing circumstances. Another great example of this is that few global pest control standards make mention of monkeys, lizards, snakes and wombats, illustrating the differences that exist among company and country practices and expectations.

Many recorded failures are instances of filth, illegal pesticides, herbicides and heavy metals, and more recently economic adulteration as evidenced by the deliberate addition of melamine to wheat gluten used in the US pet food, and only 1 year later, into milk products in China. Spices contaminated with Sudan and Para Red are another example, this time with global impact, causing market recalls to happen in many countries including the UK, the countries of mainland Europe, South Africa, Australia and China. Another global example of economic adulteration was the use of diethylene glycol (a low-cost – but frequently deadly – substitute for glycerin) in the production of toothpaste and cough syrup in China, resulting in the deaths of 51 people in Panama.

These instances serve only to underline the inadequacy of testing versus preventive approaches such as effective implementation of HACCP and prerequisite programmes across the industry. In the United States, the Food and Drug Administration (FDA) is struggling to keep up with testing the increasing number of US imports – they cannot sample everything, and many of the exporting countries are not testing before export. Even if all imports and exports could be sampled, we know very well that product testing does not work to ensure food safety (Chapter 1). No amount of additional inspectors or product testing will better enable us to ensure food safety.

A few specific considerations arise in the light of the above:

- *Difficulty in traceability at the global level.* A number of global issues have highlighted this, e.g. Sudan Red, melamine and dioxin contamination.
- *Spread of disease as a result of global trade.* Notable examples include those involving trade in live animals affected with BSE, avian influenza and *Salmonella* DT104.
- *Political and consumer reactions to global trade safety issues.* Most frequently, governments simply set up trade barriers. There is nearly always increased sampling and testing following an issue. At the consumer level, the reaction is usually fear and avoidance of the likely products associated with the issue, or sometimes any products from countries where the affected exports came from.

3.3.1 Lack of uniformity in regulations and requirements

The success of HACCP as an industry food safety management programme has led some national governments to promulgate HACCP-based regulations. There are differing opinions as to the best approach here. In Europe, the broad requirement in EC Regulation 852/2004 on the hygiene of foodstuffs (EC, 2004) to use HACCP principles for food safety management across all elements of the food industry has been fairly successful over the past several years, despite the absence of definitive controls by the industry sector. However, in other countries, views differ to the effect that regulations are sometimes described as having been incompletely developed. For example, as discussed in Chapter 1, the US HACCP regulations for raw meat and poultry products and juice products do not require definitive process controls that can be managed as critical control points (CCPs) to control identified microbiological hazards. Moreover, many small producers can be exempted from processing controls (Sperber, 2005b). Such regulations can undermine support and understanding of legitimate preventive food safety management procedures,

Table 3.8   Regulated food allergens.

Food allergen	United States	Europe	Australia and New Zealand
Celery		✓	
Cereals – with gluten	✓	✓	✓
Crustacea/shellfish	✓	✓	✓
Eggs	✓	✓	✓
Fish	✓	✓	✓
Milk	✓	✓	✓
Mustard		✓	
Peanuts	✓	✓	✓
Sesame		✓	✓
Soya	✓	✓	✓
Sulphite		✓	>10 mg/kg
Tree nuts	✓	✓	✓
Others		Lupin	Bee pollen
		Mollusc	Propolis
			Royal jelly

Source: Adapted from Higgs and Fielding (2007).

typically expected in a HACCP system. However, there are recent examples of satisfactory and effective HACCP-based regulations. One example is the US rule for *L. monocytogenes* control in ready-to-eat meat and poultry products (CFR, 2003).

Lack of uniformity in regulations around the world and the promulgation of poor regulations put an unnecessary burden and lead to additional unnecessary costs for the food industry and ultimately to the consumers A few additional examples are described below:

- *Allergens.* The differences in regulations lead to the need for multiple labelling and additional allergen control measures in companies that export around the world. It is not just about whether an ingredient is an allergen but also whether there is a recognised threshold level. In Australia and New Zealand, there is government guidance (Higgs, 2007) on the thresholds for action, unlike most other countries which in some cases have zero tolerance. Differences regarding which allergens are regulated can be very confusing if you are a food exporter (Table 3.8).
- *Mycotoxins.* The regulatory expectation related to mycotoxins in cereal and food crops is somewhat clearer. Of the several thousands of mycotoxins that have been identified, five are of principal concern in the food and feed supply; regulatory limits have been established in many countries for one or more of these mycotoxins (Table 3.9). However, there is variation among countries as to the types of mycotoxins, the types of food and feedstuffs affected and regulatory limits and testing requirements that can complicate trade for both exporters and importers. The reader is referred to Chapter 5 for additional information on mycotoxins (Table 5.17), as well as persistent organic pollutants and heavy metal contamination.
- *Other chemical contaminants.* Some of the emerging chemical contaminants such as cyanuric acid, an analogue of melamine, and melamine itself are fairly new as contaminants in food. Their example serves to indicate that any chemical which should not be there is an adulterant and that scientific knowledge is needed to help the industry figure out how to deal with these situations. Some governments have published guidelines or regulations. However, differences exist in what is considered acceptable and often more research and guidance are needed to help industry manage these hazards.

Table 3.9 Principal mycotoxins of regulatory concern in grains and nuts.

Mycotoxin	Grain or nuts	EU limits (ppb), 2007	
		Wheat	Maize
Deoxynivalenol	Barley, wheat	1250	1750
Zearalenone	Barley, maize	100	350
Fumonisin	Maize	None	4000
Aflatoxin B1	Maize, groundnuts	2	5
Aflatoxin, total		4	10
Ochratoxin	Barley	5	5

- *Counterfeiting.* There are some real concerns arising from countries which are known for their ability to counterfeit consumer and industrial products. Clearly, the counterfeiters are interested in commercial gain rather than food safety. Companies should always buy from reputable companies.
- *Microbiological criteria.* Differences exist in the use of microbiological criteria around the world. There are also differences of opinion at the public level, surrounding what these should be (e.g. zero tolerance for *L. monocytogenes* in the United States), and sometimes differences in required test protocols (e.g. Japan). It has been proposed that microbiological criteria or standards are not necessary for foods that are produced in a supply chain in which HACCP and prerequisite programmes are applied. Rather, microbiological guidelines can be used to verify the safe and sanitary operating conditions of production environments (Sperber and NAMA, 2007).

3.3.2 Lack of uniformity in audit requirements

Differing requirements, particularly amongst multinationals combined with the plethora of independent and national audit standards and auditors, can make this very difficult if you are on the receiving end. A few years ago, this was how it was. We still have challenges, but the framework is developing through the work of the Global Food Safety Initiative (GFSI) and others (see Section 3.4.2 and Chapter 14).

3.4 STRATEGIC-LEVEL RESPONSES

At a high level, strategies are needed which are preventive. Failure can be a catalyst for change but, predictably, what often emerges, particularly as an immediate response to failure, is a reactive strategy involving additional testing, more inspections and regulations. This does at least ensure that issues remain in the spotlight until more preventive work can be initiated, but that is often harder to do.

3.4.1 Government communications systems

National governments that have food safety responsibilities spread among several departments or agencies may encounter communications difficulties. That difficulty may be expected in the United States, which currently has its food safety accountabilities spread among five departments (Agriculture, Health and Human Services, Commerce, Defense and Homeland Security) and numerous agencies. Nonetheless, several excellent communication systems and databases have been established by the US Centers for Disease Control and Prevention, which are housed in the Department of Health and Human Services (http://www.cdc.gov). The Foodborne Diseases Active Surveillance Network (FoodNet) tracks foodborne illnesses to assist epidemiologists in the determination of foodborne illness outbreaks. PulseNet is a network of laboratories that share information from standardised molecular subtyping procedures, such as pulsed-field gel electrophoresis, that very specifically identify foodborne pathogens (with a specificity analogous to that obtained by fingerprinting humans). FoodNet and PulseNet enable the detection of low-incidence, widely spread illness outbreaks that could not have been previously identified by conventional microbiological and epidemiological fieldwork. PulseNet is being implemented as a collaborative system that will soon link public health laboratories of all continents. The National Antimicrobial Resistance Monitoring System for enteric bacteria assists in the identification and evaluation of the spread of antimicrobial-resistant pathogens in the food supply and human and animal populations.

Similarly, Codex Alimentarius Commission (CAC) procedures promote communications between intergovernmental organisations. CAC was created in 1963 by FAO and WHO to develop food standards, guidelines and recommended codes of practice under the Joint FAO/WHO Food Standards Programme. The purposes of this programme are to protect the health of consumers, ensure fair practices in food trade and promote coordination of all the development of food standards undertaken by intergovernmental organisations and non-governmental organisations (http://www.codexalimentarius.net). The standards, codes, etc., have the force of law in trade among United Nations member states who are signatories to the World Trade Organization. Additionally, CAC coordinates several continuing expert consultations: the Joint FAO/WHO Committee on Food Additives, the Joint FAO/WHO Meetings on Pesticide Residues, and the Joint FAO/WHO Meetings on Microbiological Risk Assessment.

3.4.2 Global Food Safety Audit Standards

This is improving and largely due to the GFSI rather than the government level activity. This is also a fairly recent development and one that is also driving auditor calibration.

GFSI was established in May 2000 and was made up of representatives from international retailers coordinated by CIES (Comite International d'Enterprises a Succursales, established in 1953) – the Food Business Forum. The objectives are, firstly, to promote convergence of food safety standards through the maintenance of a benchmarking process for (existing or new) food safety management schemes; secondly, to improve cost efficiency throughout the global supply chain through the common acceptance of GFSI standards by retailers from around the world; and thirdly, to provide a unique international stakeholder platform for networking, knowledge exchange and sharing of the best food safety practices and information. The GFSI Guidance Document was written and approved by the group, and the already existing British Retail Consortium (BRC) Global Standard for Food Safety (now in version 5; BRC, 2008) was the first standard to be benchmarked and recognised. This was followed by the International Food Standard, Dutch HACCP, Safe Quality Food, and New Zealand GAP (Good Agricultural Practice). FSSC 22000, incorporating ISO 22000 with supplementary prerequisite GMP standards via PAS220, was recognised in 2010.

The GFSI benchmarked schemes have been adopted by many retailers, food service establishments and manufacturers, notably the top three global retailers – Wal-Mart, Carrefour and Tesco.

Currently, the GFSI mission is stated as 'continuous improvement in food safety management systems to ensure confidence in the delivery of food to consumers'. Who could fail to support such a mission. All the benchmarked schemes are certification schemes which mean that they require certification bodies, who are themselves accredited by national standard institutes, to audit and certify. This requires that they adhere to certain standards regarding protocols of operation including those related to auditor competency.

The GFSI has the potential to be a significant milestone though it has taken 10 years to gain global acceptance across the industry and it needs continued support.

3.4.3   Auditor competency

It is not just harmonisation of third-party audit standards with which we need be concerned. There must also be standards of competency for auditors to give us confidence in their ability. If we can help support getting the certification bodies to a common standard for audit operations then that may truly help the industry. In Australia, a government initiative was set up in 2006 to develop auditor competency standards within a national food safety audit framework. This had some modest success. Currently, GFSI is working on this same topic. The International Register of Certificated Auditors (IRCA; http://www.irca.org) has operated an international recognition scheme for auditor competence for some years. This covers a range of different auditing fields, including quality (ISO 9001:2008) and food safety (ISO 22000:2005) auditors. Smaller national schemes have also developed such as the UK Institute of Food Science and Technology's Register of Professional Food Auditors and Mentors (http://www.ifst.org). The challenge is to bring auditor skills competency and practice standards together for global acceptance.

3.4.4   Public–private partnerships

The formation of public–private partnerships (PPPs) is a recent trend that will enhance our ability to ensure food safety by fostering collaboration between different types of organisations that ordinarily had not worked together. The participants in PPPs can be from global food companies, academia and non-governmental, national and intergovernmental organisations. A further hallmark of PPPs is that they focus more broadly on issues that affect food safety, public health and animal health, especially by focusing on the interfaces that might be overlooked by the individual disciplines.

The Safe Supply of Affordable Food Everywhere, Inc. (http://ssafe-food.org), was incorporated in 2006. Its initial projects focused on the interface of human and animal health by building veterinary capacity in developing countries and participating in the One World One Health project of the Wildlife Conservation Society (http://wcs.org) to better understand and control the threat of avian influenza. The Global Initiative for Food Systems Leadership (http://foodsystemsleadership.org) was organised in 2008 to build capacity in food safety leadership. Its initial projects to train national leaders in China and India will help to strengthen food safety in these major parts of the global supply chain.

3.5   TACTICAL LEVEL RESPONSES

What can you do as a manufacturer who is inevitably caught up in the global supply chain? A number of things can be done.

3.5.1 Supplier audits and approvals

Given everything described, to find and approve a good global supplier in these circumstances takes a lot longer than it does when working with someone locally, and at a tactical level different approaches are required. A single supplier 'audit' is insufficient in most cases, and the skills required to undertake this work are different to that which would be needed in a traditional local purchase setting. Having said that, given the number of issues that have arisen in the developed world recently, perhaps we need a different approach wherever the supplier is located.

In addition to having a deep food safety technical knowledge, enabling valid risk evaluation, supplier auditors need the softer skills to be able to educate, motivate and negotiate. This is a different skill set requirement to that of a certification body auditor whose role is solely to assess compliance. With many supplier audits, problems cannot just be identified. Practical solutions (short- and long-term) will usually need to be discussed, and training and education of suppliers must often be delivered at all levels of their organisation. This starts with motivation of the senior team to want to make the changes needed, followed by education at a cross-functional management level on food safety hazards, and training within the manufacturing environment to affect behaviour change. Training and coaching are also essential during implementation of the HACCP process, which requires deep technical knowledge and judgement, not only during the initial study but ongoing. To support suppliers in the development of a sustainable programme takes a lot of resource, but it is important to set a good example to the emerging global food supplier base. Auditors should not take the HACCP plan at face value but need to dig deeper on content (i.e. assess the scientific basis for decision-making, validation data and change management). To do this properly, it is not sufficient to visit once and then every 2–3 years. Building a relationship with the suppliers and having a frequent and ongoing presence are *critical* for success, and it is something which the private sector is able to do perhaps more effectively than the government in its enforcement role.

Where language barriers exist, the process becomes more difficult and there are also important differences in cultural understanding around the globe. For example, in some parts of the world, the culture is such that people are very polite and will usually wait to be told the requirements even if they already know them. In a company with the right attitude, once these are understood, corrective actions will be done almost immediately and they will wait to be told the next thing on the list. This in contrast with some other cultures, where although there may be an excellent knowledge of the theory in terms of what is required, there may be less action orientation in terms of implementation.

In the past, many companies would plan on only auditing their higher risk food suppliers. This was usually based on the microbiological risk, but with recent events we will have to reconsider supplier approval requirements. Sourcing strategies will have to change. Buying through brokers, even for minor ingredients, is going to require careful consideration, which will no doubt involve a much more in-depth investigation into their supply chain. A 'seeing for ourselves' mentality will increasingly apply back up through the supply chain. Focusing efforts on developing our suppliers will almost certainly include a more thorough evaluation of their ingredient supplier management programme (if it exists) than previous.

This approach is clearly resource-intensive and probably not realistic given the resources that many of us have, so a risk-based strategy is essential. Criteria should be considered such as food safety history within the food product category and whether the supplier is already exporting and/or working with multinational companies who have similar requirements. The use of third-party audits and certifications as a screen and even approval mechanism for low food safety risk suppliers can be helpful, but at the other end of the scale, hiring and training local resources to oversee production on a continuous basis can sometimes be the best or only option if your supplier base justifies this.

There are increasingly some very good and enlightened suppliers in developing countries. A number of multinationals are setting up or acquiring overseas manufacturing plants. There is heavy investment by the local food industry, and a hunger for knowledge within the regulatory enforcement arena as well as industrial workplace. Developing countries often have plentiful natural food resources, inexpensive labour and are catching up fast, aided by western companies who are sharing their knowledge. But given the complexity of the supply chain, it is unlikely that we will be able to sit back and feel confident that all issues are covered – not for many years to come.

3.5.2 Shared audits and approved supplier lists (with other companies)

There are legal issues to consider here, but in theory this makes sense. The idea is to calibrate across companies and coordinate on auditing and approvals. There are challenges, such as, which audit standard would be used – Company A or Company B? None of it is really proprietary information, but Company A might require something unique based on how the ingredient is used. As the GFSI benchmarked audit schemes develop, having access to databases of certified producers could negate the need for a separate in-company standard or calibration of in-house auditors.

There are consortium schemes emerging in industries that are on the edge of the food industry such as animal feed. TrusQ (http://www.trusq.nl), formed in 2003 in the Netherlands, is a cooperative alliance regarding food safety. Its members are Dutch animal feed producers whose objective is maximum food safety assurance. Participants share their knowledge on topics such as crisis management, track and trace systems and monitoring programmes. They have made firm agreements on choices of raw materials and suppliers and work together to do comprehensive risk assessment from point of origin up to raw material delivery. TrusQ screen suppliers back to the country of origin and auditors allocate a red, amber or green rating. Red meaning that no dealings with the suppler will occur. TrusQ is one model, and there may be others but it does offer another practical approach to small and large companies alike.

3.5.3 Business continuity planning

Challenges are many and varied, meaning that there is no one single risk mitigation strategy – flexibility is the key.

To be flexible, you need to plan ahead. Use a risk evaluation to understand not only which are your high food safety risk ingredients, but also which ingredients pose a high risk due to the following:

- Sole supplier situation.
- Unique functional role.
- Characterising ingredient (e.g. pecans in pecan pie).
- Instability of the country – economically and politically.
- Variability in currency exchange rates.
- Commodity items are vulnerable to weather or other factors that could affect crop yields.

Each of the criteria (including the food safety criteria) should be rated and assessed in terms of likelihood of occurrence and severity of effect (including impact to the business). Consider all raw materials, including those used by contract manufacturers and the contract manufacturers themselves. From a cost perspective, the trend some years ago was to go to the sole supplier. In a global market, it is preferable to understand your risks and have a dual supplier strategy to enable flexibility.

3.5.4 Sharing technology

It is a natural instinct for global food companies to protect their brands and production and trade practices in order to gain and maintain a competitive advantage over their rivals. However, in the matter of food safety management, many progressive global food companies have adopted corporate policies, declaring that advances in food safety procedures and systems will not be used as a competitive advantage. Rather, the advances will be shared with the industry through publications, vendors and trade associations. The Pillsbury Company perhaps established this trend in the 1970s by openly sharing its early ideas about HACCP and in continuing to develop HACCP for another 25 years. Cargill Inc. openly shared its otherwise proprietary knowledge in 1996 in which steam pasteurisation of beef carcasses was used to reduce the hazard of *Escherichia coli* O157:H7 in raw ground beef. Similarly, Cargill was also an early developer and continuing participant in trade association training programmes to eliminate the hazard of *L. monocytogenes* in cooked, refrigerated, ready-to-eat meat and poultry products.

3.6 CONCLUSIONS

The world is changing quickly and the speed of change is accelerating. Technology, ease of travel and distribution, the low cost of labour in the developing countries and environmental changes are just a few of the factors that are considered in this chapter. One thing is certain – that a world-class food safety programme needs the agility to anticipate and adapt to change. It is complicated. As the developing countries continue to develop their food safety programmes based on what worked in more developed countries, we need to be aware that what worked before and in a different environment may not be the best approach now. We have to be mindful of the basic concepts and open to new ideas from anywhere in the world.

Global food trading is no longer a choice. It is now a way of life. At a tactical level we have to find approaches to manage the various challenges and risks. There are varying standards and a lack of knowledge which requires a different approach to not only approving a supplier but also supporting their development. Sourcing management strategies must be risk-based and flexible: the use of third-party audit schemes, consultants and laboratories as well as going to see for yourself and carrying out detailed audits, training and follow-ups are options, and of course lead to a different cost structure for supplier approval.

Ironically, as we look at the recent major food safety outbreaks, for example *Salmonella* in chocolate and peanut butter and *E. coli* O157:H7 in salad crops, it is interesting to reflect that these occurred in the established food production arena of the West where we should have the knowledge to prevent this from happening. The lesson here is that we can never feel complacent that all issues are under control – anywhere in the world. New issues arise and we have to have the flexibility to respond and strengthen programmes as needed. As we have said already, some of these recent issues arguably could have been prevented with stronger implementation of prerequisite and HACCP programmes (chocolate and peanut butter), but some perhaps required new research into control mechanisms (salad crops).

Problems can be seen as opportunities. The US FDA developed HACCP-based regulations for canned foods in 1973 as a result of food safety failures. India set up the spice board as a result of Sudan and Para Red contamination. The UK now probably has one of the safest meat processing operations in the world as a result of bovine spongiform encephalopathy (BSE), and the Chinese government is working hard to rebuild the country's food safety systems and to deal with the economic adulteration issues that were all too visible to the western world.

It should be easy to follow best practice standards and to act with integrity, always seeking to do better and learn from others. But it is not easy. Through the lack of global coordination (e.g. of standards, regulations, training and test protocols), we make it difficult.

4 The future of food safety and HACCP in a changing world

4.1 INTRODUCTION

Change is the new norm and change is all around us – in technological advances, in globalisation, in economic and environmental conditions, and in the expectations of the new generation coming into the workforce. In addition, many social changes have impacted the food industry, for example:

- Changing lifestyles – more people eating out than cooking at home, leading to an increase in food service establishments and a decrease in domestic cooking and in-home food preparation skills and knowledge.
- More women working outside the home and an increased reliance on convenience foods – producing a further decrease in food preparation skills.
- Increased mass production of foods and globalisation of the supply chain means that more people can be affected if there is a food safety failure. Such failures are exacerbated when communications are hindered by difficulties in tracing products in distribution.
- Increased travel and tourism means that people are exposed to foodborne hazards from many countries. It also drives the consumer desire for more global food choice, which is a driver for globalised food supply chains.
- Evolution of consumer eating patterns – the demand for shorter ingredients lists, and fresh refrigerated foods with shorter shelf life.
- Attempts to improve public health, for example through reduction of salt in processed foods, reduction fat products and removal of trans fats.
- Ageing populations in many countries, together with the increase in the types and numbers of foodborne pathogens, mean that a higher number of people are susceptible to foodborne illness.

Against this backdrop and with continued high numbers of foodborne illnesses from both developed and developing countries, it is clear that something has to change. We cannot continue to operate as we have been and expect a better result – if anything the situation will get worse. And yet, we are asking the same questions now as we were 10 years ago. They were the right questions then, so *why has nothing changed?*

Food Safety for the 21st Century, First Edition By Carol A. Wallace, William H. Sperber and Sara E. Mortimore
© 2011 Carol A. Wallace, William H. Sperber and Sara E. Mortimore

The European Commissioner for Health and Consumer Protection said in 2001 (Byrne, 2001) that consumers fundamentally expected safe food. This is a basic human right. The events leading up to his speech had included bovine spongiform encephalopathy (BSE), dioxin, contaminated olive oil, and then foot and mouth disease. Consumers were shocked, but the complexity of modern food production methods became more transparent to them as a result.

This chapter looks at some possible ways forward but begins by briefly cataloguing some additional potential changes in food safety and technology that need to be taken into account.

4.2 FOOD SAFETY ISSUES

4.2.1 Emerging pathogens

Food safety professionals in the early 1960s were confronted with the need to control only four recognised foodborne pathogens: *Clostridium botulinum*, *Clostridium perfringens*, *Salmonella* and *Staphylococcus aureus*. Fifty years later, we can easily compile a list of about 20 microbial pathogens that include bacteria such as *Escherichia coli* O157:H7, *Listeria monocytogenes*, and *Vibrio vulnificus*; parasites such as *Cryptosporidium* and *Cyclospora*; viruses such as hepatitis A, and even prions associated with BSE.

One might think that this fourfold expansion of microbial foodborne hazards would signal that we have finally identified all the pathogens that will require specific control measures. That is not likely. Epidemiologists suggest that we can expect many additional foodborne pathogens in the future (Woolhouse and Gowtage-Sequeria, 2005). They have identified 1407 species of human pathogens, of which 816 are zoonotic. About 130 species are classified as emerging or re-emerging bacteria or viruses. The ability of microorganisms to rapidly mutate or adapt to changing conditions ensures that some of these will inevitably enter the food supply chain. We will need to remain vigilant to establish effective control measures as new foodborne pathogens emerge.

4.2.2 Changes in distribution of pathogens

The ability of pathogens to move globally due to increased trade and travel has been proved. It is now possible for food materials, people, animals and pathogens to move around the entire world in just 1 day (Osterholm, 2006).

The range of pathogens and vectors is expanding as a consequence of climate change and changing ecosystems. Many social factors contribute to the spread of pathogens and complicate our efforts to provide safe food and protect the public health. These include poverty, war, famine, social inequalities, inadequate political structures and the threat of bioterrorism.

4.2.3 Additional control measures

Control measures that we have in place for pathogens that we are aware of today may not be effective against emerging pathogens. This may necessitate changes in food consumption habits if adequate control measures cannot be developed. For example, the global distribution of fresh produce provides many opportunities for pathogen contamination and human illnesses. It may be necessary for immunocompromised people, for example, to consume only cooked produce to avoid a serious potential illness.

4.2.4 Antibiotic-resistant pathogens

A consumer concern in milk and animal production is that the use of antibiotics as growth promoters or to treat animal illnesses may lead to a rise in foodborne illnesses caused by antibiotic-resistant pathogens.

4.2.5 Allergens

As discussed earlier, what is considered an allergen does vary slightly from country to country. There are eight allergens common through Codex recommendations, but with several local variations (see Table 3.8 in Chapter 3). At the time of this writing, some countries have regulations on allergen labelling and some do not.

It is also anticipated that, in addition to any current local allergen patterns, similar allergen problems to those seen in the West may emerge in other parts of the world as western diets become more prevalent. Overall, there needs to be greater awareness of allergen control measures. The good news may come from the research that is underway regarding treatments. These centre on desensitising therapies and appear to give grounds for hope for the parents of children who have nut allergies. There is also ongoing research to better understand threshold levels.

4.2.6 Other chemical hazards

Melamine and its analogues such as cyanuric acid were never included in a Hazard Analysis and Critical Control Point (HACCP) plan before 2008. Hopefully, there will be no repeat of that particular issue, but there will no doubt be other chemical hazards that emerge as a concern. It is not possible to predict such hazards when they occur as the result of deliberate economic adulteration; however, it is essential that processors remain up-to-date with knowledge on chemical hazards and amend their control systems accordingly.

4.2.7 Physical hazards

In the past 10 years, advances in technology have brought improvements in physical hazard control – improved metal detector capability, x-ray detection and vision sorters. Improvements will continue and businesses should be on the alert for new improved equipments. It is hoped that there will also be advances in other areas such as sifter screens which do not break up – or have an alarm to alert the operator if they do.

With increased understanding of new or changing foodborne hazards, there may be opportunities to develop new control measures.

4.3 TECHNOLOGY ADVANCEMENTS

Examples of these are as follows:

- Robotics has given us the opportunity to reduce one of our greatest hazards from food manufacturing plants, i.e. people and the manual handling of food. We may also see developments in robotics for sanitation, thus eliminating the risk that comes when someone is wielding a pressure hose.
- The increased use of newer technologies for process control, e.g. pulsed electric fields and ultra-high pressure treatments.

- Electronic, real-time and continuous critical control point (CCP) monitoring with verification by human monitors.
- Laboratory detection equipment for chemical analysis – advances are being made all the time, which means that we will have choices to make regarding minimum detectable levels versus safe levels. Often, these decisions will be made on the basis of regulatory or legal, rather than on the basis of technical or public health, considerations.
- Rapid methods for microbiological testing – improvements in reducing detection time will be a huge help to the industry and public health laboratories.
- Communication systems – rapid communication capability exists today and is used during incidents to some effect. However, there must be opportunities to harness this for more proactive means such as best practice information sharing during normal circumstances and better coordination of information flow and consistency of messages when there is a crisis.

4.4   FOOD SAFETY MANAGEMENT

4.4.1   HACCP preliminary steps and principles

As we saw in Chapter 1, there were only three HACCP principles when Pillsbury first started using the system, but now there are seven. It is reasonable to think that there may be additional changes to come, either in the principles themselves, in the guidelines for their application, or by the addition of other food safety management tools. The following are our suggestions of where additional principles and guidelines could strengthen the HACCP system and maintain its consistent use throughout the world.

The need for a HACCP principle on validation

The way validation appears in the Codex document currently is in the guidelines on verification activities. This is somewhat confusing and almost certainly one of the reasons why validation sometimes gets missed as a requirement by those new to the concept. Validation could usefully be established as a principle in its own right. It currently plays a role in Principle 3 in the validation of critical limits and in Principle 6 to verify the accuracy of the HACCP plan (Sperber, 1999).

The need for a HACCP principle on maintenance

This would do more for ensuring that HACCP is kept up-to-date than any amount of training. History has shown the need to call this out as a basic principle of ensuring continued food safety. The need to *constantly* be on the alert for any change that could lead to a change to the HACCP plan is often lacking. Changes in plant layout, equipments, new ingredients, processes, emerging hazard data and industry issues – anything that is different to the original basis on which the plan was built – should trigger a maintenance activity.

The need for additional guidance on HACCP principle 1 – hazard analysis

This is a really important element of HACCP and yet is the area that new users struggle with the most. Practical guidance is needed not only on how to identify hazards but also on how to do a qualitative risk evaluation in terms of assessing the likelihood of occurrence and the severity of effect. Plant-based personnel, in both small and large companies, often lack access to the necessary information and competence to be able to do this properly (Wallace, 2009).

A documented risk evaluation is now a requirement of many of the third-party audit schemes, including ISO 22000. Overall, this may be helpful as in the past many companies did not formally discuss and certainly did not document their decision-making processes.

The need for additional guidance on implementation activities

Implementation of the results of the HACCP study as documented in the HACCP plan needs to be emphasised and guidance should be provided. Most companies do understand that once a CCP is identified, they have to manage it, but guidance on implementation would be helpful to drive a consistent approach. It could include the need for training, possible additional equipment, and documentation such as forms and record-keeping.

The need for additional guidance on the relationship with HACCP and need for prerequisite programmes

Since food safety can only be assured by proper implementation of both HACCP and prerequisite programmes, it seems reasonable to include the latter as a formal part of the HACCP system. Guidance is needed not so much as to what PRPs are, because that is covered in Codex General Principles of Food Hygiene (Codex, 2009), but clarification on how they enable the HACCP system to focus on the specific critical areas of control for food safety, i.e. those which manage significant hazards. A PRP decision tree such as that proposed by Campden BRI (Gaze, 2009) could be introduced into Codex guidelines (see Chapter 10; Figure 10.9).

The need for additional guidance on training

Training is such an essential component of all food safety programmes that it may well warrant inclusion as a HACCP principle in its own right. Training is currently in the guidelines but in general terms. In the future this could be enhanced to be more specific towards both training and education for the actual use of the principles, including education on hazards, how to carry out risk evaluation, training requirements for the HACCP team versus the CCP monitors and their supervisors and training for ongoing maintenance of the system. The requirement for validation and verification of the training could also be included as a requirement. Guidance on knowledge needs by HACCP principle, i.e. a standard curriculum, could dramatically aid uniformity of application in the future.

These are just a few ideas for consideration, and there will no doubt be additional thoughts and comments from other experienced practitioners. Further discussion is needed at the level of the Codex Committee on Food Hygiene to gain international recognition and acceptance. The implementation of additional preliminary steps, HACCP principles and guidance will strengthen HACCP as a food safety management tool. More importantly, it will strengthen the fundamental understanding that food safety cannot be assured by HACCP alone but is more effectively assured by the combined use of HACCP and prerequisite programmes.

4.4.2 Additions to current prerequisite programmes (Codex Principles of Food Hygiene)

- *Allergens* are included in most industry PRP documents but do not get mentioned in Codex Hygiene Principles (2003). They should be included at the next revision.
- *Training* is included in the Codex document but could be strengthened not least by adding that supervisors of food handlers should have more in-depth training to enable them to

reinforce and enforce appropriate hygienic behaviours. HACCP training has already been mentioned, but including the requirement for HACCP training (HACCP team, CCP monitors and HACCP auditors) within the Food Hygiene document would be an additional way of ensuring closer ties between HACCP and PRPs.

4.4.3   Global food safety assurance

In reviewing the numerous incidents that have occurred over the past few years, it appears that the future involves a back-to-basics approach. We have not yet mastered the basics of food safety. We have learned a lot in terms of scientific advances, but in some ways all that has done is to emphasise how much further we have to go to get control of our food supply. Consumers have a right to safe food regardless of who makes it – small producer, large producer, in whichever country – it should not matter. Consumers expect their governments to deliver on this. When we take a look at root cause, there are a number of themes:

- The human factor
- Oversight and harmonisation
- Enforcement

Despite calls for a new intergovernmental agency to coordinate on standards, we are putting people at the top of the list.

The human factor

Food safety events occur at manufacturing or food service establishments. They do not occur in the following:

- Government offices (though failure root cause may include inadequate regulation and enforcement)
- Corporate offices (though failure root cause may include inadequate policies, deployment and culture)
- Customer offices (though failure root cause may include inadequate supplier requirements and specifications)

Every scenario and every decision cannot be anticipated and documented ahead of time. We need trained, educated people *close to the action* – people who understand concepts and can think on their own, people with the knowledge to do things right and the integrity to do the right thing, and people who are held accountable and who take responsibility.

Just because global standards exist, it does not mean they will be followed, so we have to think bigger. The 2009 peanut butter incident in the United States, in which over 700 people were made ill and 9 people died, is a great example. The company is reported (USA Today, 2009) to have cleaned up just prior to their third-party customer audit and reverted to their normal (dirty) state after. This is about both trust and culture. Trust does not exist if there is a 'they/them' versus 'us' mentality. It has to be about 'we'.

Despite depressing headlines, the calls for action seem to have stayed the same over the past couple of decades. What has changed, and what is perhaps the real hope, is the recognition that food safety is very much a human as well as microbiological problem (Griffiths, 2009). Cross-contamination is still one of the major root causes of most food safety incidents. Getting consumers as well as food handlers to behave hygienically and implement food safety practices is crucially important. It needs to start in schools (which, for example, in the UK it does – see http://curriculum.qcda.gov.uk), and be a part of a global public education ongoing initiative, but with the goal that food safety is one of the life skills that is just 'known' – like crossing the road safely. This places major emphasis on effective training if this type of behaviour change is to be successful and it will not be the food scientists who lead the charge, or if they do then they will certainly not be on their own – the behavioural scientists and human resource professionals will be right beside them.

Training, education and leadership. Most authors of HACCP texts, including ourselves, highlight the need for training (Mortimore and Wallace, 1994, 1998, 2001; Scott and Stevenson, 2006; Gaze, 2009). For many other authors, however, training gets a mention but in almost a 'check the box' manner. In a couple of recent articles (Motarjemi and Gorris, 2009), the authors raise the topic to another level in which the recognition of the critical role the human factor plays in food safety assurance is finally gaining momentum. Analysis of outbreaks of foodborne illness shows that food safety incidents can often be prevented. That is why we implement HACCP and PRPs. However, unless people have the knowledge enabling them to have the belief and understanding necessary to want to work hygienically and follow procedures, then there will always be a higher probability of failure.

The quality of training can be improved. Many industrial consultant trainers have limited knowledge of the theory of adult learning and even less concerning human psychology. Some may also have limited grasp of their food safety topic. Many trainers do have the facts but often fail to make the conceptual leap between presentation of factual information and being able to use that information for risk-based decision-making (Mortimore and Smith, 1998). There is also variable awareness that, like any PRP, training needs to be both validated and verified (Wallace and Williams 2001). There is still a shortage of good-quality global resources for training and education. It is important to make the distinction between the two as both are needed:

Training: Aims to provide specific task-related skills and is often practical. Objectives are usually expressed in behavioural terms.

Education: Aims to provide theoretical and conceptual material which stimulates the learners' critical and analytical faculties.

Topics can cross over between the two, for example:

- HACCP training for operators as part of giving them the skills they need to monitor CCPs.
- HACCP training for HACCP team members as a way of thinking and working within the overall food safety hazard management programme. This would include an educational element.

We need a food safety human resource strategy for the industry that utilises both training and education and, as added value, strives to develop leadership across all levels. Figure 4.1 illustrates how the industry needs a mix of both in order to develop a trained and educated workforce. Higher education can fill the gap with up-to-date technical knowledge and thinking. Industry personnel, at all levels in the organisation, should be encouraged to take advantage of this and to continually improve skills and knowledge through both industrial experiences and academic input. The Global Initiative on Food Systems Leadership has been mentioned already (Chapter 3; Section 3.4.4) as one example.

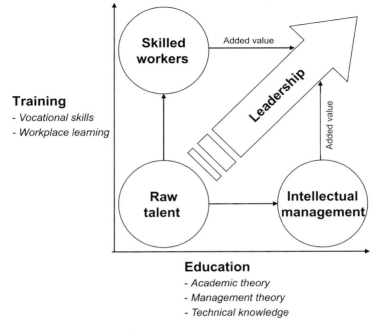

Fig. 4.1   Trained and educated workforce.

This must be a coordinated approach with oversight across countries. The Industry Council for Development's (http://www.icd-online.org) mission has been to partner with intergovernmental organisations and to jointly contribute to public health through training, education and building knowledge capacity in both food safety and nutrition. But it is not enough. Training and education need to be raised to a higher level, recognised as an integral part of food safety management and not a 'box that has to be checked'.

Availability of food safety professionals. The food industry is one of the biggest employers globally. For example, in the UK, there are an estimated 650 000 employees in 7300 companies (2006). Yet, there is a shortage of higher level skilled workers as well as blue-collar workers coming into the industry. Student interest in science-based subjects has decreased significantly and the food industry does not exactly have a glamorous image when compared with some of the newer and more fashionable professions, such as those related to the media, biotechnology or environmental sustainability. This is a challenge for the future of the food industry, and where, again, a coordinated effort across borders may be needed.

Knowledge resources. Linked to above areas, there is a need now and in the future for a trustworthy knowledge base. Many books, internet sites and academic papers exist and many are excellent, but some are not peer-reviewed or written by acknowledged experts. It is difficult for those still learning to differentiate between those that are good and those that may mislead and confuse. In addition to this, global food corporations are the major repositories of information about food safety hazards and their means of control but may treat it as proprietary knowledge. We need to consider how to make more of their knowledge available publicly.

Oversight and harmonisation

Improving food safety oversight and harmonisation depends on an accurate recognition of the roles and responsibilities of the principal stakeholders in their efforts to ensure and verify food safety. Without such recognition of distinct responsibilities, individual parties may work at cross-purposes. The resulting confusion or ignorance may lead to a lack of adequate oversight, as demonstrated by the 2009 issues with *Salmonella* in peanut products.

Assuring food safety is primarily the responsibility of the food industry in the broadest sense and not just the manufacturers. The industry is responsible for the safety of food materials throughout the supply chain. Food corporations have most of the knowledge and expertise necessary to identify and control foodborne hazards. In its processing facilities, the food industry is responsible to implement and maintain adequate HACCP and prerequisite programmes, a responsibility that is literally borne 24 hours/day. The primary role of governmental bodies is to verify that food safety practices in the industry are acceptable and in place. This responsibility is most effectively discharged through the audits of food processors' documentation of its food safety programmes. The effectiveness of current regulatory inspections has been challenged. The UK has a recent example of this, where a number of children fell ill and one boy died as a result of *E. coli* O157:H7 contaminated food in South Wales (Pennington, 2009). Governmental agencies can develop guidance documents, lead risk assessments of newly detected hazards, and promulgate reasonable and practical rules and regulations when necessary.

Accepted food safety practices vary – between countries and within countries. It is really driven home by seeing some of the Global Food Safety Initiative (GFSI) benchmarked standard owners try to find a way to reach a common understanding with regards to expectations. Codex establishes the principles, but interpretation of these varies. Interestingly, HACCP interpretation varies a lot less than PRPs, possibly because it is a much narrower subject, but problems remain despite that. The time could be right for the creation of a new organisation that could provide coordination of the various efforts, provide oversight and provide guidance to governments regarding enforcement. As more and more countries regulate in the area of food safety, the differences between them is likely to increase. It is essential that the food industry and governmental bodies collaborate more effectively to fulfil their individual responsibilities for food safety.

Multinational food companies have worldwide operations that enable them to fulfil their individual responsibilities for food safety. They have developed the HACCP system and spread it worldwide within their supply chains. They are ready to participate in a more effective and consistent global food safety programme that will require them to continually challenge their own systems as well as sharing non-competitive knowledge with others in the supply chain. In particular, better practices must be established and maintained at the myriad points of origination of food commodities and ingredients that enter global food commerce.

There is however no comparable array of individual regulatory and health or intergovernmental agencies with the global connections and authority to support necessary improvements in global food protection. Much of the global influence for food safety resides in the United Nations (UN) organisations. The Food and Agriculture Organization (FAO) is principally involved in food security and some elements of food safety, while the World Health Organization (WHO) is principally involved in public health. Both are supported in their missions by the Codex Alimentarius Commission (CAC), which develops food standards, guidelines and codes of practice to assist in protecting public health and ensuring fair trade. Likewise, the World Organization for Animal Health (OIE) is responsible for animal health and food safety for products of animal origin.

No single organisation, however, has the necessary accountability, scope or focus to enhance and ensure global food protection – a situation further complicated by the complex setting of critical public health, animal health and environmental protection issues in which it must operate. Furthermore, current trends in population growth, climate change and resource availability will make it even more difficult to protect our food supply. The necessary leadership to make progress, in spite of these trends, can be provided by a new global food industry–intergovernmental collaboration.

It has been proposed that a new intergovernmental organisation be created to complement and expand the industry food protection efforts (Sperber, 1998). It could be placed within the UN parallel to WHO and FAO, supported by CAC and OIE, with its sole emphasis on food protection. Named the 'Food Protection Organization (FPO),' its programme would include the many elements necessary to ensure food protection, including, but not limited to, the following:

- Promoting global understanding, implementation and verification of enhanced food protection measures.
- Establishing incentives for intergovernmental collaboration.
- Providing farm-to-table coverage with a focus on points of origination.
- Requiring the use of HACCP and prerequisite programmes.
- Advancing uniform audit procedures with industry collaboration.
- Establishing traceability systems for food ingredients and products.

The FPO could be considerably more proactive and cost-efficient than the current fragmented system, ultimately enabling the effective use of all applicable resources to enhance food protection for all consumers. Standards set by this organisation must be outcome-based to ensure a degree of flexibility and foster development of new best practices. It needs to provide guidance that has scientific foundation but is risk-based and practical.

The industry, government, academia, non-governmental organisations and consumers need to collaborate openly. Leadership can come from any of these areas (HACCP and GFSI came from industry), but it must inspire, model (the future) and coach in order to combat cultural resistance and the 'it won't work here' attitudes that abound. Globalisation is a reality, and that will not change. We need a way to communicate and collaborate across borders.

There may not be a need for new standards. Plenty of them already exist (Figure 4.2), but the necessary coordination does not.

Enforcement

In the United States, it is no secret that, the FDA has had insufficient resources for enforcement in the field. Other countries have their own particular challenges. Here are a few thoughts on this as we consider the future of food safety and HACCP:

- There is always a benefit in having someone else take a look at your systems providing that they are competent. This can:
  - provide a learning opportunity,
  - identify areas for improvement,
  - enable sharing of best/different practices, and
  - validate that existing practices and programmes are appropriate.

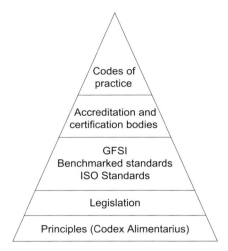

Fig. 4.2   The relationship of standards. (Adapted from Robach, April 2009.)

- A partnership between government enforcement agencies, corporate auditors or customer auditors (all have an enforcement role) is far more effective than a police mentality.
- An effective and efficient future might look like this:
  ○ Third-party auditing across *all* businesses by highly skilled experienced (in the food sector) auditors.
  ○ Carried out at the request of the auditee (company being audited).
  ○ Shared widely with anyone who wants to see the report (transparent).
  ○ Audits and inspections should always be unannounced, i.e. should be a true reflection of the day-to-day condition of the company.

The *benefits* of this new style third-party auditing are worth having. It would negate the need for the masses of customer-supplier audits that still occur each year, which will allow suppliers and customers to focus on more specific issues relating to their business transactions (product-specific food safety, sensory, functionality, productivity initiatives). However, third-party audit firms must work from the common standards (perhaps GFSI) and be overseen by the national accreditation bodies who in turn have some additional (not in place today) oversight and accountability.

Combined approach

Often heard at the plant level is the request for a combined review of programmes – food safety and quality, environmental health and safety, and now a plea to include sustainability and corporate social responsibility. There are benefits in terms of ensuring that all managers are on site during the audits, but the downside is dilution of effect – inability to review in detail and the impact is not as great when reviewing opportunity areas for each programme. That said, there is a certain amount of overlap and, of course, competing resources. Certainly, at the governmental level, having oversight of issues such as the following could be beneficial:

- Public health
- Animal health

- Environmental sustainability
- Food safety
- Food security
- Food defence

Combining this list with concerns over the following list (which are discussed in more detail in Chapter 3), it becomes clear that a more collaborative approach between the food industry, academia, national, non-governmental and intergovernmental organisations will be needed to improve food safety and public health protection:

- Increasing human population. While already at 7 billion people, policy makers seem to calmly accept that the population will increase another 2–3 billion in the next 40 years. How can we feed 9–10 billion people, and what exactly is expected to happen after 2050?
- Increase in the number of immunocompromised people. It is estimated that about 20% of the human population is immunocompromised by age, pregnancy (in which case the fetus is at greater risk than the mother) or chronic illness. This proportion is expected to increase as human life expectancy increases (Gerba *et al.*, 1996).
- Pressure to use land for biofuel production, which does not require the same food safety standards and reduces flexibility of use if non-food-grade grain is produced.
- Decreasing availability of water, arable land and fossil energy sources.
- Climate change is reducing crop production by droughts and by rising sea levels that will flood coastal crop land.
- Food demand is increased by the improving economic status in developing countries.
- Political instability in developing countries and political inaction in developed countries are major obstacles to progress in maintaining an adequate and safe food supply.
- Changes in technology and agricultural practices to improve yields.
- Loss or lack of knowledge.

4.5 CHANGES IN THINKING/POLICY MAKING

4.5.1 Food safety objectives

Not exactly new thinking but not yet widely used in industry, food safety objectives (FSOs) can be used by government (or industry) to communicate food safety targets (ICMSF, 2005). FSOs have been defined as:

> Statements based on a risk analysis process which includes an expression of the level of a hazard in food that is tolerable in relation to an appropriate level of consumer protection. When justified by the risk assessment, the FSO should include expression of the level of the hazard as a maximum tolerable concentration and/or frequency (Hathaway, 1999).

Thus, they form a bridge between the application of control measures (i.e. HACCP and prerequisite programmes) and the acceptable level of risk to the consumer population.

FSOs are distinct levels of foodborne hazards, often microbiological, that cannot be exceeded, but are different to setting microbiological criteria which are used for acceptance and rejection. FSOs can be used for measuring improvement of food safety controls. They can be set at the point of consumption or higher up in the supply chain. The principle is that HACCP and PRPs are outcome-based and FSOs are a metric that can be used to assess success. In doing it this way, different control measures, processing technologies, formulations, etc., can be compared for their ability to meet or exceed the objective.

4.5.2 End product testing

As pointed out in Chapter 1, the food processing industry after World War II relied heavily on finished product testing to verify food safety. However, product testing is not practical to detect public health hazards that occur at a low incidence. This dilemma led to the emergence of the HACCP system in the 1960s. Even though companies using HACCP do not usually need to conduct product testing and rely on lot acceptance criteria, many customers still require a wide variety of microbiological tests to be performed on finished products. This is a logical disconnect that seems to have its origins in the 1960s publications by the International Commission on Microbiological Specifications for Foods (ICMSF, 1974) and the National Academy of Sciences (National Academy of Sciences, 1960; Food Protection Committee, 1964; Foster, 1971).

These publications detail hazard and risk categories, sampling plans and microbiological specifications. Together, these can lead to extensive and expensive programmes for product testing. It must be pointed out that these programmes were developed for the analysis of materials of unknown origin and unknown means of control, to be applied anywhere in the global supply chain. They are also necessary to investigate acute problems, such as *Salmonella* in dried eggs in the 1960s or *Salmonella* in nuts in 2009. The logical disconnect that carries over to this day is that many food processors and their customers expect that the same procedures should be applied to all food production, the vast majority of which is of known origin and produced under adequate controls. Such products do not require stringent testing, and in many cases, any testing at all.

Food suppliers and processors operating under normal controlled conditions are beginning to use microbiological monitoring guidelines rather than microbiological specifications and lot acceptance criteria to verify the safety and quality of their products. Periodic monitoring of the production environment and product build-ups, using indicator tests such as aerobic plate counts and mould counts, can suffice to verify compliance with sanitation and HACCP requirements. The monitoring results can be shared with customers. Finished products should not need to be tested (Sperber and NAMA, 2007). A substantial effort is still required to educate the food industry and other stakeholders to reduce or eliminate the use of unwarranted microbiological specifications and testing requirements.

4.5.3 Hazard analysis versus risk assessment

The processes of hazard analysis and risk assessment share a common step – hazard identification. That common step has confused some food professionals to think that hazard analysis and risk assessment are essentially identical, to the point that the risk assessment process could replace hazard analysis in the formation of a HACCP plan (Sperber, 2001). We want to strongly dispel this misguided notion as it has the potential to confuse and undermine the established success and practicality of HACCP systems.

Hazard analysis in Codex terms is the process of identifying hazards at each process step together with determination of which are of such significance that they need to be controlled. It also includes the consideration of effective control measures. It is a qualitative, local process conducted by the facility HACCP team, over a period of several weeks or months. Because of the unique nature of manufacturing facilities, ingredient supply chains and product formulations, each food processing facility will have a unique HACCP plan. Qualitative databases such as sensitive ingredient lists for bacterial pathogens, mycotoxins, allergens and foreign materials are useful in this process.

Brainstorming by the multidisciplinary HACCP team is an important means to identify potential hazards associated with new products, processes, facilities, markets and regulations, thus allowing their evaluation. In the industrial setting (manufacturing plant or food service operations), many third-party audit standards and organisations are using the term 'risk assessment' to prompt an evaluation of likelihood and severity during hazard analysis. Also, it is used extensively for the evaluation of appropriate controls within a formalised PRP, and these are very good things to be doing.

However, the term 'risk assessment' has (perhaps unfortunately) been hijacked by Codex (1999) which, in stark contrast to hazard analysis, describes a formal risk assessment, which is a quantitative, global process by which a particular hazard can be numerically quantified for a particular food category. Risk assessments, requiring several months or years to complete, are conducted by a large, sometimes global group that includes food safety experts and risk assessors from academic, industrial, public health and regulatory entities. Prominent examples of risk assessments include *L. monocytogenes* in ready-to-eat foods (FAO/WHO, 2002) and *E. coli* O157:H7 in ground beef (FSIS, 2001). One such risk assessment can serve to guide the deliberations of countless HACCP teams around the world for many years.

Hazard analysis and risk assessment in this sense are distinctly different processes. Each should be valued and used for its distinctive purpose.

4.6   CONCLUSIONS

The future of food safety and HACCP could be very exciting. It could also be bleak and disappointing if we do not have the courage to take some bold steps. We need to address the urgent need for *knowledge and leadership* across the global food industry (every single person), we need to continue the work aimed at having *common standards* and science-based regulations, and we need a global *infrastructure* to provide global strategy and oversight.

We cannot let another decade go by and *still* be asking the same questions and calling for change.

Part Two
Foodborne Hazards and Their Control

5 Recognising food safety hazards

5.1 INTRODUCTION

A great many hazards of different types may enter the food supply, making the food potentially harmful when consumed. Product development teams, food safety managers, and the Hazard Analysis and Critical Control Point (HACCP) teams must be aware of these hazards when developing products and processes and when conducting hazard analyses so that proper control measures can be established as necessary.

5.1.1 What is a food safety hazard?

A foodborne hazard is 'a biological, chemical, or physical agent in, or condition of food with the potential to cause an adverse health effect' (Codex, 2003). The definition is focused very sharply on food safety considerations. This chapter describes and explains foodborne hazards according to this globally accepted Codex definition.

Biological hazards include pathogenic bacteria, fungi, viruses, prions, protozoans and helminthic parasites. Manifestations of these hazards typically involve foodborne illnesses with symptoms including gastrointestinal distress, diarrhoea, vomiting and sometimes death.

Chemical hazards include allergens, mycotoxins, heavy metals, pesticides and cleaning and sanitation chemicals. When ingested, these may cause gastrointestinal distress, organ damage and immunological reactions that may result in death. The long-term ingestion of foods containing toxic chemicals can lead to chronic effects, including cancer.

Physical hazards typically include materials that enter the food throughout its production chain, such as extraneous vegetable material, stones, bone fragments, wire pieces, broken glass and wood splinters. Their presence in food may result in choking, or oral or internal cuts, but rarely result in death.

5.1.2 What is not a food safety hazard?

Many types of quality and regulatory defects that occur during food processing are not considered to be food safety hazards because they would not produce an adverse health effect if such foods were consumed. Therefore, these defects are not identified as significant hazards during a hazard analysis, and they are not included in the HACCP plan. Rather, they are controlled by the use of prerequisite programmes, as described in Chapter 10, as well as specifications and quality control mechanisms. Spoilage microorganisms may produce flavour, odour and visual defects without making the food harmful for consumption. Souring of milk, putrefaction of meats, gassing of liquid products and the appearance of microbial colonies on the surface of foods are examples of quality defects. Regulatory defects occur when a food does not conform to the requirements of particular regulations. The presence of certain foreign materials, undeclared non-hazardous ingredients, extraneous vegetable material or otherwise mislabelled product containers may violate regulations without representing an overt health hazard.

5.2 BIOLOGICAL HAZARDS

5.2.1 Epidemiology and morbidity statistics

Epidemiological data are reported up to several years after the occurrences of illnesses and outbreaks. Therefore, it is somewhat difficult to estimate the current number of foodborne illnesses. Nonetheless, it is estimated that 1.8 million people died of diarrhoeal diseases in 2005, most of which could be attributed to contaminated food or water (WHO, 2007a). The US Centers for Disease Control and Prevention (CDC) estimated that the annual foodborne illness burden in the United States was responsible for 76 million illnesses, 325 000 hospitalisations and 5000 deaths. About 20% of the illnesses were caused by known pathogens, the remainder by unknown agents (Mead *et al.*, 1999). A great disparity exists between these estimates and surveillance data because many foodborne illnesses are mild and not reported to public health agencies. The CDC's Foodborne Diseases Active Surveillance Network (FoodNet) recorded 40 foodborne illnesses and infections per 100 000 population in 2005 (Table 5.1). Extrapolated to the entire US population, FoodNet would have reported 120 000 illnesses for that year, versus the 76 million illnesses estimated by CDC. Notifiable diseases not included in the FoodNet programme – botulism, trichinellosis, giardiasis and hepatitis A – account for additional 20 000 cases per year, many of which are food- or waterborne (Tables 5.2 and 5.3).

The Center for Science in the Public Interest compiled data on the US outbreaks and cases of foodborne illness as attributed to specific food categories for the period 1990–2003 (Table 5.4). Another compilation of illness outbreaks that implicated fresh produce covered the period

Table 5.1 The incidence of bacterial and parasitic infections and haemolytic uremic syndrome in 2005.

Pathogen/condition	Incidence/100 000 population
Campylobacter	12.71
Listeria	0.31
Salmonella	14.81
Shigella	6.09
STEC^a O157	1.31
STEC non-O157	0.46
HUS	1.63
Vibrio	0.34
Yersinia	0.35
Cryptosporidium	1.91
Cyclospora	0.09
Total	40.01

Source: FoodNet surveillance data (CDC, 2007a).
[a]Shiga Toxin-producing *E. coli*.

Table 5.2 Provisional cases of infrequently reported notifiable diseases, United States, 2002–2006, annualised.

Disease	Number of cases/year
Botulism	
Foodborne	21
Infant	83
Wound and unspecified	33
Listeriosis	*777*
Trichinellosis	11

Source: CDC (2007b).

Table 5.3 Provisional cases of selected notifiable diseases, United States, 2006.

Disease	Number of cases
Giardiasis	16 919
Hepatitis A	3 263
Salmonellosis	41 924
Shiga toxin-producing *E. coli*	3 199
Shigellosis	13 660

Source: CDC (2007c).

Table 5.4 Food categories and number of associated outbreaks and cases, 1990–2003.

Food category	Outbreaks	Cases
Beverages	66	2 643
Breads and bakery	116	3 493
Dairy	153	5 156
Eggs	329	10 849
Game	25	182
Multi-ingredient	812	23 126
Produce	554	28 315
Seafood	899	9 312
Beef	438	12 702
Luncheon meat	145	5 287
Pork	170	5 859
Poultry	476	14 729
Totals	4 183	121 653

Source: Smith de Waal *et al.* (2006).

Table 5.5   Fresh produce implicated in the US outbreaks of foodborne illness, – 1973–1997.

Produce item	Number of outbreaks
Mixed salad	76
Mixed fruit	22
Mixed vegetables	7
Lettuce	25
Melon	13
Seed sprouts	11
Apple or orange juice	11
Berry	9
Tomato	3
Green onion	3
Carrot	2
Eight other items	8

Source: Sivapalasingam *et al.* (2004).

1973–1997 (Table 5.5). Nearly half the outbreaks in the latter report were caused by unidentified aetiologic agents (Table 5.6).

The numbers and types of foodborne illnesses are not well defined in many of the developing regions of the world. Those regions and countries with modern surveillance and reporting systems, including Canada, Europe, Australia and New Zealand, report statistics similar to those cited earlier from the United States (Notermans and Borgdorff, 1997; Todd, 1997). In contrast, several island nations of the southeast Pacific report a much higher proportion of foodborne illnesses caused by *Vibrio* spp. (Su *et al.*, 2005; Azanza, 2006).

It has long been generally known that most foodborne illnesses occur as a result of food handling and food preparation errors in food service operations and in the consumers' homes. Epidemiological data from the OECD member states (Organization for Economic Cooperation

Table 5.6   Aetiologic agents implicated in the US outbreaks of produce-borne illness, 1973–1997.

Aetiologic agent	Number of outbreaks
Salmonella	30
Escherichia coli O157:H7	13
Non-O157 *E. coli*	2
Shigella	10
Campylobacter	4
Bacillus cereus	1
Yersinia enterocolitica	1
Staphylococcus aureus	1
Hepatitis A virus	12
Norovirus	9
Cyclospora cayetanensis	8
Giardia lamblia	5
Cryptosporidium parvum	3
Chemical or toxin	4
Unidentified	87
Total	190

Source: Sivapalasingam *et al.* (2004).

and Development; representing North America, most of Europe and four Asia–Pacific countries) document this long-established belief. In an examination of 7191 foodborne illness outbreaks in the period 1998–2001, it was learned that 42% occurred in food service establishments and 31% occurred in private homes. Only 3% of the outbreaks were attributed to food manufacturers or retailers. The remaining 24% were attributed to unknown or 'other' locations. The principal contributing factors to these illness outbreaks were determined to be (in descending order of frequency) time/temperature abuse, cross-contamination, improper storage, raw foods, infected persons and inadequate food-handling practices (Rocourt *et al.*, 2003).

Another review of foodborne illness outbreaks in the same geographical regions over the past 80 years revealed that 816 outbreaks could be attributed to errors by food handlers throughout the food service industry. The principal aetiological agents in the outbreaks were norovirus (41%), *Salmonella enterica* (19%) and hepatitis A virus (10%) (Greig *et al.*, 2007).

5.2.2 Characteristics of foodborne illnesses

Types of illness

Foodborne illnesses can be grouped according to the mechanism of pathogen–host interaction. *Intoxications* (earlier called 'food poisoning') occur when a pathogen produces toxin(s) while growing in a food. Upon consumption of a sufficient quantity of 'poisoned' food, the host becomes ill. Botulism and staphylococcal food poisoning are the best-known foodborne intoxications. *Infections* are caused when viable pathogens in a food survive passage through the host's stomach into the intestine. Enterotoxins may be produced in the intestine, often causing diarrhoea and/or vomiting. Some pathogens may invade the intestinal wall and cause septicaemia or meningitis. Salmonellosis and listeriosis typify these two infective mechanisms, respectively. *Opportunistic* pathogens sometimes cause foodborne *illness* when a host is exposed to very large numbers of a microbe that normally is incapable of causing illness. Often, the host is severely immunocompromised. *Cronobacter* (formerly *Enterobacter*) *sakazakii* is a recent example of an opportunistic pathogen. It is occasionally implicated in infections of prematurely born infants.

Predisposition to illness

Humans elicit a wide range of responses when exposed to foods that might be capable of causing illness. Healthy persons with a strong immune system often fend off infectious or toxic doses that would cause illness in less healthy individuals. It is well established that very young (<1 year old) or elderly (>70 years old) humans are more vulnerable to foodborne illnesses than are humans of intermediate age. Furthermore, humans of any age may be immunocompromised by pre-existing conditions such as autoimmune diseases, chronic illnesses or immune-suppressive drugs. Cancer, AIDS and organ transplant patients, for example, are highly immunocompromised. It is estimated that about 20% of the human population is immunocompromised by age, pregnancy (in which case the fetus is at greater risk than the mother) or chronic illness. This proportion is expected to increase as human life expectancy increases (Gerba *et al.*, 1996). This consideration illustrates the importance of making all foods safe for all people.

Infectious or toxic dose

The number of pathogenic microbes or quantity of a toxin required to cause illness depends on the concentration of pathogens or toxins in a particular food, the nature and amount of food

Table 5.7 Characteristics of common foodborne illnesses.

Microorganism	Incubation period (range)	Fever	Diarrhoea	Vomiting
		Symptoms		
Salmonella spp.	12 hours (6–48)	+	+	+
Staphylococcus aureus	2 hours (0.5–8)	−	+	+
Bacillus cereus[a]	1 hours (0.5–6)	−	−	+
Clostridium perfringens	12 hours (9–15)	−	+	−
Shigella spp.	24 hours (12–48)	+	+	−
Hepatitis A virus	28 days (15–50)	+	−	−

[a]Emetic toxin-producing strains.

consumed, the virulence of the pathogen or toxin, and the health status of the consumer. In some cases, less than 100 pathogenic cells are sufficient to cause illness. In other cases, more than 1 million pathogenic cells will not cause illness. In the case of foodborne intoxications, the mere presence of a toxigenic pathogen in a food is not sufficient to cause illness. The toxigenic pathogen must first grow in the food and produce sufficient toxin to cause illness. For example, for staphylococcal food poisoning to occur in a healthy adult, *Staphylococcus aureus* must grow to a minimum population of 3 million cells/g of food and the individual must consume about 100 g of the toxic food, the toxic dose being about 1 μg of enterotoxin.

Incubation period

The elapsed time between the consumption of an implicated food and the onset of illness symptoms is the incubation period. These are typically short (hours or days) for bacterial illnesses and range to weeks, months and even years for parasitic infections. Foodborne intoxications usually have shorter incubation periods because the toxin is preformed in the food. Growth of the pathogen in the host's intestinal tract is not required, as in the case of foodborne infections.

Illness symptoms

The most common foodborne illnesses typically cause fever, diarrhoea and/or vomiting. Considered with the incubation periods of the illnesses, these symptoms provide a quick estimation as to the type of foodborne illness (Table 5.7). The less common foodborne illnesses elicit many additional key symptoms (Table 5.8).

Under-reporting of foodborne illnesses

While it may appear simple to determine the cause of a foodborne illness from its incubation period and key symptoms, a great deal of effort is required to confirm the illness and report it

Table 5.8 Characteristics of less common foodborne illnesses.

Microorganism	Incubation period (range)	Key symptoms
Escherichia coli O157:H7	4 days (3–9)	Bloody diarrhoea, possible HUS
Listeria monocytogenes	7 days (3–21)	Septicaemia, meningitis
Clostridium botulinum	24 hours (12–40)	Double vision, difficulty swallowing, possible respiratory paralysis
Yersinia enterocolitica	24 hours (18–36)	Diarrhoea, vomiting, severe abdominal pain
Vibrio spp.	12 hours (4–30)	Fever, diarrhoea, vomiting

Table 5.9 Sources of bacterial pathogens involved in foodborne infections.

Pathogen	Natural habitat	Associated food sources
Salmonella spp.	Animal intestine	Raw meat, poultry and eggs
	Process environments	Multiple dry foods
Campylobacter jejuni	Animal intestine, soil	Raw poultry and milk
Escherichia coli O157:H7	Ruminant intestine	Raw beef, milk, water
		Farm animal contact
Listeria monocytogenes	Soil, animal intestine, moist processing areas	Raw milk and meat, soft cheeses, ready-to-eat deli products
Shigella spp.	Animal intestine	Freshly prepared foods
	Human carriers	Food service operations
Vibrio spp.	Marine water	Raw shellfish and seafood
Yersinia enterocolitica	Animal intestine, water	Raw milk, water

to public health offices. Most cases of foodborne illness are mild. The victims often recover in one or several days without seeking medical care. When severe cases require medical attention and even hospitalisation, proof of foodborne illness usually requires isolation and identification of the pathogen or toxin from the implicated food and/or from clinical specimens taken from the patient. It is estimated that less than 1% of foodborne illnesses are officially reported. Therefore, the 120 000 extrapolated cases of foodborne illness reported in the United States each year would represent more than 12 million actual cases of these particular illnesses (Table 5.1; CDC, 2007a).

Principal types and sources of foodborne pathogens

In much of this chapter, the authors present a great deal of information without tediously documenting every specific data point. The data are accumulated from our own research and experience in the food industry, covering nearly five decades, from several outstanding compilations devoted almost solely to the topic of foodborne pathogens (Doyle, 1989; CAST, 1994; ICMSF, 1996; Lund *et al.*, 2000), and from the indicated references.

Foodborne infections are often zoonotic diseases, meaning that the pathogen can infect both humans and one or more other species of animals. Therefore, the human illnesses are usually linked to raw animal products or the environments in which the animal was raised or processed (Table 5.9). Foodborne intoxications are less directly linked to an animal source. They are often linked to soil and other environmental sources (Table 5.10).

The ability of bacterial pathogens to cause foodborne illness most often depends on their ability to grow in the implicated food. The growth requirements of the principal foodborne pathogens as related to oxygen, temperature, pH and water activity of the food are summarised in Table 5.11.

Table 5.10 Sources of bacterial pathogens involved in foodborne intoxications.

Pathogen	Natural habitat	Associated food sources
Staphylococcus aureus	Human and animal skin	Raw meat and poultry
		Fermented sausage and cheese
Clostridium botulinum	Soil, water	Raw vegetables, fish
Bacillus cereus	Soil, root and cereal crops	Cooked rice and potatoes
Clostridium perfringens	Soil, animal intestines	Steam table meat and poultry
		Stuffed poultry

Table 5.11 Growth limits of foodborne bacterial pathogens.

Pathogen	Temperature (°C) Minimum	Temperature (°C) Maximum	pH Minimum	pH Maximum	Water activity Minimum
Salmonella spp.	6	46	4.0	9.5	0.94
Campylobacter jejuni	27	45	4.9	9.0	0.98
Escherichia coli O157:H7	8	43	4.4	9.0	0.95
Listeria monocytogenes	0	44	4.6	9.2	0.93
Shigella spp.	7	46	4.9	9.3	0.98
Vibrio parahaemolyticus	3	44	4.8	11.0	0.95
Yersinia enterocolitica	0	44	4.4	9.6	0.97
Staphylococcus aureus	7	48	4.2	9.0	0.85[a]
					0.92[b]
Clostridium botulinum					
Proteolytic	10	49	4.8	8.5	0.94
Non-proteolytic	3	45	5.0	8.5	0.97
Clostridium perfringens	15	50	5.0	8.9	0.95
Bacillus cereus	5	50	4.4	9.3	0.93

[a] Limit for aerobic growth.
[b] Limit for enterotoxin production.

Emerging pathogens

At the beginning of 'modern' food microbiology more than 60 years ago, there were just a handful of pathogens that commanded the attention of the food industry and public health agencies. Today, as evidenced by the remainder of this chapter, dozens of pathogens command the same attention. Epidemiologists predict that food safety and public health professionals will be confronted by an ever-increasing number of pathogens. An evaluation of 1407 human pathogens determined that 177 are currently 'emerging', most often from a zoonotic source (Table 5.12).

It has long been recognised that human pathogens may emerge from animal pathogens, the microflora of children and opportunistic infections of immunocompromised people (M.T. Osterholm, personal communication). The emergence of pathogens is accelerated today by the increased trade of food and raw materials throughout the global food supply chain (Chapter 3), the global trade of live animals, global human travel, microbial evolution, increased susceptibility of the human population, and by the consumer demands for more convenient foods, thereby increasing exposure to food handlers.

Table 5.12 Numbers of emerging and re-emerging human pathogens.

Pathogen group	Known pathogens	Emerging pathogens
Bacteria	538	54
Fungi	317	22
Viruses and prions	208	77
Helminths	287	10
Protozoa	57	14
Total	1407	177

Source: Woolhouse and Gowtage-Sequeria (2005).

The matter of emerging pathogens cannot be taken lightly. Just in the past 25 years, a number of important foodborne pathogens have emerged, including *Listeria monocytogenes*, *Escherichia coli* O157:H7, *Cyclospora cayetenensis* and variant Creutzfeld–Jakob disease.

5.2.3 Bacterial pathogens – special considerations and features

Spore-forming bacterial pathogens

Several genera of bacteria, particularly *Clostridium* and *Bacillus*, are able to produce endospores that are resistant to physical and chemical factors such as heat, dehydration, and acidification. They can survive in the environment for a very long time. Upon introduction into a food, the spores may germinate, enabling growth and/or toxin production by the pathogen. In addition, spores can survive heat treatments, grow and produce toxin in a suitable food (Chapter 6).

Clostridium botulinum. This obligately anaerobic (i.e. cannot grow in the presence of oxygen) organism consists of seven serotypes, from type A through type G, with one or more subtypes. All strains of type A and most strains of type B are proteolytic and mesophilic, and produce very heat-resistant spores. The remaining strains of type B, along with types E, F, and G, are non-proteolytic and psychrotrophic, and produce spores with little heat resistance. Types C and D are pathogenic for animals, but not for humans. Botulinal neurotoxin is the most potent known biological toxin. Less than one ng/kg body weight is sufficient to kill a human (Schecter and Arnon, 2000). Death is caused by paralysis of the respiratory muscles. Botulinal toxins are quite heat-labile and are easily inactivated by pasteurisation and cooking procedures (Sperber, 1982). Infant botulism has been recognised as one cause of sudden infant death syndrome. Originally believed to be caused by *C. botulinum* spore contamination of honey and other sweeteners used in infant formula, it is now thought that many cases of infant botulism are caused by ingestion of household dirt and dust (Nevas *et al.*, 2005).

Clostridium perfringens. This anaerobic organism, usually found in animal intestines and soil, is commonly associated with hot-held (e.g. steam table/hot buffet) foods such as meats and gravies. It is slightly thermophilic and grows very rapidly without sporulation in the food. After consumption, the cells sporulate in the human small intestine, liberating an enterotoxin during lysis. Because a high number of cells must be ingested to cause the later infection, this illness is known as a toxico-infection. Diarrhoea is the principal symptom of the mild illness caused by *C. perfringens*. The same organism is often involved in wound infections and gas gangrene.

Bacillus cereus. Strains of this facultative pathogen can produce one of two general types of toxins. Heat-stable emetic toxin is produced by strains that grow in farinaceous foods such as cooked rice and potatoes. Illnesses associated with these foods can be avoided by adequate refrigeration of the cooked foods. Heat-labile enterotoxin is produced by other strains, some of which are psychrotrophic, that grow in proteinaceous foods.

Non-spore-forming bacterial pathogens

While illnesses caused by non-spore-forming bacterial pathogens usually result from the ingestion of large numbers of cells, several notable exceptions are described below. This group of pathogens is easily inactivated by heat, yet several of its members are hardy survivors in food processing environments, and one member produces a very heat-stable enterotoxin. All are facultative aerobes.

***Salmonella* spp.** Salmonellosis is one of the most common bacterial foodborne illnesses (Table 5.1). Control of salmonellae in the food processing environment commands more attention than any other pathogen from the food industry. Long recognised as being ubiquitous, salmonellae survive and grow very well outside their normal habitat, the animal intestine. In particular, salmonellae survive very well in dry environments. They have often contaminated food products that are processed and/or packaged in dry environments, such as dried dairy products, dried eggs products, chocolate, soy flour, peanut butter, pet foods and animal feed. Such contamination incidents are often caused by the unwise or inadvertent introduction of moisture into otherwise dry processing areas by means of wet cleaning and sanitation procedures, condensation, roof leaks, etc. Typically, salmonellosis is caused in healthy persons by the ingestion of millions of cells. Much lower numbers of cells, about 100 or less, can infect immunocompromised persons. Healthy individuals can also be infected by such low numbers if the salmonellae are protected during passage through the stomach, particularly by high-fat foods such as chocolate or peanut butter, or if the pH of the stomach is increased by antacid therapy or a large quantity of food. A global outbreak of illnesses that began about 20 years ago has been associated with shell eggs in which transovarian infection by *Salmonella* Enteritidis leads to internal contamination of the egg yolk; however, the risk of this occurring has been reduced in some countries by flock vaccination.

Campylobacter jejuni. Campylobacters are another very common cause of foodborne illnesses, with about 99% of the cases being caused by *C. jejuni*. This pathogen is associated with raw poultry, and sometimes raw milk. Illnesses are often caused when cross-contamination between these sources and ready-to-eat (RTE) foods, such as salads, occurs during food preparation. Campylobacters are often linked to cases of Guillain–Barré syndrome. This microorganism is more difficult than most to grow *in vitro*, requiring a headspace containing 2–10% carbon dioxide. It does not survive readily in processed foods.

***Escherichia coli* O157:H7.** First implicated in outbreaks of foodborne illness in 1982, *E. coli* O157:H7 has been found to be a particularly virulent pathogen. The infectious dose is as low as 10–100 cells. Death can be caused, particularly in children, by haemolytic uremic syndrome (HUS), a condition that may lead to kidney failure. Many serotypes of *E. coli* produce Shiga toxins that are capable of causing HUS, but *E. coli* O157:H7 is the principal serotype associated with foodborne transmission. Originally associated with raw or undercooked ground beef, illnesses caused by this pathogen have more recently been associated not only with meat products, but also with unpasteurised milk and juices, sprouts and other produce, water, and contact with animals on the farm or in petting zoos. The original source of *E. coli* O157:H7 was found to be cattle, in which this pathogen colonises the rectoanal junction (Lim *et al.*, 2007). Currently, environmental sources are considered to be equally or more important than cattle or raw ground beef as a source of contamination (Strachan *et al.*, 2006).

Listeria monocytogenes. Long known to be a disease of animals or humans closely associated with farm animals, listeriosis was first documented as a foodborne illness in 1981. This microorganism is a relatively weak pathogen; however, it justifiably requires a great deal of attention in the food industry. Although high numbers of cells are required to infect even immunocompromised individuals, the consequences of infections in this group can be drastic. The infections often proceed to septicaemia and/or meningitis and can lead to spontaneous abortion, stillbirth and death, with a mortality rate of about 30%. *L. monocytogenes* is psychrotrophic. It can grow to infectious numbers in cooked RTE refrigerated foods of extended shelf life that were contaminated at some point between the cooking and packaging operations, or during multiple uses in food service operations or the home. For this reason, control of *L. monocytogenes* receives a great deal of attention in food processing plants that produce these types of food products and in retail establishments that sell or serve them.

***Shigella* spp.** Most shigellae foodborne illnesses are caused by *Shigella flexneri*. Similarly to *E. coli* O157:H7, this pathogen produces a Shiga toxin that can cause HUS. It does not survive very well in processed foods and is transmitted principally by food handlers via the faecal–oral route.

***Vibrio* spp.** Vibrios are halophilic microorganisms that thrive in warm marine or estuarine environments. They are a common part of the microflora of fish and seafood. *Vibrio parahaemolyticus* is most commonly associated with foodborne illness. It is psychrotrophic and often found in shellfish. *Vibrio vulnificus* is found in shellfish harvested from warm water. It causes a serious infection, with a 50% mortality rate in persons who have a liver dysfunction, such as hepatitis. *Vibrio cholera* causes cholera, which is primarily a waterborne infection.

***Yersinia enterocolitica*.** This pathogen is psychrotrophic and sometimes causes illnesses that are associated with raw milk or water. It can cause extreme abdominal pain, sometimes resulting in erroneous emergency appendectomies.

***Staphylococcus aureus*.** *S. aureus* produces many serological types of enterotoxins, classified as types A through H. Type A staphylococcal enterotoxin is most commonly associated with foodborne illness. It is very heat-stable, surviving 10 minutes at 121°C or more than 20 minutes at 100°C. The toxic dose of this toxin is about 1 μg. Historically, staphylococcal food poisoning has been associated with fermented foods such as cheeses and sausages. Slow fermentation or starter culture failure enables *S. aureus* to grow to high numbers and produce toxic levels of enterotoxin. Cream-filled bakery products have also been a common cause of staphylococcal illness. These are caused by contamination from food handlers during the filling operation.

***Mycobacterium avium* subsp. *paratuberculosis*.** This pathogen is known to cause Johne's disease in cattle. Although there is no definitive proof, it is suspected to cause a similar disease in humans, Crohn's disease (Gould *et al.*, 2005), and this has led to changes in recommendations for milk pasteurisation in some countries. Its primary food sources are raw milk, raw meat and water.

***Cronobacter* (formerly *Enterobacter*) *sakazakii*.** This microorganism is a prime example of an opportunistic pathogen. A common coliform bacterium, it can be detected in a great many food and environmental sources, but it rarely causes illness (Friedman, 2007). It has been on very rare occasions found to be the cause of infection and death of premature infants. These rare infections are caused by mishandling (extreme temperature abuse) of rehydrated infant formula in the clinical setting (Gurtler *et al.*, 2005). Nonetheless, a heavy burden to eliminate *C. sakazakii* in dried ingredients is being placed on food processors that supply the producers of dehydrated infant formulas.

5.2.4 Viral pathogens

Foodborne viral pathogens are obligate parasites that grow only in human cells. Therefore, the primary source of these viruses is human faeces. Viruses are transmitted hand-to-mouth or by

Table 5.13 Characteristics of viral foodborne pathogens.

Virus	Incubation period in days (range)	Key symptoms
Hepatitis A	28 (15–50)	Hepatitis, jaundice
Norovirus	1 (1–2)	Projectile vomiting
Rotavirus	2 (1–6)	Infects primarily very young people

food handler contamination of prepared food, facts that have led to the 'golden rule of food safety', wash your hands. The major symptoms of viral foodborne illnesses are gastroenteritis and hepatitis (Table 5.13).

Several zoonotic viruses are receiving attention in the food safety arena because of their potential to become foodborne.

Influenza viruses

Currently, receiving the most attention is the H5N1 avian influenza virus and the H1N1 swine influenza virus. At this writing several hundred humans have died after contracting H5N1 pulmonary infection directly from domestic poultry or birds. Foodborne transmission of avian influenza viruses will not occur if poultry is cooked to 70°C (Thiermann, 2007). At the end of 2009, a pandemic of the H1N1 virus is underway and its full ramifications are not known. However, in contrast to the avian influenza virus, this swine influenza virus is not contracted directly from swine; it would also be inactivated by cooking pork products.

SARS

A flare-up of sudden acute respiratory syndrome (SARS) in 2003 revealed a new zoonotic virus whose host appeared to be civet cats in Asia. While not thought to be foodborne, SARS is a prime example of an emerging zoonosis. It demonstrates that food safety and public health professionals must always be prepared to 'expect the unexpected'.

Foot and mouth disease

Foot and mouth disease (FMD) is not a zoonosis, as it does not infect humans. It is, however, of great economic importance to the food and agricultural industries. FMD is a severe and highly communicable disease of cloven-hoofed ruminants — cattle, sheep, goats and deer — and swine (Haley, 2001). FMD outbreaks are controlled by quarantine zones and the destruction of sometimes millions of animals, whether infected or not. In a world of increasingly limited resources, procedures to protect the food use of these animals could be developed, as the FMD virus does not infect humans and is inactivated by cooking.

5.2.5 Prions

Originally thought to be 'slow virus diseases' because of their very long incubation periods, the causative agent has been discovered to be not a virus, but a misshapen normal cellular protein called a prion (Table 5.14). The infectious prion causes disease by catalysing a

Table 5.14   Features of prion diseases of animals and humans.

Disease	Host	First detection	Incubation period (years)
Scrapie	Sheep, goats	1732	2–5
BSE	Ruminants, felines	1986	4–5
vCJD	Humans	1996	10–15

Source: Hueston and Bryant (2005).

change from normal to infectious state and the crystallisation of prions in brain cells. With the destruction of sufficient brain cells, microscopic holes appear in the brain, hence the term 'transmissible spongiform encephalopathy' (TSE). Creutzfeld–Jakob disease (CJD) is a TSE that regularly occurs throughout the human population at an incidence of about one case per million people each year. Scrapie is a TSE that has been known for several centuries to infect sheep.

A new TSE was identified in 1986 when bovine spongiform encephalopathy (BSE or 'mad cow disease') was detected in the UK cattle. Eventually, it was determined that BSE had originated from sheep scrapie prions. In 1996, it was determined that a new human disease, inaptly called 'variant Creutzfeld–Jakob disease, (vCJD)', had emerged from BSE. Although it is unrelated to conventional CJD, each reported case of CJD tends to cast suspicion on beef products because of the vCJD misnomer.

BSE was found to be transmitted between cattle by the consumption of contaminated meat and bone meal. Lack of adequate feed controls enabled the spread of BSE by 2003 to nearly every European country and Japan, with two cases detected in North America. Ultimately, nearly 200 000 heads of cattle contracted BSE. Millions more were incinerated as a precautionary measure. At this writing, about 200 humans have died of vCJD.

BSE can be transmitted to humans and other animals by the consumption of brain or spinal cord material from infected cattle. The infectious agent is extremely stable and cannot be denatured by existing food processes. Therefore, stringent control of infected animals, prevention of over 20-month-old animals entering the food chain, and complete removal of all tissues likely to contain prions – specified risk material – from cattle, sheep and goats became the main control mechanisms in the UK. Derogation of the 'over 30-month rule' has since been granted, but only for slaughterhouses operating to a 'Required Method of Operation' that has been approved by the Secretary of State (UK Statutory Instrument, 2008).

The important lesson from this outbreak was that what had begun as a long-recognised disease of sheep, scrapie, in the span of 15 years became a serious public health and food safety matter. The emergence and initial mismanagement of the BSE epidemic emphasise the importance of managing food safety from farm-to-table and throughout the global food supply chain.

5.2.6   Protozoan parasites

A number of protozoan parasites have been associated with human illnesses that were caused by the consumption of contaminated raw foods or water (Table 5.15). Typical symptoms of protozoan infections, which can persist for months, include diarrhoea, fever and enteric distress.

Table 5.15 Protozoan parasites of humans associated with raw foods or water.

Protozoan	Incubation period (days)
Toxoplasma gondii	10–23
Cryptosporidium parvum	1–12
Cyclospora cayetanensis	7
Giardia lamblia	5–25
Entamoeba histolytica	14–28

5.2.7 Parasitic worms

Numerous tapeworms and roundworms are involved in foodborne infections (Table 5.16). Most have long incubation periods and are associated with the consumption of raw or undercooked meat or seafood, and raw fruits and vegetables, principally soil crops.

Table 5.16 Parasitic worms of humans associated with food consumption.

Parasite	Incubation period	Typical food source
Tapeworms		
Taenia solium	8 weeks to 10 years	Undercooked pork
Taenia saginata	10–14 weeks	Undercooked beef
Diphyllobothrium latum	3–6 weeks	Undercooked fish
Roundworms		
Ascaris lumbricoides	2–8 weeks	Raw fruits and vegetables, soil
Trichinella spiralis	8–15 days	Undercooked meat
Anisakis simplex	1–14 days	Undercooked marine foods

5.3 CHEMICAL HAZARDS

A very wide variety of chemical hazards may appear in food products either by natural occurrence in a raw material or by deliberate or unintentional addition during processing. The health effects of the chemical hazards can range from rather benign, e.g. residual cleaning compounds, to acutely toxic or carcinogenic, e.g. some mycotoxins or persistent organic pollutants. While chemical hazards require control because of their overt food safety risks, food producers must also contend with two additional consequences of the presence of unwanted chemicals – regulatory non-compliance and trade disruptions. This situation has become more complex because of different chemical residue limits around the world (Chapter 3) and the tendency to 'chase zero' when knowledge of threshold tolerances in not available.

5.3.1 Allergens

Allergens are naturally occurring proteins to which some persons develop a hypersensitivity or immunological response. Concentrations of allergenic material as low as 1 ppm can induce responses in a matter of several minutes or less. Allergens affect up to 2% of adults and 7% of children. Symptoms range from rashes and nausea to anaphylaxis and death.

Food intolerances or sensitivities are often confused with allergic responses. These are, however, non-immunological responses and are associated with non-proteinaceous compounds (Timbo *et al.*, 2004). Food intolerances may be caused, for example, by sulphites at concentrations of more than10 ppm, monosodium glutamate, food colours such as yellow #5 (tartrazine), lactose and histamine (Section 5.3.3).

The principal global foodborne allergens are associated with groundnuts (peanuts), tree nuts, crustaceans, fish, eggs, milk, soybeans and wheat (Hefle and Taylor, 1999). Regional allergens are associated with, for example, celery (the European Union), buckwheat (Southeast Asia) and rice (Japan) and numerous others. Minor allergens include the seeds of cotton, sesame, sunflower and poppy, as well as legumes and molluscs, and cross-reactivity with other allergens, for example reaction to apples in individuals sensitive to birch pollen, can cause further problems for sensitive consumers. Properly refined oil produced from allergenic seeds is not itself allergenic, as all the proteins have been removed. Allergen hazards may be introduced during food processing by poor control of reworked materials, addition of the wrong ingredient to a food or mislabelling of the consumer product. They may also be introduced by cross-contamination during food preparation in the home or food service operations.

5.3.2 Mycotoxins

Ergotism is an illness that has been known for millennia, long before awareness of mycotoxins emerged in 1961. Ergot is produced during the growth of *Claviceps purpurea* in grassy cereal groups such as rye, oats, wheat and barley. It is contained in sclerotia that range in size from mouse droppings to several centimetres. The threat of ergotism can easily be avoided by proper grain-handling practices.

Thousands of mycotoxins are produced as secondary metabolites during mould growth. They can enter the food supply when substantial mould growth occurs in field crops. Most mycotoxins are produced by only three mould genera — *Aspergillus*, *Fusarium*, and *Penicillium* (Murphy *et al.*, 2006).

Aflatoxin, named after *A. flavus*, was discovered after contaminated feed caused the deaths of turkeys in the UK in 1961. It is a potent liver carcinogen. Aflatoxin is a normal hazard in peanut crops during wet years and in corn (maize) crops during dry years when drought-stressed plants are vulnerable to mould infestation. Four serological types – B_1, B_2, G_1 and G_2 – of aflatoxin can be produced. Aflatoxin B_1 can be altered in the digestive tract of ruminant animals, appearing as aflatoxin M_1 in their milk.

Patulin can be produced in damaged fruits, particularly apples. Therefore, it is of some concern in products such as apple juice.

Ochratoxin can be produced during mould infestations of wheat, corn or oats. It can contaminate the meat or milk derived from animals that consumed ochratoxin-contaminated grains. It is of most concern in Africa and the EU.

Zearalenone, sometimes produced in grain crops, is a mycoestrogen that can disrupt steroidal hormone functions.

Deoxynivalenol, also called 'vomitoxin', is one of 180 trichothecene mycotoxins. Typically, produced in wheat and barley during wet years, it is often responsible for feed refusal in animals fed contaminated grain. It is a protein synthesis inhibitor.

Fumonisin was first detected in 1990 after the unexplained deaths of horses. Unlike other mycotoxins, it possesses no aromatic ring structure and is highly water-soluble. Its three

Table 5.17 Examples of regulatory action levels on mycotoxins in food and feedstuffs in several countries or regions.

Mycotoxin	Food/feedstuff	Action level		
		United States	**EU**	**Codex**
Aflatoxin	Peanuts, nuts			
B_1, B_2, G_1, G_2	Mouldy grains	20 ppb	2–15 ppb	15 ppb
Aflatoxin M_1	Milk	0.5 ppb	0.05 ppb	0.5 ppb
Fumonisin	Maize (corn)	2–4 ppm	—[a]	—
Patulin	Apple juice	50 ppb	50 ppb	50 ppb
Deoxynivalenol	Milled wheat products 1 ppm	—	—	
Ochratoxin A	Cereal grains	—	5 ppb	—

Source: Murphy *et al.* (2006).
[a]No action level.

serologically distinguished forms – B_1, B_2 and B_3 – are produced principally in corn (maize). Fumonisin B_1 can cause leukoencephalomalacia, a condition that causes the brains of horses to be dissolved. It is also a suspected cause of human oesophageal cancer (Chu and Li, 1994).

Of the several thousand identified mycotoxins, only two – aflatoxin B_1 and fumonisin B_1 – are known to be overt animal pathogens. These are suspected, but not proved, to be carcinogens in humans. Many mycotoxins, however, pose substantial regulatory and trade hazards to the food industry. Even in these matters, global concern is largely limited to aflatoxin and patulin (Table 5.17).

5.3.3 Marine foodborne toxins

Shellfish poisoning

Several illnesses are associated with bivalve molluscs such as mussels, clams and oysters. The molluscs filter large quantities of seawater, thereby concentrating pathogenic dinoflagellates or diatoms that produce a number of toxins. Illnesses caused by contaminated molluscs are more likely during red tides (Liston, 2000).

Paralytic shellfish poisoning. This can be caused by several genera of dinoflagellates, including *Alexandrium*, *Gymnodinium*, *Pyrodinium* and *Saxidomus*. The last produces saxitoxin a very heat-resistant and lethal toxin. The illness is characterised by tingling, nausea and potential respiratory paralysis and death. The lethal dose is 2 mg. The US Food and Drug Administration (FDA) enforces an action level of 80 μg/100 g meat (0.8 ppm).

Diarrhoetic shellfish poisoning. It results in a mild gastroenteritis, also caused by dinoflagellates.

Neurotoxic shellfish poisoning. It is a gastrointestinal illness with a low fatality rate. It is caused by brevetoxin, produced by the dinoflagellate *Gymnodinium breve*.

Amnesic shellfish poisoning. Also called as domoic acid poisoning, it begins as gastroenteritis and can proceed to neurological symptoms, coma and death. First recognised in 1987, it is caused by the genus *Pseudonitzchia*, a diatom that produces domoic acid.

Finfish poisoning

Ciguatera poisoning. This is caused by ciguatoxin produced by the dinoflagellate *Gambier discus toxicus*. It can be associated with about 400 species of tropical fish. Ciguatoxin also causes gastroenteritis and can proceed to neurological symptoms and death.

Scombroid (or histamine) poisoning. It is associated with a family of fish, involving most often tuna, mahi-mahi and mackerel, that contains high levels of free histidine in their flesh. *Proteus* spp. can grow on improperly chilled fish, decarboxylating histidine to histamine, which produces symptoms that mimic an allergenic response.

Puffer fish (fugu) poisoning. This is caused by tetrodotoxin, an often lethal neuroparalytic toxin, that is produced in the liver or internal organs of the puffer fish by several genera of Gram-negative bacteria including *Vibrio*, *Alteromonas*, *Aeromonas*, *Plesiomonas*, *Pseudomonas* and *Escherichia*. It has typically been reported in Japan and Southeast Asia due to the inadequate removal of the internal organs during food preparation.

5.3.4 Genetically modified foods

Genetic modification is the alteration of an organism by the introduction into its genome, or chromosomes, of one or more genes from a different organism. Numerous crop applications have been developed to prevent plant diseases without using chemical pesticides (BT corn) or to reduce herbicide applications (herbicide-resistant soybeans) (WHO, 2007c). Originally, theoretical public health concerns regarding the potential altered allergenicity of genetically modified (GM) crops were raised. At this point there is no indication that GM foods are unsafe. In the interests of due diligence, however, food safety assessments need to be conducted on all new GM applications. While no overt food safety issues have been identified with GM foods, major logistical and regulatory difficulties can be caused by the need to segregate GM from non-GM crops, and in the labelling of GM-containing consumer foods.

5.3.5 Antibiotics

Potential food safety and public health issues related to the presence of antibiotic residues in food are not well understood. The hypothesised concern is that the use of antibiotics in animals, either therapeutically or as growth promoters, may give rise to the occurrence of antibiotic-resistant pathogens in the food supply or in humans. Therefore, when therapeutic antibiotics are used to treat animal diseases, adequate clearance times before harvesting are enforced by veterinarians to prevent human consumption of the antibiotic.

Regulatory problems occur because antibiotics that are approved for use in some countries may not be approved in other countries, or their use may be approved at various levels in different countries. In 2003, EU member states established harmonised minimum required performance reporting limits for the detection of residues of nitrofurans at 1 ppb and chloramphenicol at 0.3 ppb (Food Standards Agency, 2003). Nevertheless, nitrofurans became a temporary, but severe, regulatory problem in 2004. At that time nitrofurans were legal for use in Thailand poultry, but not in the UK, whose regulatory limit of 1 ppb in poultry was near the limit of detectability, essentially a zero tolerance. An Irish laboratory applied a new technology that could detect nitrofurans with a much greater sensitivity (chasing zero). Although in regulatory compliance when the original testing method was used, Thai poultry was banned in the UK when it yielded positive results with the new testing method.

5.3.6 Persistent organic pollutants

Several thousand complex chemicals have been synthesised since the 1930s for use as pesticides or industrial chemicals. These are generally chlorinated or brominated aromatic compounds that are resistant to microbial or chemical degradation. Therefore, they persist for a very long time in soil and sediments. Most of these compounds are lipophilic, bioaccumulating in fatty tissues in the food chain (Jones and de Voogt, 1999). Because many of the compounds are volatile, they can be airborne for great distances from their point of use.

Persistent organic pollutants (POPs) in aquatic sediments can be accumulated in algae and plankton, which in turn are ingested by fish. Humans and other animals further accumulate POPs when eating contaminated fish. In the 1960s, birds of prey became nearly extinct because DDT accumulated in fish reduced the birds' eggshell thickness. On land, POPs carried by air can contaminate forage, which is consumed by grazing animals. The meat or milk from these animals can transfer the POP to humans. In humans, POPs may elicit chronic effects. They are known or suspected carcinogens, and can function as sex hormone, endocrine or immune system disruptors. The principal POPs are well described by Ulberth and Fielder (2000).

Pesticides

Aldrin, chlordane, DDT, dieldrin, endrin, heptachlor, mirex and toxaphene are applied to a wide variety of crops. They must be applied according to proper schedule. Crops cannot be harvested until the prescribed clearance period has elapsed.

Industrial chemicals and contaminants – hexachlorobenzene and polychlorinated biphenyls

Used as a fungicide in crop seeds such as wheat, hexachlorobenzene (HCB) causes human illness when the seed grain is mistakenly used as food, even though the treated seeds are coloured to discourage consumption. Polychlorinated biphenyls (PCBs) are used in industrial applications, such as coolants in electrical transformers. The inadvertent or deliberate contamination of feed-grade oils or fats with transformer fluid can occur (Larebeke et al., 2001). The US action level for PCBs in red meat (fat basis) is 1 ppm (FDA, 2000).

Dioxins

The principal human exposure to dioxin is through consumption of contaminated fish, meat or dairy products (Peshin et al., 2002). The contamination of fats used in animal feed with industrial oil led to a massive recall of food products because of dioxin contamination in Belgium in 1999. The EU has established action levels for dioxins in human food and animal feed ranging from 0.5 to 4.5 ppt (Larebeke et al., 2001; Commission, 2002).

POPs and organophosphate pesticides present a significant toxic hazard to pesticide applicators and workers who handle the treated crops. While POPs were widely used in developed countries during the 1950s and 1960s, their toxic effects led to diminished usage and outright bans. While POP usage in developed countries is either limited or banned, developing countries can have wider usage. For example, DDT remains an important mosquito control agent in malaria-endemic regions. Although banned in developed countries, PCBs are present in older electrical transformers and capacitors. Improper maintenance or disposal of these units can lead to contamination of the food supply chain.

5.3.7 Heavy metals

Heavy metals in the food can have a wide variety of toxic effects in humans and animals including tunnel vision, deafness, chronic brain damage, congenital defects, peripheral neuropathy, hyperkeratosis, nephrotoxicity and skin cancer (Peshin *et al.*, 2002). Heavy metals typically enter the environment from industrial effluents, coal-fired power plants and municipal garbage incinerators.

Mercury

Historically, recognised as the cause of 'mad hatter's disease', metallic mercury is transformed to methyl mercury in marine and freshwater environments. It is bioaccumulated in the food chain in the same manner as POPs. Methyl mercury-treated seed grain, when mistakenly used as food, has been responsible for foodborne illnesses. Mushrooms obtained near mercury and copper smelters often cause illness. The United States has an action level of 1.0 ppm for methyl mercury in fish, shellfish and wheat (FDA, 2000).

Lead

Lead poisoning can be caused by ceramic ware used for serving food. In addition to neurological symptoms, lead can also cause kidney damage. The action level in various types of ceramic ware ranges from 0.5 to 7.0 ppm (FDA, 2000).

Cadmium

Sometimes associated with rice, cadmium is a nephrotoxicant. It can also be associated with ceramic ware, for which action levels of 0.25–0.5 ppm have been established (FDA, 2000).

Arsenic

Used in various compounds as a rodenticide or fungicide, arsenic is a well-known human poison. Chronic exposure can result in cancer or skin lesions. Arsenic naturally found in rocks or minerals can be released into the groundwater by biological or chemical processes. While groundwater contamination with arsenic occurs globally, it is thought to be a more serious problem in Southeast Asia (Sakai, 2007).

Uranium

The heaviest of naturally occurring metals, the toxic effects of uranium are emerging as a serious public health matter. The 1986 meltdown of a nuclear reactor at Chernobyl caused widespread contamination of soil and crops in Eastern Europe with radioactive fallout. Responsible for many human deaths and congenital defects, the full effects of this tragedy have not yet been realised.

A new hazard has been associated with uranium, because of its density more than its radioactivity. During the enrichment of uranium^{235} for use in nuclear power plants, large quantities of the natural isotope, uranium^{238} (U^{238}), are left over. Massive quantities of low-radioactivity U^{238} are being used to produce 'hardened' ammunition and armour plating for military vehicles. The heat of explosions or impacts vapourises U^{238}. Contemporary war zones are becoming contaminated with U^{238} at levels that are hazardous to the indigenous populations, and will contaminate animals and food crops.

5.3.8   Chemicals used in food processing environments

A great number and variety of chemicals are used in food processing plants for the routine operation, maintenance and cleaning of the processing equipment. Care must be taken to prevent cross-contamination of food materials with the chemicals used in food processing environments. Representative chemicals include boiler additives, lubricants, refrigeration fluids, detergents, sanitizers and pesticides.

5.3.9   Chemicals used in food packaging materials

Care must be taken in the selection of food contact packaging materials, as plasticizers and additives used in their manufacture may leach into the food product. Adequate knowledge of the chemicals used and their toxicity should be obtained. For example, in 2008, research was begun to understand the public safety impact of bisphenol A that leached into water and foods packaged in polycarbonate plastic bottles or metal cans that had been manufactured with the use of bisphenol A. Testing procedures, requirements and standards for the migration packaging additives into foods have been republished in England this year (Statutory Instrument No. 205, 2009).

5.3.10   Unanticipated potential chemical hazards

The myriad of naturally occurring and synthetic chemicals can interact in foods to create derivatives or analogues that may come under suspicion as foodborne hazards. At these times food safety and public health professionals sometimes need to act quickly to assess and control the potential hazard.

Acrylamide has long been used to synthesise polyacrylamide, which is widely used for water treatment, soil conditioning, laboratory applications, and in the production of paper, textiles and cosmetics (Friedman, 2003). The WHO guideline for acrylamide in drinking water is 0.5 ppb. Also produced in cigarette smoke, acrylamide has neurotoxic effects in humans and animals. The serendipitous discovery[1] in 2002 by the Swedish National Food Authority of acrylamide in foods caused temporary alarm throughout the food industry. It was determined that acrylamide is formed when glucose and asparagine interact during the baking or frying of foods at temperatures above 120°C. It is not presently thought that acrylamide is a significant foodborne hazard (WHO, 2007d).

Other instances of potential chemical foodborne hazards can occur through deliberate adulteration of food. Chemical adulteration can change the nutritional profile and utility of a food. A 2007 outbreak of cat and dog illnesses and deaths due to kidney failure was caused by the economic adulteration of wheat flour in China with melamine and its analogues — cyanuric acid, ammelide and ammeline – so that the flour could be sold as wheat gluten. The toxic effects are greatest when melamine and cyanuric acid can interact in the host (FDA, 2007). Used in the production of fertilisers and plastics, melamine is a nitrogen-rich compound that can mimic the presence of protein in analytical tests. Cyanuric acid is commonly used in chlorine-based

---

[1]   When testing blood samples of workers in the acrylamide industry, the Swedish authority found similar levels of acrylamide in a set of control workers who had no industrial exposure, leading to its discovery in certain foods.

cleaning powders. While this outbreak of pet illnesses was quickly contained without known human exposures, it dramatically emphasised the need for effective food safety controls and verification procedures in global trade. However, just 1 year later a similar outbreak occurred in China, when at least 6 infants died when fed milk that had been adulterated with melamine (Jiang, 2008). In a separate incident in 2004, at least 13 Chinese children died of malnutrition while being fed counterfeit milk that had no nutritive value (Anonymous, 2004).

The production and trading of adulterated or counterfeit foods are criminal acts in most countries. Such acts undermine confidence in the safety of the food supply and must be handled swiftly and firmly to discourage further incidents. In November 2009, the Chinese government executed a dairy farmer and a salesman and jailed 19 others for their roles in the melamine contamination of milk.

5.4 PHYSICAL HAZARDS

Foodborne physical hazards are commonly called 'foreign materials' or 'foreign bodies' because their presence in food is unnatural. Foreign materials in food can be potentially harmful or merely undesirable.

5.4.1 Sources of foreign material

Foreign material can enter foods at almost every point of the food supply chain. The general sources of contamination are the environment, the food itself, the food processing facilities and personal objects.

Soil and stones are typical environmental contaminants during harvesting. During harvesting or by later infiltration, food may be contaminated by insects, rodents or reptiles. These may be disintegrated during harvesting or processing so that body parts or excreta are the residual evidence of contamination.

Some foreign material contamination originates with the food itself. Examples include fruit pits, stones and stems: bones or bone chips from fish and meat, pieces of corn cobs, nut shells, and hardened or crystallised sugars.

A great deal of foreign material contamination originates in food processing facilities. Metal shavings, nuts and bolts, and lubricants from the processing equipment can enter the food. Nails, cut wires and broken utility blades can be dropped into the food stream by maintenance workers. Pieces of glass, hard plastic and wood splinters can enter the food from other fixtures and utensils in the processing area.

Personal objects used or worn by maintenance and line workers and by food handlers in food service operations often fall into the food. These may include rings, pencils, papers, earrings, nose rings, buttons, thermometers, hair and gloves.

5.4.2 Injuries associated with physical hazards

Because of their relatively low occurrence and severity, foreign materials are often not considered to be significant hazards in foods. While most incidents of foreign material ingestion do not result in overt bodily harm, up to 5% of such incidents do result in injury (Hyman *et al.*, 1993; Olsen, 1998). Slender, pointed objects such as bone slivers and glass shards are the most hazardous. These can cause oral or gastrointestinal lacerations or perforations. Sharp foreign materials about 2–5 cm long are most likely to be involved in intestinal perforations. In these rare cases, subsequent abdominal infections can result in death.

It is generally considered that objects smaller than 7 mm in their longest dimension are not likely to cause injury in adults, as these can pass through the digestive tract without causing harm. However, some particles less than 7 mm may cause harm in infants. Because it is readily softened in the food or upon ingestion, paper is usually not considered to represent a risk of injury.

Many small toys and other objects are marketed with food products to increase their appeal to children and their parents. Objects that are less than about 3 cm in diameter or 6 cm long can present a choking hazard for infants and small children. Larger objects that cannot be swallowed do not present a choking hazard.

Some foods, e.g. grapes or pieces of meat, may present a choking hazard for small children.

5.5 CONCLUSIONS

A complete awareness and understanding of the potential biological, chemical and physical foodborne hazards must be maintained by product development teams, food safety managers and HACCP teams. Such knowledge is necessary to conduct product and process research and to conduct a responsible hazard analysis and determination of the appropriate control measures that will need to be implemented at critical control points and in prerequisite programmes. All food safety professionals must also remain vigilant in order to be able to quickly respond to unanticipated hazards. 'Expecting the unexpected' is an underappreciated aspect of food safety management.

6 Designing safety into a food product

6.1 INTRODUCTION

This is the first of several chapters that describe the development and application of control measures to assure the safety of food products. A great many physical, chemical and biochemical technologies can be used to control potential hazards in food.

We have reached the conclusion that, at its core, the Hazard Analysis and Critical Control Point system (HACCP) consists of two essential processes – product design and control, and process design and control. Both the product and process design requirements are typically defined by research and development groups. It is during this period that control measures must be identified, tested, validated and incorporated into the product formulation (Mortimore and Wallace, 1998; Sperber, 1999). Upon completion, the design requirements are communicated to manufacturing teams that scale up and validate the design requirements at a commercial production level. Individual plant HACCP teams complete the HACCP plan for each product and process and establish the monitoring, verification and record-keeping procedures.

This chapter describes product design and control, which is based on the adjustment and control of factors that are intrinsic to the food product. These factors include the ingredients that are included in the food and the equilibrium chemical and biochemical properties of the finished food product.

6.2 FORMULATION INTRINSIC CONTROL FACTORS

A number of intrinsic, or inherent, food properties, e.g. water activity (a_w) and pH, can be controlled to ensure food safety, as well as to inhibit microbiological spoilage and to protect product quality. These two intrinsic properties in particular have been used, perhaps crudely, since ancient times to preserve foods by salting, drying, acidification and fermentation. As professional food scientists in the 21st century, it is sobering to contemplate that some of our most effective food safety control measures have been used for millennia. In modern times these intrinsic properties are better understood and can be used more precisely to minimise microbial growth without compromising product quality.

Food Safety for the 21st Century, First Edition By Carol A. Wallace, William H. Sperber and Sara E. Mortimore
© 2011 Carol A. Wallace, William H. Sperber and Sara E. Mortimore

6.2.1 Water activity

All microorganisms, of course, require water for growth. The a_w value of a food product has a significant effect on the growth rates and types of microorganisms that can grow in the food. It provides an indication of the availability of water to support microbial growth. Some of the water in a food is unavailable to support microbial growth because it is hydrogen-bonded to constituent molecules in the food. It has long been known that the microbiological stability of a food can be improved either by removing water (e.g. dehydration) or by adding one or more solutes (e.g. salts and sugars). Early efforts to quantify and control these factors involved determination of percentage of moisture in the product and, eventually, the percentage of the principal solutes. Both of these are relatively crude measures in terms of their ability to estimate the control of particular microorganisms in a food. Modern food product developers need to maximise the inhibition of microorganisms by water activity reduction without creating organoleptic defects because of increased solute concentrations.

The modern use of a_w as a control factor in food product development was greatly facilitated by Scott (1957) as he clearly described its relationship with several physico-chemical properties that can be measured or calculated:

$$p/p_o = n_2/(n_1 + n_2) = ERH/100 = a_w$$

The earliest accurate measurements of a_w involved manometric determinations of vapour pressures and by dividing the vapour pressure of a solution or food (p) by the vapour pressure of pure water (p_o). The ideal a_w of a solution or food can be determined by dividing the moles (gram molecular weight) of solvent (n_2) by the sum of the moles of solvent and the moles of solute (n_1). For example, the addition of one mole of a salt or sugar to 1000 g (55.5 moles) of water would give a 1.0 molal solution with an ideal a_w of 55.5/56.5 = 0.982. Of course, solutes do not act ideally (ideal behaviour is conceived as the impact of one molecule of solute in an infinite volume of solvent). Therefore, a_w values for solutions with different solutes will vary from ideality (Table 6.1). Recently, it has become feasible to accurately determine a_w values by using instruments that measure the equilibrium relative humidity (ERH) of a food. The a_w of a food is simply its ERH/100 (Scott, 1957). Because ERH readings are limited to the range of 0–100%, a_w readings will range from 0 to 1.0. In laboratories today, a_w determinations are made quickly and accurately by the instrumental determination of dew point depressions.

The ability of a solute to reduce a_w is inversely proportional to its molecular or ionic weight. For this reason, food product developers favour the use of solutes of low molecular weight, as reduced quantities of a particular solute could be used to achieve the desired a_w value. For

Table 6.1 Molalities of different solutes required to provide a particular water activity value.

	Molality			
Ideal		**Actual value**		
a_w	**Value**	**NaCl**	**Sucrose**	**Glycerol**
0.980	1.13	0.61	1.03	1.11
0.940	3.54	1.77	2.72	3.32
0.900	6.17	283	4.11	5.57
0.850	9.80	4.03	5.98	8.47

Source: Scott (1957).

Table 6.2 Representative water activity values of foods and ingredients.

Food or ingredient	Water activity
Water	1.00
Fresh meat, poultry and seafood	0.99
Mayonnaise and salad dressings	0.90
Icing, frosting, pancake syrup	0.80
Dried fruit	0.65–0.75
Saturated sodium chloride	0.75
Corn syrup	0.70
Wheat flour, freshly milled	0.65
Wheat flour, 55% high-fructose corn syrup	0.60
Dry pasta, spices, milk, cocoa	0.1–0.4

Source: Sperber (1983) and Christian (2000).

example, the average ionic weight of sodium chloride is 29.25, while the molecular weight of sucrose is 342. On a theoretical basis, sodium chloride would be about 12 times more effective than sucrose in reducing a_w.

The a_w values of foods and food ingredients (Table 6.2) provide a good indication of the types of microorganisms that can spoil a particular food (Table 6.3). Microorganisms that have a low tolerance for increased osmotic pressure can only grow in high-a_w foods such as fresh meats and beverages. Highly osmotolerant microorganisms such as osmophilic yeasts can grow at a_w values as low as 0.60. No microbial growth has been reported in foods at a_w values lower than 0.60.

Table 6.3 Minimum water activity values and pH ranges that support the growth of various microorganisms in foods.

Microorganisms	Minimum a_w	pH range Minimum	pH range Maximum
Alicyclobacilli	0.98	2.0	6.0
Pseudomonads	0.97		
Enteric bacteria	0.95	4.5	9.0
Clostridium botulinum			
Non-proteolytic strains	0.97	5.0	8.5
Proteolytic strains	0.94	4.8	8.5
Clostridium perfringens	0.95	5.0	8.9
Salmonella spp.	0.94	4.0	9.5
Bacillus cereus	0.93	4.4	9.3
Listeria monocytogenes	0.93	4.6	9.2
Staphylococcus aureus			
Toxin production	0.92	4.2	9.0
Aerobic growth	0.85	—	—
Lactic acid bacteria	0.92	3.5	9.0
Moulds	—	0.5	11.0
Normal spoilage	0.84	—	—
Xerotrophic spoilage	0.62	—	—
Yeasts	—	1.5	8.5
Spoilage	0.90	—	—
Osmophilic	0.60	—	—

Source: Doyle (1989), Jay (2000) and W.H. Sperber, unpublished data.

Table 6.4   The effect of solute on the minimum water activity that will support the growth of microorganisms.

Microorganism	Water activity achieved by		
	NaCl	**Glycerol**	**Sucrose**
Pseudomonas fluorescens	0.957	0.940	ND
Clostridium sporogenes	0.945	0.935	ND
Saccharomyces cerevisiae	0.92	ND	0.89
Candida dulciaminis	0.86	ND	0.81

Source: Sperber (1983), Deak and Beuchat (1996) and Christian (2000).
ND, test not performed or not reported.

The ability of some types of microorganisms to grow at lower a_w values is often related to their ability to accumulate 'compatible' solutes such as glycerol. When a microorganism is exposed to increased osmotic pressures, the movement of water out of the cell will greatly slow or stop its metabolism and growth. Those microbes capable of growth at lower a_w values have been shown to accumulate small solute molecules or ions to restrict the outward flow of water (Sperber, 1983; Csonka, 1989). Enteric bacteria can grow at a_w values as low as 0.95 by accumulating potassium ions. Continued accumulation of potassium ions is toxic to the enteric bacteria, so growth below a_w 0.95 is not possible. Some of the more osmotolerant microorganisms can grow by accumulating non-ionic solute molecules, which permit their growth at a_w values lower than those provided by ionic solutes (Table 6.4).

6.2.2   pH

The pH value, total acidity and type of acidulant are important intrinsic factors that affect the types of microorganisms that can grow in foods. The pH value is expressed as:

$$pH = -\log_{10} 1/[H^+]$$

where $[H^+]$ is the hydrogen ion concentration. Because pH is a logarithmic function, doubling or halving the amount of acid or hydrogen ions in a food will alter the pH by 0.3 units ($\log_{10} 2 = 0.3$). Pure water is defined as having a neutral pH 7.0. Values below 7.0 are acidic, while those above 7.0 are basic or alkaline. pH values can range from 0 to 14.

Collectively, microorganisms can grow over most of the pH range, at values ranging from 0.5 to 11.0 (Table 6.3). Some pathogens and extremophiles have evolved elaborate acid tolerance responses to survive and grow in reduced pH environments. Most foodborne bacteria grow in a narrower pH range, from pH 4.5 to 9.5. Most foods are in the acidic range, below pH 7.0 (Table 6.5).

The type of acidulant(s) used in a food can have a major effect on the growth of microorganisms (Table 6.6). Some of these are short-chain organic acids, e.g. acetic acid, that impose an inhibitory effect substantially greater than the effect that would be expected from pH reduction alone.

6.2.3   Chemical food preservatives

Contrary to common consumer and media perceptions, chemical food preservatives are not harmful. Almost all those described here occur naturally in foods. Early food scientists observed

Table 6.5 Representative food pH values.

Food	Typical pH value
Carbonated beverages	2.0
Vinegar	3.0
Apple juice	3.1
Orange juice	3.6
Tomato juice	4.2
Cheddar cheese	5.2
Minced beef	6.2
Milk, bovine	6.4
Maize, peas, honeydew melons	6.5
Fresh fish	6.7
Surface-ripened cheeses	>7.0
Hominy	8.5
Nixtamalised maize	10.0

Source: Lund and Eklund (2000) and W.H. Sperber, unpublished data.

the antimicrobial activity of certain foods, isolated the antimicrobial agents, and learned how to produce and use them as food preservatives. The commercial use of chemical preservatives is often limited by organoleptic considerations, particularly flavour and odour.

Organic acids

The inhibitory effect of short-chain fatty acids on the growth of some microorganisms has led to the widespread use of sorbic acid, propionic acid, benzoic acid and parahydroxybenzoic acid (parabens) as food preservatives. Sorbic acid is found in European mountain ash berries. Propionic acid is produced in Swiss cheese by propionibacteria. Benzoic acid is found in cranberries. Parabens are synthesised by additions to benzoic acid.

Two properties of food preservatives are of utmost importance and require careful consideration during the research and development of their uses in specific foods. These are the dissociation constant (pK_a) and the partition (or distribution) coefficient (PC).

Food preservatives exert antimicrobial effects by interfering with internal metabolism, which requires their passing through the microbe's cytoplasmic membrane. Only the undissociated

Table 6.6 Influence of the type of acidulant on the minimum pH growth limit for salmonellae.

Acidulant	Minimum pH for growth
Citric, hydrochloric	4.05
Tartaric	4.10
Gluconic	4.20
Malic, fumaric	4.30
Lactic	4.40
Succinic	4.60
Glutaric	4.70
Pimelic, adipic	5.10
Acetic	5.40
Propionic	5.50

Source: Chung and Goepfert (1970).

form of organic acids can pass through the cytoplasmic membrane. The pH of a food directly affects the proportion of the preservative that can enter the cell, as described by the Henderson–Hasselbalch equation (Lund and Eklund, 2000):

$$pH = pK_a + \log_{10}[A-]/[HA]$$

where [A−] is the concentration of dissociated acid, or anionic form, and [HA] is the concentration of undissociated acid, or acid form.

When the concentrations of the anionic and acid forms are equal, pH = pK_a, meaning that 50% of the added preservative is undissociated and can be effective as a preservative. When the pH is 1.0 unit below the pK_a value, the preservative is about 91% undissociated. When 1.0 unit above the pK_a, the preservative is about 9% undissociated. Thus, the antimicrobial effectiveness of organic acids increases as the pH value of the food is lowered.

Most organic acids are lipophilic, being more soluble in fats and oils than they are in water. Therefore, fat and oil-containing foods can concentrate the acids in their fat phase, effectively blocking their antimicrobial properties. This is the most important consideration because microorganisms can grow only in the water phase of food products. In some foods that are water-in-oil emulsions, e.g. butter, microbial growth is physically restricted by the very small size of the water droplets that are encased in fat. The distribution of the preservative in the fat and water phases of the food is quantified as:

PC = proportion of compound in fat phase/proportion of compound in water phase

Higher PC values indicate that the preservative is increasingly less soluble in water. The relatively high PC values and low pK_a values of most preservatives limit their effectiveness to acidic foods (pH <5.5) and lower-fat foods (Table 6.7). There are several exceptions to this general situation. Methyl paraben has a high pK_a (8.47), meaning that it can be effective in foods of all pH values. Propionic acid is about six times more soluble in water than in fat (PC = 0.17), enabling it to be a more effective antimicrobial agent in higher-fat foods.

Organic acids are generally difficult to incorporate into the water phase of food. Therefore, they are almost always used in their salt forms, which are more easily solubilised in water, e.g. potassium sorbate, calcium propionate and sodium benzoate. Sorbic acid/potassium sorbate is the most broadly effective preservative available for food use, being inhibitory to bacteria,

Table 6.7 Chemical properties, usage and antimicrobial effectiveness of major food preservatives.

	Preservative			
Property	**Sorbic acid**	**Propionic acid**	**Benzoic acid**	**Methyl paraben**
Dissociation constant	4.76	4.87	4.20	8.47
Partition coefficient	3.0	0.17	6.1	6.0
Typical usage (% w/w)	0.1–0.3	0.2–0.8	0.1	0.1
Principal usage	Many foods, salads, syrups, beverages	Yeast-leavened bakery products	Fruit drinks, soda	Beverages
Relative effectiveness against				
Bacteria	++	−	+	+
Yeasts	++++	−	++++	++
Moulds	++++	++	++	+++

Source: Raczek (2005) and W.H. Sperber, unpublished data.
+, inhibition; −, no inhibition.

Table 6.8 Influence of partition coefficient on the mould-free shelf life of a refrigerated, high-fat, unbaked pastry product.

Ingredient (%, w/w)	Formulation 1	Formulation 2
Wheat flour	47.9	48.3
Water	20.0	25.0
Shortening or lard	30.0	25.0
Sodium chloride	2.0	1.6
Potassium sorbate	0.1	0.1
Total	100.0	100.0
Properties		
pH	5.0	5.0
Water activity	0.92	0.94
Mould-free shelf life at 7°C	70 days	>100 days

yeasts and moulds. However, as depicted in Table 6.7, food preservatives are generally much more effective against yeasts and moulds than they are against bacteria. A major exception is that calcium propionate is not inhibitory to yeast metabolism. Therefore, it is widely used as a mould inhibitor in yeast-leavened bakery products, in which yeast inhibition would be highly undesirable.

One of the authors experienced a dramatic example of the influence of the partition coefficient on preservative effectiveness (Table 6.8). A product development team had commercialised a successful refrigerated, unbaked, pastry product, represented by Formulation 1 in this example. Several potential product improvements were investigated in Formulation 2, in which the sodium chloride and shortening concentrations were decreased and the water content was increased. The increase in moisture content raised the a_w of Formulation 2 from 0.92 to 0.94, but its pH and concentration of potassium sorbate were unchanged.

Given the increased moisture content and a_w value of Formulation 2, even an experienced food product developer would have predicted that Formulation 1 would have a longer mould-free shelf life than Formulation 2. Quite surprisingly, the mould-free shelf life of Formulation 2 proved consistently to be about 50% longer than that of Formulation 1. This counterintuitive result could be explained only by an altered distribution of potassium sorbate (PC = 3) between the product's fat and water phases. The reduction of fat and increase in water contents in Formulation 2 allowed more of the potassium sorbate to remain in the water phase and provide greater mould inhibition. The calculated 43% increase of potassium sorbate in the water phase of Formulation 2 closely correlated to the observed increase in product shelf life.

The preservatives discussed earlier are not universally inhibitory to fungi. Sorbate-resistant *Penicillium* spp. (Marth *et al.*, 1966) and benzoate-resistant yeasts (Pitt, 1974) have been reported.

Seemingly, countless research papers have been published about a great many chemical preservatives. Comprehensive reviews have been published by Foegeding and Busta (1991) and Lund and Eklund (2000). Some of the putative preservatives, e.g. antioxidants, have not proved practical for commercial use as antimicrobial agents in foods. They are not included in this chapter. Several additional commercially practical examples are described here.

Sulphur compounds

Sulphur compounds have been known to inhibit microbial growth since early civilizations burnt sulphur in barrels to preserve wine quality. Sodium bisulphite is used today to prevent yeast spoilage of dried fruits and wine. It is also used to prevent the growth of bacterial spore formers during the production of products such as dehydrated potatoes. A thorough description of sulphur chemistry has been provided by Block (1991).

Sodium nitrite

Sodium nitrite is the major antimicrobial component of curing salts that are used in many meat, poultry and seafood products. In addition to inhibiting the germination and outgrowth of bacterial spores, it has desirable organoleptic properties, particularly colour stabilisation and a characteristic flavour. It is sometimes required by regulation because of its ability to reduce the risk of botulism in cured products.

Nisin

Nisin is the primary example of bacteriocins, small proteins that are usually produced by lactic acid bacteria. Much like sodium nitrite, it inhibits the germination and outgrowth of bacterial spores. The use of nisin and other bacteriocins is subject to regulatory limitations. Nisin has been approved in the United States for use in pasteurised cheese spreads and liquid pasteurised eggs.

Carbon dioxide

Carbon dioxide is an effective antimicrobial inhibitor, though it is usually not considered in discussions of chemical preservatives. A feedback inhibitor of aerobic respiration, carbon dioxide, at sufficient concentrations, inhibits the respiration and growth of obligately aerobic organisms such as pseudomonads, moulds and humans (its use in food production environments can pose an occupational safety hazard). It is frequently used as a component of headspace gases in modified atmosphere packaging of bakery and meat products (see Chapter 7). It also provides a secondary preservative benefit in carbonated beverages, where the active moiety is quite likely carbonic acid, because of the low beverage pH (J. I. Pitt, 1982, personal communication).

Essential oils

Essential oils from spices, herbs and other plants have been found to have antifungal properties (López-Malo *et al.*, 2005). Along with other naturally present flavour compounds, e.g. diacetyl and smoke, essential oils may contribute to effective microbial inhibition in food products through interactions and hurdle effects.

6.2.4 Oxidation–reduction potential

The types of microorganisms that can grow in food are sometimes influenced by the food's oxidation–reduction potential, also referred to as the O/R potential, redox potential or Eh. The electro motive force in a food (Eh) can be directly measured. It ranges from a minimum value of −421 mV (highly reduced) to a maximum value of +816 mV (highly oxidised). Eh values in food are largely 'poised' or buffered by food constituents. The surface Eh may be raised by exposure to atmospheric oxygen (Jay, 2000; Morris, 2000).

Obligate aerobic microbes are favoured to grow in foods with positive Eh values, while obligate anaerobic microbes require highly reduced microenvironments for growth. Most foods are naturally poised at a negative Eh value. Therefore, the growth of obligate aerobes is usually restricted to the surface of a food that is exposed to the atmosphere. The interior portions of low-acid canned foods when opened and exposed to the atmosphere can support the growth of obligate anaerobes such as *Clostridium botulinum* because of their low internal Eh value.

Fresh or cooked foods that are typically exposed to the atmosphere can support the growth of obligate anaerobes, e.g. *C. botulinum*, if access to oxygen is eliminated, or if the interior of the food sustains a sufficiently low Eh. A potential botulism hazard was detected in the early 1970s when fresh mushrooms for commercial sale were packaged with oxygen-impermeable film. Respiration of the mushrooms removed oxygen in the package's headspace, providing anaerobic conditions suitable for the growth of *C. botulinum*. This potential hazard is easily eliminated by creating one or more small holes (3–6 mm) in the packaging material (Sugiyama and Yang, 1975).

An outbreak of botulism, consisting of 28 cases and one death, was caused by sautéed onions that were held on the side of a restaurant grill for long, but indeterminate times. The onions were used throughout the day in the restaurant's popular grilled sandwiches, and were possibly kept on the grill for one or more days. The interior of a mound of sautéed onions in margarine was shown to readily support growth and toxin production by *C. botulinum*. It is possible that the fat content of the margarine increased the botulism hazard by blocking the access of oxygen to the onion surfaces (Solomon and Kautter, 1986). One of the authors and his family ate at the implicated restaurant during the period of the outbreak in 1983. Fortunately, all had ordered grilled sandwiches *without* onions. This curious fact remains of some importance to the surviving author.

6.2.5 Interactions between preservative factors

Many preservative factors, e.g. pH, a_w, chemical preservatives and temperature, can be used to control microbial growth. Microorganisms also vary in their resistance to the inhibitory effects of individual preservative factors (Figure 6.1a). In some instances, the required degree of food safety and quality assurance can be achieved simply by increasing the amount of a preservative factor (Figure 6.1b).

Of course, no food is microbiologically stabilised by a single preservative factor, as every food has its characteristic pH, a_w and storage temperature. However, the additive effects of the individual preservative factors can sometimes suffice to protect the food during its expected shelf life (Figure 6.1c). The use of multiple preservative factors can permit the development of a food in which no single preservative factor imparts an undesirable organoleptic property. For example, the use of multiple preservative factors could be used to stabilise a food with only a slight reduction in a_w, rather than a more drastic dehydration of the food. Modern technology enables food scientists to incorporate multiple preservative factors at the appropriate levels to optimise the functional performance and organoleptic properties of the food. The use of multiple preservative factors to inhibit microbial growth is commonly known as 'hurdle technology' (Leistner and Gould, 2002).

The additive effects of preservative factors to control microbial growth have long been understood. Various combinations of pH and a_w can be used to prevent the growth of *C. botulinum* without resorting to extreme reductions in either factor (Table 6.9). Guynot *et al.* (2005) showed that pH, a_w and chemical preservatives could interact to inhibit mould spoilage

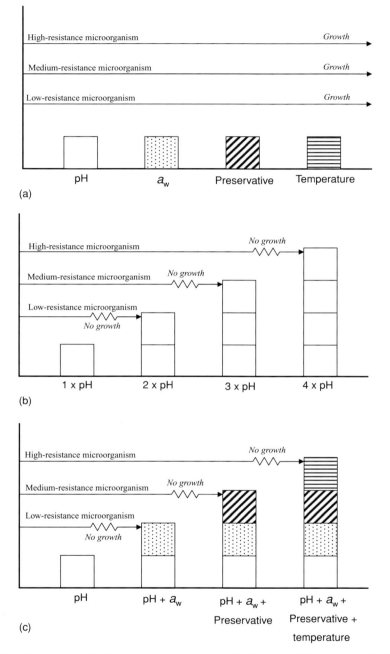

Fig. 6.1 The effect of various factors on microorganism growth: (a) effect of single factor, (b) effect of single factor with increased concentration, (c) additive effects of combined factors and (d) interactive effects.

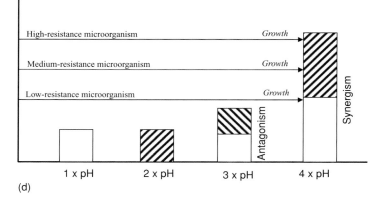

Fig. 6.1 *(Continued)*

of bakery products. Properly designed research and analysis can be used to demonstrate intricate interactions between preservative factors. Antagonistic interactions can occur when two or more preservative factors interact to diminish the combined effectiveness expected from observations of the individual effectiveness of the factors. Synergistic interactions occur when the combined preservative factors produce a greater inhibitory effect than that expected from a simple additive effect of the individual factors (Figure 6.1d).

Figures 6.1c and 6.1d differ from those commonly used to explain hurdle technology. A microorganism introduced into a particular food encounters all the food's preservative factors simultaneously, rather than sequentially (as depicted in Leistner and Gould, 2002, p. 19).

6.3   USE OF EXPERIMENTAL DESIGN AND ANALYSIS

6.3.1   Challenge testing

During the research and development of food products, it is usually necessary to conduct challenge testing in which the food is inoculated with one or more types of microorganisms, incubated and periodically evaluated to learn the fate of the target microbes. Challenge testing can be conducted for one or more of the following purposes: to develop and validate food

Table 6.9   Combined effects of water activity and pH on the growth of *Clostridium botulinum* type B.

	pH				
a_w	**7.0**	**6.0**	**5.5**	**5.3**	**5.0**
0.997	+	+	+	+	+
0.99	+	+	+	+	−
0.98	+	+	+	+	−
0.97	+	+	+	−	−
0.96	+	−	−	−	−
0.95	−	−	−	−	−

Source: Baird-Parker and Freame (1967).
+, growth; −, no growth.

safety control measures, to ascertain potential product quality and shelf life issues related to microbial spoilage, to optimise new product formulations, and to verify changes made in existing commercial formulations.

Care must be given to the selection of the microorganisms used to challenge a product. Generally, a minimum of two or three strains of each microorganism is used in order to account for potential response variability between strains (Doyle, 1991). The purity and identity of each culture should be confirmed before it is used in a challenge test. This important task is typically performed regularly by the curator of the laboratory's culture collection. Important strains of particular pathogenic microorganisms can be obtained from commercial culture collections, e.g. the American Type Culture Collection. Strains of spoilage microorganisms are best obtained from spoiled samples of the product being evaluated, or from a very similar food product. In this case, the spoilage microorganism should be isolated, identified and shown to be responsible for the observed spoilage defect before it is used in a challenge test.

Individual strains of vegetative microorganisms need to be cultured by conventional procedures, usually for 24 hours, to produce a predictable number of viable stationary-phase microbes. As a guide to the number of bacteria in a culture medium, it is useful to remember that the maximum population of vegetative bacteria in a mature culture is about 10^9 cells/mL. The strains can be used individually, but are generally pooled to facilitate inoculation of multiple strains into the food. The actual population density of the pooled inocula and the inoculated food is usually determined at the beginning of the challenge test. Dormant bacterial spores are often prepared and maintained for weeks or months at known population levels under refrigerated or frozen conditions. Depending on the type of product, the spores may be heat shocked before being inoculated into the food. If a heating step is involved in the food's processing after spore inoculation, this step would serve as the heat shock treatment.

It is important to challenge food products with a reasonable microbial load. After processing, the typical microbial load of a commercial food product, whether in the spore or vegetative forms, is quite low. Therefore, an initial inoculation level of about 10^2 colony forming units (cfu)/mL or gram of food is recommended in situations where the growth potential of the target microbe is being evaluated. Higher inoculation levels could overwhelm the food's preservative system, giving a false indication of spoilage or a food safety hazard in the commercial product, which would never be naturally contaminated at such high levels. In situations in which destruction of the target microbe by heat, acidification or other process steps is being evaluated, the target inoculation level should be at least 10^6 cfu/g of food so that a quantifiable reduction in population can be readily determined. In certain situations in which the growth potential of a particular microorganism is completely unknown, an intermediate inoculation level of 10^4 cfu/g of food can be used, affording the opportunity to readily observe growth, death or stasis of the target population.

The challenge test should replicate the conditions of commercial production as closely as practical. A food manufacturing company struggled in the 1960s to identify the spoilage microorganism that was responsible for a large outbreak of spoilage in salad dressing at the retail level. The spoiled dressing readily yielded a large population of a particular yeast which was duly isolated and used for subsequent challenge studies. None of the salad dressing formulations inoculated with the putative spoilage yeast spoiled during months of storage. Drummed starch paste had been used in the manufacturing plant to provide a desirable texture to the salad dressing. Interviews with production personnel revealed that several months before the spoilage outbreak, one drum of starch paste had a 'funny' odour. Rather than 'waste' a drum of starch paste, the production operator portioned small amounts of spoiled starch paste, along with the required balance of unspoiled starch paste, into 19 consecutive days of salad dressing production. Within several months, about one million jars of salad dressing had spoiled. Armed with this knowledge, company microbiologists then inoculated starch paste with the putative spoilage yeast and incubated it for several days before using the paste to produce salad dressing. All the salad dressing samples spoiled quickly. It seemed that the starch paste provided conditions that enabled the spoilage yeast to adjust to the harsh (acetic acid) environment of the salad dressing (Sperber, 2009b). Food researchers need to be alert to such possibilities when designing challenge studies for new food products.

The above example will be mentioned in later chapters as it superbly demonstrates the need for adequate training and awareness on the part of all plant personnel involved in processing operations. This example also illustrates a potential very important product safety hazard. While yeast spoilage might not seem to be a food safety hazard, it must be pointed out that the salad dressing was packaged in glass jars with screw-on lids. Vigorous yeast fermentation in the affected products caused many jars to explode, the consequences of which could have been serious to any person near an exploding jar. Many food products formerly packaged in glass containers are now packaged in plastic containers. Nonetheless, the potential hazard of exploding glass jars cannot be ignored for the remaining products packaged in glass containers. Of course, the desired remedy for such a problem would be proactive attention to product design and process controls so that such spoilage outbreaks would not happen.

Careful attention must also be given to the method of product inoculation in challenge studies. It is vital that product characteristics such as a_w and pH not be altered in the inoculated products. Another earlier incident, this one from the 1970s, readily proved this point. Soon after the 1973 adoption of canned foods regulations in the United States, some food safety officials questioned the potential safety of pasteurised process cheese spreads. The canned foods regulations required that any shelf-stable canned food above a_w 0.85 or pH 4.6 be autoclaved to ensure control of the potential botulism hazard. Pasteurised process cheese spreads typically have an a_w value around 0.93 and a pH value around 5.5. Unlike retort-processed low-acid canned foods, the cheese spreads were simply hot-packed into glass jars and distributed without additional processing. Despite the disparity between the cheese spread a_w and pH and the regulatory requirements for low-acid shelf-stable canned foods, no cases of botulism and negligible spoilage of commercial shelf-stable cheese spreads had been noted during several decades of widespread distribution and consumption. It had been assumed that the interaction between the a_w and pH of the cheese spread was sufficient to control the growth of *C. botulinum* and other bacterial spore formers that could survive the cooking and hot-filling process.

Nonetheless, the food safety officials quite correctly decided to validate the safety of shelf-stable cheese spreads by conducting challenge tests. Fifty jars each of five commercial cheese spreads were injected with 0.1 mL aqueous *C. botulinum* spore preparations. After incubation nearly 100% of the samples of two of the cheese spreads tested positive for botulinal toxin (Kautter *et al.*, 1979). Additional research was then conducted by others to verify the above results, which were not in harmony with the observed experience of several decades of commercial production and consumption. In these projects, 11 formulations of cheese spreads were inoculated with *C. botulinum* spores during mixing of the product, which was then cooked and hot-packed into glass jars, steps which closely matched the commercial process. None of the incubated samples in these trials was found to contain botulinal toxin (Tanaka, 1982). It was concluded that the original demonstration of botulinal toxin production was made possible by the means of inoculation by injection. At least some of the water in the spore inoculum remained in the injection site and was sufficient to increase the a_w in microenvironments to permit toxin production.

In many challenge tests, it is possible to mix the test inoculum thoroughly into the product and subject it to relevant process steps before packaging and incubation. Sometimes, it is necessary to inoculate the product surface in order to evaluate spoilage. In these cases, it is necessary to prepare the inoculum in a substrate that will not alter the product's properties. For example, a suspension of mould spores can be prepared in dry flour or starch, precisely quantified, and verified for stability before sprinkling onto a product surface. It is also possible to use an aqueous suspension of any microorganism and transfer it to a product surface with a negligible transfer of water by using small sterilised squares of a bristly painting pad (R. B. Ferguson, personal communication).

The evaluation of microbial growth or death in challenge tests can be accomplished with the use of a wide variety of non-selective and selective microbiological media to quantify, isolate and identify the test microbes as necessary (Downes and Ito, 2001).

6.3.2 Accelerated shelf life testing

Challenge test product samples should be stored at the commercial distribution temperature and evaluated over the entire expected product shelf life, or until the product exhibits quality defects or fails to meet food safety requirements. When the primary purpose of the challenge test is to validate the safety of the product, e.g. a *C. botulinum* challenge test in which toxin production represents a failure, the test should be continued for 1.5 times the anticipated shelf life, or until toxin production is demonstrated (Doyle, 1991).

Quite often, in research and development projects and in commercial plant production, it is impractical to store products at the expected commercial distribution temperature for the full shelf life to observe whether a food spoilage defect or potential food safety hazard will develop. It is, however, generally possible to store products at temperatures higher than the commercial distribution temperature for shorter periods. Within practical ranges, chemical reaction rates and microbial growth rates in foods will increase as temperature is increased (Labuza and Schmidl, 1985), and this facilitates accelerated shelf life testing (ASLT), where samples are periodically evaluated for types of product failure over a shorter period.

In the case of research and development ASLT tests, it is usually important to monitor the samples for many types of product failure. In the case of commercial production samples, the ASLT conditions can be tailored to accurately predict the potential occurrence of the product's principal failure mechanism before the end of shelf life. The example in Table 6.10 describes the ASLT conditions that can predict potential product failure caused by excessive growth of lactic acid bacteria. The expected commercial shelf life of this product category is 90 days at refrigerator temperatures (4°C). Any of the other sets of temperature and times in Table 6.10 can be used in this company's laboratories to evaluate the growth of lactic acid bacteria. For example, evaluation of samples stored for 2 days at 25°C or 4 days at 20°C can accurately

Table 6.10 The practical example of ASLT conditions to predict the shelf life of a refrigerated food.

Storage temperature (°C)	Storage time (days)
4	90
7	28
15	10
20	4
25	2

Source: Adapted from Sperber (2009a).

predict potential lactic spoilage of this particular product category within the 90-day expected shelf life at 4°C. When premature product failure is predicted by analysis of the ASLT samples, the manufacturer can take action to hasten the distribution cycle of the product or to prevent product distribution.

6.3.3 Predictive microbiology and mathematical modelling

The use of conventional research and development methods, including challenge tests, can consume months or even years before a safe product of high quality is commercialised. Food researchers have been increasingly assisted in the past several decades by the development and use of computer-based programmes for experimental design, analysis and prediction (Ross and McMeekin, 1994).

Before the revolution in information technology, experimental design at its best consisted of the production of a large number of samples in which individual preservative factors would be tested at various levels. In complex foods, up to hundreds of formulations would be prepared and evaluated separately. The advent of simple factorial designs permitted the simultaneous testing of two or more preservative factors, providing a modest increase in information yield.

Today, computer-based programmes such as response surface methodology, predictive microbiology and advanced mathematical computations can provide far more information with far fewer samples (Whiting, 1995). These technologies permit the simultaneous evaluation of multiple intrinsic and extrinsic preservative factors in a single product formulation. Quantitative estimates of the effects of each individual preservative factor as well as all possible combinations of two or more factors are produced. The relative speed of these technologies permits ready optimisation of product formulations, as well as the identification of antagonisms or synergisms between preservative factors that would likely have not been detected by conventional procedures.

Mathematical models are the latest advance to assist the food product development team. These are strictly a computer-based activity in which relevant product information and microorganism characteristics are entered into the database. The programme calculates the desired output, e.g. amount of microbial growth or destruction, and the time for toxin production or the end of shelf life. A detailed description of the use of a mathematical model has been provided by McMeekin *et al.* (1997). A model for the growth of *Pseudomonas* spp. in raw poultry has been published by Dominguez and Schaffner (2007).

The US Department of Agriculture (USDA) has for some years developed, continually improved, and made freely available its Pathogen Modelling Program (ARS, 2007). Its current version (7.0) will predict growth rates of 11 pathogens, cooling and growth rates of 2 pathogens in cooked meat products, survival of 4 pathogens, thermal inactivation of 3 pathogens and radiation inactivation of 2 pathogens.

In 2003, the USDA joined the Food Standards Agency of the UK and the Norwich-based Institute of Food Research to establish the ComBase Consortium, a combined database for predictive microbiology. ComBase can generate graphs and predict the rates of growth, survival or inactivation of a very wide array of spoilage and pathogenic microorganisms. ComBase is accessible online without charge at http://www.combase.cc.

6.3.4 Theory versus reality

Mathematical modelling permits the researcher to conduct a great deal of research on the computer, virtually at the speed of light. One simply inputs the product and microbial characteristics and gets a quick estimate of preservative efficacy, the impact of changing one or more preservative factors in a food, inactivation rates during heat treatments, etc.

Researchers must be aware that such powerful tools should be used alongside expert judgement and interpretation and cannot be used as the ultimate assurance of food safety. Ultimately, challenge tests with appropriate pathogenic or spoilage microorganisms must be conducted to validate product safety, quality and shelf life. Predictive modelling programmes can be used advantageously to sort through a great many factors before final development and validation is confirmed in the 'real' world. Keep in mind that, useful and powerful as they have proved to be, predictive modelling programmes are no substitute for laboratory confirmation. Ultimately, food safety is not a computer game. Food safety is, rather, ultimately assured by the common sense application and validation of necessary product and process control measures.

6.4 INGREDIENT CONSIDERATIONS

The selection, handling and use of ingredients require careful consideration both in product design and process control activities. Relatively small quantities of ingredients used in the production of research samples can present the same food safety and public health hazards that can be presented when very large quantities are used in commercial food production. Therefore, research laboratories must provide the same assurance of the safety of test samples that manufacturing facilities must provide for consumer food products.

6.4.1 High-risk ingredients

While the term 'sensitive ingredient' has long been used in the United States, much of the world refers to these as 'high-risk ingredients'. We have used the latter term throughout this book. High-risk ingredients have a history of potential contamination with pathogenic microorganisms, mycotoxins, etc. Therefore, measures are taken to control high-risk ingredients so that contamination of food products does not occur.

Salmonella was the first foodborne pathogen to merit the description, 'ubiquitous'. While its natural habitat is the animal intestine, *Salmonella* grows well in food processing environments and it survives well in dry environments. In the past 50 years, it has received more attention from the food industry than any other pathogen. A number of ingredient categories became known as '*Salmonella* high-risk ingredients'. Chief among these are cooked meat and poultry products, dried milk and egg products, and soy products.

Aflatoxin, the first identified mycotoxin, has also received a great deal of attention from the food industry because of its high toxicity and suspected role as a liver carcinogen. Peanut and corn products receive the most attention because aflatoxin can be produced in the peanut and corn field crops and, unless properly controlled, persist into a wide variety of food products. Ingredients receiving lesser attention include tree nuts, dried coconut, tapioca flour, cottonseed meal and figs.

Staphylococcus aureus, one of the earliest detected foodborne pathogens, had in earlier years been a regular problem in fermented sausages and cheeses. In the event of poor starter culture performance, *S. aureus* could grow to very high numbers, producing heat-stable enterotoxins. These problems have largely been eliminated and can be controlled simply by monitoring and verifying a timely pH drop during fermentation. Other outbreaks of staphylococcal food poisoning have been traced to food service products which were stored or transported without adequate temperature controls, enabling the growth of *S. aureus*. In these cases, the source

Table 6.11   Pathogenic microorganisms that are associated with high-risk food ingredients.

Microorganism	Food ingredient
Salmonella spp.	Eggs and egg products
	Dried dairy products
	Milk chocolate
	Soy flour
Staphylococcus aureus	Fermented cheese above pH 5.4
	Fermented sausages above pH 5.4
Listeria monocytogenes	Ready-to-eat perishable delicatessen meat and poultry products
	Soft, fresh cheeses
	Surface-ripened cheeses
Escherichia coli O157:H7	Raw, ground beef
	Raw milk
	Fresh vegetables

of contamination can be food handlers or cross-contamination from ingredients such as hand-deboned meat and poultry products.

Listeria monocytogenes has relatively recently been identified as a potential hazard in refrigerated, ready-to-eat foods of extended shelf life that will support its growth. Refrigerated foods typically implicated in listeriosis outbreaks include soft, ripened cheeses; cooked meat, poultry and seafood products; delicatessen meats and fluid milk. While these foods do not easily transfer a listeriosis hazard to processed compound food products, the product development team must be alert to the potential listeriosis hazard whenever ingredients of this type are used.

Bacillus cereus has long been recognised as a risk in starchy foods. Much like staphylococcal enterotoxin, the emetic toxin of *B. cereus* is heat-stable. *B. cereus* spores can also survive cooking processes. Therefore, its presence can pose a risk in cooked or dried potatoes and in cooked rice that is not stored under adequate refrigeration before preparation as refried rice.

A summary of some of the key microbiological and chemical hazards associated with high-risk ingredients is provided in Tables 6.11 and 6.12. Extensive information on high-risk and other ingredient control measures is provided in Chapter 7.

Table 6.12   Chemical agents that are associated with high-risk food ingredients.

Chemical agent	Food ingredient
Mycotoxins	
Aflatoxin	Peanuts (groundnuts) and peanut products
	Dry corn products
	Tree nuts
Fumonisin	Dry corn products
Deoxynivalenol	Barley, wheat
Ochratoxin	Barley
Zearalenone	Barley, corn
Ergot	Rye
Allergens	Peanuts and peanut products
	Tree nuts
	Crustaceans
	Fish
	Eggs and egg products
	Milk and dairy products
	Soybeans and soy products
	Wheat and wheat products

As detailed in Chapter 11, all food prototypes must be safe for consumption. This requirement pertains to samples tasted in the test kitchen, or evaluated by internal sensory panels and during external central location consumer tests, home-use tests, and market trials before product commercialisation. It is vital that the product development team coordinate its development activities with the product safety assessment team so that the safety of all sensory analysts and consumers is assured.

6.5 CONCLUSIONS

The design of product formulations by the product development team is the most important first step in the production of a safe food product. If appropriate control measures based on the product's intrinsic factors are not researched and validated at this step, significant time may be lost to correct the early mistakes. Worse, a product that is potentially unsafe or of reduced quality and shelf life might be produced and distributed.

7 Designing a safe food process

7.1 INTRODUCTION

As indicated at the beginning of Chapter 6, we have reached the conclusion that, at its core, the Hazard Analysis and Critical Control Point system (HACCP) consists of two essential processes – product design and control, and process design and control. Both the product and process design requirements are typically defined by research and development groups. It is during this period that control measures must be tested, validated and incorporated into the product formulation (Mortimore and Wallace, 1998; Sperber, 1999). As is the case for product design, the process design activity is best performed by an experienced team that typically includes food safety specialists, microbiologists, food scientists, process engineers and packaging engineers. Upon completion, the process design requirements are communicated to manufacturing teams that scale up and validate the design requirements at the commercial production level. Individual plant HACCP teams complete the HACCP plan for each product and establish the monitoring, verification and record-keeping procedures.

This chapter describes a complete range of practical food processing technologies that are applied extrinsically to the food product in order to control identified foodborne hazards. The product development team should consider that the extrinsic control factors described in this chapter might interact with the intrinsic control factors described in Chapter 6. For example, the effectiveness of chemical preservatives may be enhanced at reduced storage temperatures. While many of the control measures described in this chapter are most appropriate for the control of microbiological hazards, similar measures and considerations of controls must be undertaken by the HACCP team for the control of chemical and physical hazards.

Some educators refer to the 'three Ks' as a comprehensive programme of microbiological hazard control in food production, those being, 'kill them', 'keep them from growing' and 'keep them out'. Described more scientifically, these three procedures are:

1. *Destruction of microorganisms.* Many well-established and several novel procedures are available to kill microorganisms. These include thermal processes such as cooking, pasteurisation and sterilisation, and non-thermal processes as such irradiation, high hydrostatic pressure and pulsed electric fields.
2. *Prevention of microbial growth.* The primary extrinsic factors used to control microbial growth are refrigeration, freezing and drying. The intrinsic factors used to control microbial growth are described in Chapter 6.

Food Safety for the 21st Century, First Edition By Carol A. Wallace, William H. Sperber and Sara E. Mortimore
© 2011 Carol A. Wallace, William H. Sperber and Sara E. Mortimore

3. *Prevention of contamination.* Many potential microbiological hazards can be avoided by preventing cross-contamination from raw materials and the processing environment to processed foods. Cleaning and sanitation procedures and personnel practices used in food processing facilities are most important in this regard.

Chemical and physical hazards are generally controlled by prerequisite programmes and by the use of detection devices. Programmes must be in place to avoid contamination of the processed food with chemical hazards such as undeclared allergens and physical hazards such as insects, metal, wood and glass fragments.

7.2 PROCESS CONTROL OF MICROBIOLOGICAL HAZARDS

7.2.1 Destruction of microorganisms

Microorganisms in food materials can be killed by thermal processes and non-thermal processes, including chemical disinfection. Because the reduction of microbial populations occurs logarithmically, several terms have come into common usage to easily compare the lethal effects of various treatments (Joslyn, 1991). The '*D* value', or decimal reduction time, is the amount of time at a particular temperature required to reduce a microbial population by 90%, or one $\log_{10}$ unit. The '*z* value' indicates the amount of change in temperature (°C) that is required to shift the *D* value by 90%, or one $\log_{10}$ unit. For example, if a microorganism has $D_{110°C} = 5.0$ minutes and $z = 10°C$ (a value typical of bacterial spores), it would have $D_{120°C} = 0.5$ minute and $D_{100°C} = 50.0$ minutes.

Thermal processes

The importance of proper thermal processing for food safety led to a number of regulations in the United States. First written in 1923, the Pasteurized Milk Ordinance (PMO) deals with the pasteurisation of milk (FDA, 2006). In 1973, Food and Drug Administration (FDA) promulgated regulations for low-acid canned foods (CFR, 2008a) and acidified foods (CFR, 2008b). All these are HACCP-based regulations. They deal with time and temperature process controls, sanitary design and sanitation requirements, and administrative requirements, including record-keeping. As explained in Chapter 1, the PMO preceded the origin and evolution of the HACCP system by about 50 years. The canned foods regulations, a collaborative effort in which the Pillsbury Company assisted the FDA, sprang directly from the early HACCP developments (Chapter 1).

Pasteurisation. In conventional usage, the term pasteurisation refers to the destruction of vegetative microbial cells and viruses in food products. Several official definitions are more extensive and specific. Pasteurisation is: 'Any process, treatment, or combination thereof that is applied to food to reduce the most resistant microorganism(s) of public health significance to a level that is not likely to present a public health risk under normal conditions of distribution and storage' (NACMCF, 2006). A more specific definition is: 'Pasteurization is a microbiocidal heat treatment aimed at reducing the number of any pathogenic microorganisms in milk and liquid milk products, if present, to a level at which they do not constitute a significant health hazard' (Codex, 2009). Pasteurisation conditions are designed to effectively destroy the organisms

Table 7.1 Thermal resistance properties of vegetative bacterial cells.

Microorganism	°C	D (minutes)	z
Pseudomonas fluorescens	55	1–2	—
Escherichia coli	55	4	—
E. coli O157:H7[a]	60	0.75	—
	71	0.01	—
Salmonella spp.[a]	70	0.05–0.5	—
Salmonella senftenberg (775 W)[a]	60	10.8	6
Staphylococcus aureus[a]	60	7.8	4.5
Listeria monocytogenes[a]	70	0.1–0.3	

Source: Derived from Doyle (1989), Lewis and Heppell (2000) and Farkas (2001).
[a]Pathogens.

Mycobacterium tuberculosis and *Coxiella burnettii* (Codex, 2009). Thermal resistance values of representative vegetative microorganisms are presented in Table 7.1.

Most often, pasteurisation involves a cooking or heating procedure conducted at atmospheric pressure. It is used to protect the public health by killing pathogenic microorganisms and to extend product shelf life by killing spoilage microorganisms. Pasteurised products are not sterile. They must be refrigerated during further distribution unless they are otherwise preserved, for example, by water activity or pH reduction.

The pasteurisation of some foods has such prominent public health significance that the time and temperature requirements are the subject of regulations (Table 7.2). It is important to note that both time and temperature must be considered. As the pasteurisation temperature is increased, the required processing time is reduced. The tabulated parameters are the minimum time and temperature that must be applied to the respective products. In practice, food processors typically use somewhat higher temperatures and/or heating times (operational limits – see Chapter 12) to provide a margin of safety, both for food safety and for regulatory compliance considerations. It is difficult, but not impossible, to pasteurise dried materials. The treatment of dried egg albumen at 54.4°C for 7 days is known as 'hot-boxing'. Storage of the albumen, packed in approximately 20 kg boxes, under such conditions will usually kill residual salmonellae that may have contaminated the albumen during packaging. Similar hot-box treatments can be used for milled cereal grains, starches, etc., that are to be used in infant or geriatric formulations to provide a greater degree of assurance of the absence of vegetative microbial pathogens. A

Table 7.2 Examples of pasteurisation requirements in the United States.

Product	Time	Temperature (°C)	References
Milk	30 minutes	63	FDA (2006)
	15 seconds	72	FDA (2006)
	1 second	89	FDA (2006)
Ice cream mix	30 minutes	69	FDA (2006)
	25 second	80	FDA (2006)
Liquid eggs	3.5 minutes	60	CFR (2008)
Salted eggs or yolks	3.5 minutes	63.3	CFR (2008c)
Spray-dried egg albumen	7 days	54.4	CFR (2008d)
Blue crab meat	1.0 minute	85	Ward et al. (1984)
	4.2 minutes[a]	85	Cockey and Tatro (1974)

[a]For 12-log kill of *Clostridium botulinum* type E spores.

comparison of the markedly longer times required to pasteurise dried eggs versus the short times required to pasteurise liquid eggs (days vs minutes) indicates the difficulty in inactivating microorganisms with dry heat.

Most pasteurised products require refrigeration to prevent the growth of spoilage microorganisms and pathogenic spore formers whose spores survive pasteurisation. However, products with a sufficiently low water activity (usually below 0.85, such as sugar syrups) or pH value (usually below 4.0, such as acidified foods) will be microbiologically stable at ambient temperatures if properly handled and packaged. Such products are usually cooked and hot-filled at 80–100°C into the final consumer package. The residual heat in the product is sufficient to kill vegetative microorganisms that may have been in the container, as well as airborne microorganisms that may have entered during the filling operation. Bottled products usually need to be inverted for 2 minutes so that the bottle neck and cap interior are adequately pasteurised. The product's reduced water activity and/or pH is sufficient to prevent the germination and growth of bacterial spores.

Bakery products are pasteurised during the baking process. However, during the cooling period before packaging, the surface of the bakery product will become contaminated with airborne mould spores, which can spoil the product before consumption. Such products can be further heat-treated in the final sealed consumer package to prevent spoilage and to extend shelf. Several heat sources such as convection ovens, microwaves or infrared bulbs can be used to heat the product inside its commercial package. The internal generation of steam will kill mould spores that have contaminated the product surface or package interior (Bouyer, 1970; Richardson and Hans, 1978). Upon cooling, the generated moisture is absorbed by the product. The expansion of the package during heating verifies the integrity of the package seals. This technology can also be applied to other products, such as irregularly shaped cooked meats, for the surface destruction of *Listeria monocytogenes*.

Many consumers cook shell eggs for a short time so that the yolk remains soft. This process is adequate to kill surface microorganisms, including salmonellae, which are common contaminants of raw poultry and eggs. The interior of the egg, particularly the yolk, is usually sterile. Therefore, soft-cooked shell eggs had been considered safe for consumption. However, over the past several decades a 'pandemic' of salmonellosis was linked to the consumption of soft-cooked shell eggs. Investigators soon discovered that a new strain of *Salmonella* Enteritidis infected laying hens and was deposited in the ovaries into the newly produced egg yolk. Protected during the soft-cooking of shell eggs, *S.* Enteritidis would survive to cause illness upon consumption. Procedures have been developed by egg producers to heat shell eggs before distribution for consumption. Shell eggs are heated in warm water at times and temperatures adequate to produce a 5-log reduction of salmonellae in the egg yolk without cooking the egg. Treated in this manner, soft-cooked shell eggs are considered safe for consumption (USDA, 1997). Widespread vaccination of laying flocks has significantly reduced egg infection by *S.* Enteritidis, eliminating the need for specialised heat treatments. Vaccine trials conducted in the United States in 1997–1999 found that shell eggs from 8.1% of non-vaccinated flocks contained *S.* Enteritidis. None of the eggs tested from 93 vaccinated flocks contained *S.* Enteritidis (Mirandé, 2000).

Sterilisation

Practical sterilisation procedures for foods involve high-temperature thermal processes. Many foods to be sterilised are packaged into metal, glass or plastic retail containers, hermetically

Table 7.3 Thermal property values of representative bacterial spores.

Bacterial spores	°C	D (minutes)	z
Bacillus coagulans	121	0.01–0.1	–
Bacillus stearothermophilus	121	4–4.5	7
Clostridium sporogenes	121	0.1–1.5	9–13
Clostridium botulinum			
Proteolytic types A and B[a]	100	25	10
	121	0.2	10
	141	0.0025	10
Non-proteolytic types B, E and F[a]	100	0.05	–
Bacillus cereus[a]	100	3	–
Clostridium perfringens[a]	90	1–9	–
	110	0.5–1.5	–

Source: Derived from Doyle (1989), Lewis and Heppell (2000) and Farkas (2001).
[a]Pathogens.

sealed and processed under pressure with steam at 121°C or higher. Some foods are sterilised by ultra-high temperature (UHT) procedures and filled into separately sterilised containers. Such processes are designed to kill bacterial spores that could otherwise cause product spoilage or foodborne illness upon consumption of the food. The thermal properties of the relevant bacterial spores are summarised in Table 7.3. Sterilised foods are processed to provide 'commercial sterility'; that is, they are not absolutely sterile. Quite likely, spores of obligately thermophilic bacteria such as *Bacillus stearothermophilus* will be present in sterilised foods. They are, however, incapable of growth in sterilised foods under normal conditions of storage and transportation at ambient temperatures. An important consideration is that such foods might spoil if stored at very high temperatures, e.g. more than 50°C, but they would not be capable of causing illness. Obligately thermophilic bacterial spore formers are not pathogenic. Commercial sterilisation processes are not adequate to inactivate prions, the infectious proteins that are responsible for spongiform encephalopathies in many mammals, including humans (Hueston, 2003).

Canning processes. Low-acid canned foods require a minimum process to assure a 12-D 'botulinum cook', to ensure the absence of spores of *Clostridium botulinum*. As can be calculated from the data in Table 7.3, the 12-D botulinum cook would be at least 2.4 minutes at 121°C. In practice, longer times than the minimum botulinum cook are used by canning operators to provide a margin of safety and to ensure the destruction of mesophilic spore formers that are more heat-resistant than *C. botulinum* and could spoil the canned foods during storage and distribution. Many additional factors are controlled to assure the safety and stability of sterilised canned foods, including, but not limited to, product processing conditions, product temperature before can filling, product viscosity, can headspace, can seam integrity and chlorination of can cooling water.

A commercial sterilisation treatment as applied to low-acid canned foods is not necessary for the production of acidified canned foods. Acidified canned foods either have a natural pH below 4.6, or are acidified so that the pH is below 4.6, or have a water activity value below 0.85. Under these conditions, surviving bacterial spores cannot germinate and grow. Acidified canned foods are usually heated to about 100°C. They may be heated in the retail package, or they may be hot-filled as described earlier in Pasteurisation section. In the United States, canning processes are defined by regulations (CFR, 2008a, 2008b).

Table 7.4 Time and temperature requirements for sterilisation by hot-air treatments.

Temperature (°C)	Time (minutes)
170	60
160	120
150	150
140	180
121	'Overnight'

Source: Joslyn (1991).

UHT processes. Foods may be sterilised at ultra-high temperatures (UHT) for a very short time, e.g. 140–150°C for several seconds. In an enclosed system, the UHT-treated foods are aseptically packed into packages that have been separately sterilised by chemical sterilants such as hydrogen peroxide, flame sterilisation, superheated steam or high-intensity UV irradiation (Lewis and Heppell, 2000). While microbiologically stable, UHT foods might be susceptible to spoilage by food enzymes, some of which are very resistant to UHT treatment.

Dry heat processes. Dry hot air can be used instead of steam to sterilise materials; however, very long times are required (Table 7.4). While seldom used directly for food sterilisation, dry heat is commonly used for the sterilisation of laboratory glassware and sampling devices.

Non-thermal processes

The thermal processes described earlier can be generally applied to foods that are cooked, canned, baked, fried, etc. Non-thermal processes can be used in specific applications intended to minimise organoleptic changes that are caused by thermal processes. Non-thermal processes tend to be more costly and less effective in reducing microbial populations than are thermal processes. Therefore, they are more suitable for pasteurisation rather than sterilisation processes. Furthermore, the extensive research and development expenses and the costs of commercial applications make many of the non-thermal processes impractical for commercial use. For these reasons, processes such as pulsed light, non-thermal plasma, oscillating magnetic fields and ohmic heating are not discussed here. Some non-thermal processes are practical for particular applications; these are discussed below.

Filtration. Filtration can be used to remove microbes, particles and some chemicals from clear liquids and gases (Levy and Leahy, 1991). Microbes can be removed from liquids to produce sterile products when filters of an effective pore size of 0.22–0.45 μm are used. Similar filters can be used to filter the incoming air supply for food production areas, thereby minimising product contamination during production and packaging. Many organic compounds can be removed from liquids and gases by filtration through columns of activated carbon. A UK dairy filters pasteurised milk to extend its shelf life.

Chemical disinfectants. Chemical disinfectants can be used to reduce microbial loads in liquid or dry food materials (Parisi and Young, 1991). Chlorine compounds and ozone are often used to sanitise water that may be used in the dipping of fruits and vegetables, as ingredient water, and in the cleaning of food processing equipment. Gaseous disinfectants, including ethylene and propylene oxides, chlorine dioxide, β-propiolactone and formaldehyde can be used to disinfect production rooms and packaging materials. Ethylene and propylene oxides were formerly used to kill mould spores in nuts, dried fruits and cocoa powder. However, ethylene oxide has fallen into disuse because of its carcinogenic and mutagenic properties. Propylene oxide is little used because of its relative ineffectiveness in killing mould spores.

Ultraviolet light. Ultraviolet (UV) light has a number of food safety and public health applications, even though its antimicrobial effectiveness is diminished by its low penetrating ability and by the shadowing effects of particles in air or liquids (Schechmeister, 1991). Major uses of UV light for disinfection involve arrays of high-intensity UV bulbs over which liquids flow. Very large arrays are used to disinfect municipal water supplies; smaller arrays are used in food processing plants to disinfect recycled flume water. UV arrays can also be used in ventilation systems to disinfect air that is supplied to food production areas.

Ionising irradiation. Ionising irradiation can be used to pasteurise some food materials. While theoretically possible, it is not usually used to sterilise foods with irradiation, as the very high doses can create organoleptic defects. The typical ionising radiations in food applications are gamma rays from radioactive isotopes such as cobalt[60] or high-energy electron beams (Silverman, 1991). The irradiation doses used in food processing are measured in kilograys (kGy). One kGy equals 10^5 rads; one rad equals 100 erg/g. The irradiation D values presented in Table 7.5 illustrate several important points. The lethal effect of ionising irradiation is caused by its ionisation of molecules it happens to strike. The ionised molecules in turn can react with and denature important cellular molecules such as DNA. Therefore, large targets such as parasitic worms are much easier to inactivate with irradiation than are smaller targets such as bacteria and viruses. Bacterial spores are more difficult to kill with irradiation than are vegetative cells. Regulatory approvals have been granted for the use of irradiation to treat a wide variety of foods; for example, in the United States this list includes milled cereal grains, fruits, vegetables, spices and meat and poultry products. The legitimate commercial use of irradiation processes is greatly limited by consumer scepticism about their safety due to misinformation and lack of education.

High hydrostatic pressure. High hydrostatic pressure (HHP) treatments at pressures up to 1000 megapascals (MPa) for several seconds to several minutes can be used to reduce the

Table 7.5 D values of representative microorganisms treated with ionising irradiation.

Organism/molecule	D (kGy)
Clostridium botulinum[a]	3.3
Bacillus subtilis	0.6
Enterococcus faecium	2.8
Salmonella Typhimurium[a]	0.2
Pseudomonas spp.	0.06
Aspergillus niger	0.5
Saccharomyces cerevisiae	0.5
Foot and mouth virus	13.0
Complete inactivation of	
Enzymes	20–100
Insects	1.5
Trichinella spiralis	0.2–0.5

Source: Silverman (1991).
[a]Pathogens.

microbial populations in packaged foods. One MPa is equal to 10 atmospheres of pressure, or 150 pounds per square inch. HHP is an adiabatic process; the product temperature at the end of treatment is the same as its initial temperature. However, at very high pressures the product temperature during the pressure treatment will increase about 10°C for each 100 MPa increase in pressure. Therefore, some of the lethal effects during HHP processing are due to the increase in temperature during the process. A major benefit of this process is that it does not alter product texture and other organoleptic properties, yielding products of higher quality and extended shelf life. It is favoured for the treatment of high-value, perishable, refrigerated products that would be altered by thermal processes. The process can be applied to moist foods in which the high pressure is transferred uniformly throughout the food product. It cannot be used to treat dried foods. The high-pressure treatment alters the conformation of proteins and nucleic acids by disrupting non-covalent bonds, such as hydrogen bonds, thereby killing microbial cells (Jay, 2000; Ross *et al.*, 2003). A limiting factor in the commercial application of HHP is that it must be used in batch operations.

Pulsed electric fields. Pulsed electric fields can be similarly used to pasteurise liquid foods, such as milk, eggs and juices, by passage through a high-voltage electric field, up to 80 kV/cm, that is pulsed at microsecond intervals. Commercial applications have been limited by difficulties in scalability (Ross *et al.*, 2003).

7.2.2 Prevention of microbial growth

The principal process controls to prevent the growth of microorganisms in foods include refrigeration, freezing, hot-holding, modified atmospheres and moisture control.

Refrigeration

The widespread availability of mechanical refrigeration for the distribution and storage of foods in commerce and in homes enables consumers to have a wide variety of foods available throughout the year. Despite the shelf life extension provided by refrigeration, some spoilage and pathogenic microorganisms are able to grow in refrigerated foods. Each microorganism has a distinct growth range as related to temperature. The optimum growth temperature is much closer to the maximum growth temperature than it is to the minimum growth temperature (Table 7.6). As the growth temperature is decreased, intracellular metabolism slows. Below its minimum growth temperature, a microorganism is no longer able to grow. The generally accepted temperature for optimum refrigeration is 5°C or lower. Five of the pathogens in Table 7.6 can grow, albeit slowly, below this temperature.

It is important to remember that there can often be interactions between preservative factors that can sometimes be used for commercial or organoleptic advantage. For example, at 27°C osmophilic moulds can grow at a minimum water activity value of 0.65, while at 7°C they cannot grow below water activity value of 0.83. Therefore, a product developed for distribution at 27°C could be reformulated to a higher water activity value for distribution at 7°C (W.H. Sperber, unpublished data).

The shelf life of many refrigerated foods ranges from several weeks to several months. Eventually, almost every food will be spoiled by microorganisms if the storage temperature is

Table 7.6 Temperature growth parameters of bacterial pathogens.

Microorganism	Growth temperature (°C)		
	Minimum	Optimum	Maximum
Listeria monocytogenes	0	37	44
Yersinia enterocolitica	0	33	44
Vibrio parahaemolyticus	3	37	44
Clostridium botulinum (non-proteolytic)	3	30	45
Bacillus cereus	5	30	50
Salmonella spp.	6	37	46
Staphylococcus aureus	7	37	48
Escherichia coli O157:H7	8	37	43
C. botulinum (proteolytic)	10	40	49
Clostridium perfringens	15	44	50
Campylobacter jejuni	27	42	45

Source: Doyle (1989) and Herbert and Sutherland (2000).

too high, the storage time too long, or if the food has a higher than normal initial load of spoilage microorganisms.

Freezing

The shelf life of refrigerated perishable foods can often be extended by frozen storage. Commercially produced frozen foods are usually stored at −18°C, a temperature that prevents the growth of all foodborne microorganisms. Some microbial growth is possible in foods stored at temperatures above −18°C, but below 0°C, as food solutes prevent some water from freezing at the normal freezing point of pure water. Bacterial growth has been detected in foods at −3°C; mould growth has been found in foods at −8°C (Lund, 2000). Microorganisms can generally grow more quickly in thawed meat and produce products than in the fresh counterparts because of the release of intracellular nutrients when frozen meat and produce cells are thawed.

Some degradative enzymes remain active after vegetables are frozen. Therefore, vegetables are blanched in steam or hot water at temperatures about 90°C to inactivate the enzymes. Blanching, of course, also serves to pasteurise the vegetables.

Hot-holding

'Keep cold foods cold and hot foods hot' is a wise saying in public health circles. 'Cold' is generally defined as 4°C or cooler and 'hot' as 60°C or higher; regulatory requirements may differ slightly in various countries. Foods held at temperatures between these limits should not be held for more than 6 hours without prompt refrigeration or heating (FDA, 2005b). The hot-holding temperature provides a margin of safety; none of the foodborne pathogens are capable of growth above 50°C (Table 7.6).

Widely used in food service operations, hot-holding is sometimes a necessary procedure for food processors. When food products or ingredients are cooked, they are sometimes stored in a holding tank until final processing or packaging. It is important that the temperature of such materials not falls below 60°C. Leftover hot-held materials may be 'toted off' and placed into refrigerated storage; however, the centre of large containers filled with hot food will take several days to reach refrigerated temperature. Therefore, it is important to rapidly chill the food before placing into a large container, or to portion the food into small containers so that the food will chill to optimum refrigeration temperature within 6 hours.

Modified atmosphere and vacuum packaging

Some perishable food products are packaged in containers with a headspace of air under atmospheric pressure. Other factors permitting, aerobic microorganisms can grow freely in such products. Their growth can be inhibited or prevented by the removal of headspace oxygen (vacuum packaging) or the addition of inhibitory gases (modified atmosphere packaging). In each case it is important that the packaging material be impermeable to the appropriate gases and that the final package be hermetically sealed to prevent the entry of oxygen or the escape of inhibitory gases.

The most practical benefit of modified atmosphere or vacuum packaging is the inhibition or prevention of mould growth. Packaging products in such a manner could create a potential hazard by providing opportunities for the growth of anaerobic pathogens, as previously described for fresh mushrooms (Section 6.2.4 in Chapter 6). The opportunities for such a hazard must be prevented by other means, such as refrigeration.

In vacuum packaging, it is very difficult to remove all the air and oxygen. The growth of some moulds can occur at oxygen levels as low as 0.4%. Some commercial applications have been developed in which residual oxygen in vacuum packages is removed by oxygen scavengers (Smith *et al.*, 1986). It has long been recognised that a similar phenomenon occurs in fat- or oil-containing foods that are hermetically packaged with a small headspace volume, e.g. salad dressings, mayonnaise and refrigerated ground beef. Residual oxygen combines readily with vegetable oil, thereby preventing the growth of aerobic yeasts and moulds in salad dressings and mayonnaise (Sperber, 2009b). Similarly, absorption of residual oxygen by the fat in refrigerated ground beef will prevent the growth of aerobic microorganisms, including psychrophilic spoilage bacteria. The elimination of atmospheric oxygen in this case further selects for the growth of lactic acid bacteria, which in turn provides additional protection against the growth of spoilage and pathogenic microorganisms (Frazier, 1958).

Carbon dioxide gas is often used in modified atmosphere packaging to prevent the growth of aerobic microorganisms. It can be used in combination with other gases that are used to back-flush a product headspace after a vacuum is drawn. Shelf life of fresh fruits and vegetables can be extended with gas mixtures of 8–10% carbon dioxide, 2–5% oxygen and the remainder as nitrogen (Jay, 2000; Farkas, 2001). As expected with most preservative factors, there is an interaction between carbon dioxide levels and temperature to prevent microbial growth. The growth of mould on bakery products can be retarded at 5°C by a mixture of 30% carbon dioxide and 70% nitrogen. However, at ambient temperatures a mixture of 70% carbon dioxide and 30% nitrogen is required to inhibit mould growth (Cook and Johnson, 2009).

7.2.3 Prevention of contamination

The third type of process controls to enhance the protection of food products against microbial defects is the prevention of contamination by relevant spoilage and pathogenic microorganisms.

Several types of effective controls to keep undesirable microorganisms out of food products are well understood, but sometimes overlooked by food processors.

High-risk ingredient control

The first of these is the control of high-risk ingredients, as briefly introduced in Chapter 6. In principle, high-risk ingredients are more readily addressed as an extrinsic hazard than as an intrinsic hazard. High-risk ingredients are those that have been historically associated with particular microbiological or chemical hazards, e.g. *Salmonella* contamination of dried egg products or aflatoxin contamination of peanut products. Over the years, many participants in the global food supply chain have agreed informally on the major types of high-risk ingredients (Tables 6.11 and 6.12 in Chapter 6). Validation and verification of supplier capability is the most effective means of high-risk ingredient control.

Beginning in the 1960s, the potential hazards in high-risk ingredients were controlled by quarantine and laboratory testing. These ingredients were released from quarantine only after an agreed sampling plan was followed and negative results were obtained for the identified hazard. Similarly, finished products manufactured with the sensitive ingredients were often quarantined and tested before being released into commerce.

Obviously, the extensive ingredient and product testing plans did not fit well with the modern product design and process control features of HACCP programmes, which focus on real-time observations and measurements. Nor did the extensive quarantine period (often several weeks) fit well with the modern global system of just-in-time manufacturing. Therefore, since the 1990s in the food industry, the mode of ingredient control has shifted from quarantine and testing to validation and verification of supplier capabilities.

In those rare circumstances that may require ingredient or product testing, responsible specifications, as detailed by the National Research Council (1985) and responsible sampling plans, as developed by the ICMSF (2002) must be followed. Microbiological testing has been found to be usually an impractical means to ensure the safety and quality of ingredients and food products. Whenever possible, the use of specifications, lot acceptance criteria and ingredient or product tests should be replaced by the use of microbiological monitoring guidelines in the food production environment (Sperber and NAMA, 2007).

Allergenic ingredient control

A great many allergenic ingredients can be used in food products. The major allergens are peanuts (groundnuts), tree nuts, crustaceans, fish, eggs, milk, soy and wheat. Minor or regional allergens include celery, buckwheat, rice, legumes, molluscan shellfish, and the seeds of cotton, sesame, sunflower and poppy.

A key control measure to protect allergen-sensitive consumers is adequate product labelling so that the consumer is aware of the real or potential presence of an allergen. Many operational measures can be taken to minimise the use of known allergens or to prevent the contamination of foods that are not expected to contain allergens. A non-allergenic ingredient can be substituted for an allergen when feasible. Many food products can be reformulated to eliminate the use of minor amounts of an allergenic ingredient that is not essential to maintain functional or organoleptic properties of the food. It is always necessary to prevent cross-contamination with allergens in product development kitchens, pilot plants, sensory testing areas and food manufacturing facilities. Allergen control measures are presented in greater detail in Section 7.3.1.

Aqueous ingredient control

Numerous aqueous ingredients or subassemblies are used in food products. These are often minor ingredients that are used in small quantities. Therefore, the ingredient or stock solution may be used over long periods during its storage at ambient or refrigerator temperatures. If such materials were contaminated with a pathogen or toxic substance, it could contaminate finished products for several days, weeks or even months. Many flavours, colours, preservatives and processing aids could be used in this manner.

A prominent example of this potential hazard was detected and controlled before it could develop into a public health problem. Routine testing of one company's research samples produced for sensory evaluation revealed the presence of enteric bacteria. Further evaluation showed that a colour solution stored in a 2 L bottle at ambient temperature supported the ready growth of salmonellae to a level of about 10 million cells/mL. The use of this colour solution at a 0.1% level in the product would have yielded 10 000 salmonellae/g in the product. Used in the company's manufacturing facilities, the same colour solution was prepared and stored at ambient temperature in a 2000 L tank, a quantity sufficient to support, and potentially contaminate, 4 weeks of production. Had salmonellae grown in this storage tank, a major illness outbreak could have occurred. This particular colour solution was stabilised against the growth of all bacteria (including salmonellae) by the incorporation of 15% propylene glycol (W.H. Sperber, unpublished data).

After this incident we learned quickly that many aqueous-based ingredients or subassemblies used in food production can present similar potential hazards. It is incumbent upon the product development team to be aware of such hazards, to evaluate ingredients that could present such a hazard, and to implement effective control measures when necessary. The following control measures for aqueous-based ingredients have proved effective:

1. Control a_w at or below 0.85.
2. Control pH at or below 4.0.
3. Use approved preservatives, e.g. propylene glycol.
4. Use validated combinations of the above three measures.

If none of the above measures can be used, the aqueous subassembly must be used on the same day it is prepared. It should not be stored, even under refrigeration, for further use.

Sanitary design and sanitation

Other principal means to prevent food product contamination are imbedded in prerequisite programmes (Chapter 10), particularly sanitary design of food processing equipment and cleaning and sanitation procedures. We have observed that food safety and food quality failures are often associated with a lack of adequate cleaning and sanitation procedures. In turn, the inadequacy of such procedures is sometimes associated with the improper sanitary design of food processing equipment and facilities. The matters of sanitary design and proper sanitation are of utmost importance as microorganisms can grow in substrates that are similar, and sometimes identical, to the food product that must be protected against such microbes. For example, the spoilage of chemically leavened refrigerated dough products was a major commercial problem in past decades. Some of the dough-handling equipment was designed in such a manner that it could not be properly cleaned and sanitised. Dough accumulations in the bearings and crevices of such equipment served as continual incubators for lactic acid bacteria, the principal spoilage microorganism of the dough products. Eventually, spoilage was essentially eliminated because of the use of new equipment whose design permitted easy and effective cleaning and sanitation. Similar types of product quality defects were caused in yeast-leavened doughs simply because dough mixers were cleaned only once per week, rather than daily.

Moisture control

The inadvertent and unsuspected presence of moisture can contribute to food safety and food spoilage incidents. Bakery mix products produced in a seven-level tower facility were found to be contaminated with *Salmonella*. According to facility managers, the facility was completely dry and could not support the growth of *Salmonella*. Thorough evaluation of the bakery tower by a friendly microbiologist revealed 43 sites of moisture contamination, including air control valves, open windows, hand-wash sinks, floor mop pails and condensation on cold exterior walls. Several sites were found to harbour salmonellae. These could likely have contributed to the products' contamination.

Condensation in ventilation systems has been shown to be a source of *L. monocytogenes* that contaminated dairy products during packaging. Even blast freezers operated at −40°C can become a source of unsuspected microbial contamination. Some foods are blast frozen before packaging. Blast freezers are usually operated and kept cold during the production week and cleaned on weekends. During production, loose product material can be blown onto the conveyor tracks and accumulate on the floor of the blast freezer. During the daylong thawing procedure before cleaning, the accumulated food debris becomes an incubator for microbial growth. It is difficult to remove all the food debris and to adequately clean and sanitise all the equipment and utility surfaces of the blast freezer. Unless these steps are satisfactorily completed, the microbes will become airborne when the blast freezer is put back into operation and can contaminate new foods that enter the freezer. Strict attention to sanitary design and adequate cleaning and sanitation procedures can prevent many potential microbiological safety and quality problems (Troller, 1993).

7.3 PROCESS CONTROL OF CHEMICAL HAZARDS

Food manufacturing facilities should maintain a chemical control plan to prevent contamination of its products with allergens, mislabelled or adulterated ingredients, and cleaning and maintenance chemicals. As with all food safety and quality practices, employee training and awareness is an essential factor in minimising the risk of chemical hazards in foods.

7.3.1 Allergen control

In recent years, a great deal of regulatory attention has been given to the presence of undeclared allergenic ingredients in food products. Food processors must enact effective control measures to prevent the occurrence of this regulatory and potential public health hazard (Taylor and Hefle, 2005; Jackson *et al.*, 2008). Some of the necessary measures are taken during product design and commercialisation, as described in Chapter 11. These include verification of the accuracy of the ingredient declarations on product labels, and the implementation of suitable prerequisite controls for the receipt, storage and use of high-risk ingredients.

Additional allergen control measures need to be applied in the food production facility. Allergen-containing ingredients need to be clearly labelled and stored separately from non-allergen-containing ingredients. Dedicated storage bins, utensils and conveying equipment are often used for specific allergenic ingredients, such as peanuts. Sometimes, dedicated food processing equipment is used. Large corporations can dedicate a production line or even an entire production facility solely to the production of a specific allergen-containing or allergen-free food. Production sequencing and scheduling can be a useful separation technique when multiple foods are produced on a single production line. In a given production run, all non-allergen-containing foods can be produced before those that contain allergens, and thorough cleaning as well as verification of allergen absence will be necessary at the end.

The use of product rework can be a source of contamination with food allergens. For example, in the era before foodborne allergens became a public health concern, it was common practice for ice cream producers to rework leftover ice cream from all flavours into chocolate ice cream, whose colour and flavour would tend to mask the presence of ingredients normally foreign to chocolate ice cream. Of course, some of the foreign ingredients were significant allergens such as peanut butter and tree nuts. While it remains an economic necessity to rework leftover ice cream, responsible producers today have a strict 'like into like' rework policy in which leftover peanut butter ice cream can be reworked only into peanut butter ice cream etc. Similarly, in the past, leftover enrobing chocolate and remelted chocolate bars would be reworked without consideration that they may have been contaminated by nuts.

Adequate cleaning and sanitation procedures are essential to prevent the contamination of non-allergen-containing foods with residual allergens from foods produced on the same equipment. Complete wet-cleaning and sanitation is the best way to remove residual allergens. When dry-cleaning procedures must be used, wiping, vacuuming or rinsing with vegetable oils can be done. Air pressure hoses should never be used in place of vacuum hoses, as the former will simply spread dry allergenic material to other production equipment. Longer production runs with allergen-containing foods can be used to minimise the number of cleaning and sanitation periods. After cleaning, the absence of specific allergens should be verified by using one of various enzyme-linked immunosorbent assay (ELISA) or dipstick test procedures that are commercially available. These tests can detect the presence of allergens at or near 1 ppm, the threshold level considered necessary to cause an allergic response in a sensitive individual for some allergens.

7.3.2 White powder control

Many food processors attempt to implement 'white powder control' procedures. Hundreds of food ingredients – salts, leavening agents, preservatives, acidulants, sugars, flours, starches and proteins – are white powders – as are cleaning agents and other non-food chemicals. Several control steps are important to be certain that the white powders are used correctly. Upon receipt at the facility, a sample of the white powder should be tested by visual, organoleptic or chemical means to verify its identity. The accuracy of the ingredient labels should be confirmed when placed into storage. Quite often, minor ingredients are weighed, combined and blended in a separate area and later front-loaded into product mixers. Quality assurance and production personnel must verify that the ingredients are added correctly to product mixers or preblend operations. The mistaken use or omission of a particular 'white powder' could lead to a significant quality or product safety defect.

7.3.3   Cleaning, sanitation and maintenance chemicals

Each food processing facility should establish a chemical control plan to organise control and monitoring procedures to prevent food product contamination with chemicals that are used throughout the facility, but are not intended for use in foods. Chief among these are many cleaning and sanitation chemicals, and pesticides. All such chemicals need to be stored in a confined, locked area and not be used during periods of food production. When food production is halted, all food materials must be properly stored before these chemicals can be removed from storage and used. It is essential that ingredient or product containers are never used to store or handle non-food chemicals such as cleaning agents, lubricants and pesticides that are used in the food processing facility. Occasional contamination of consumer food products with floor cleaners, hydraulic fluids, etc., has occurred when this restriction is not in place and enforced.

7.4   PROCESS CONTROL OF PHYSICAL HAZARDS

Many types of foreign materials may contaminate food products during processing and packaging. Recent major recalls of food products in the United States because of foreign material contamination were caused by metal, wire, glass and hard plastic fragments (Peariso, 2007). The primary causes of contamination were attributed to the following:

- Inadequate maintenance of processing equipment and facilities
- Lack of supplier management systems
- Lack of HACCP programme
- Flawed hazard analysis

Foreign materials can also include extraneous vegetable material, e.g. nut shells, and insects, rodent hair and droppings, bird feathers and small animals. As with the control of biological and chemical hazards, many elements of prerequisite programmes are essential for the control of foodborne physical hazards.

There are three principal means to control physical hazards in foods:

1. *Exclusion* – including programmes for glass, wood, personnel practices and pest control
2. *Removal* – by the use of devices such as magnets, sifters, screens and stone traps
3. *Detection* – by using instruments such as metal detectors, x-rays and optical sorters

7.4.1   Exclusion techniques

Control of glass and brittle plastic contamination

Most food processing facilities maintain a strict prohibition on the use of glass or brittle plastic instruments, utensils or food storage and handling vessels to avoid the possible entry of glass fragments into the food product. Necessary light bulbs – incandescent, fluorescent and ultraviolet – must be constructed with shatterproof glass or installed in shatterproof fixtures.

Control of wood contamination

Once a major problem in food processing facilities, the contamination of food with wood splinters, has largely been eliminated by the exclusion of wooden pallets, and wooden handles on tools and maintenance equipment used in all production areas. Contamination with wood (and other foreign material) may remain a problem in some developing countries, emphasising the need to implement effective control measures throughout the global supply chain.

Personnel practices

Implementation and training in personnel practices are essential to eliminate the hazard of items falling into the product stream. Employees should not wear items of jewellery (except, usually, for a solid wedding ring). Employee uniforms and hair/beard covers are usually required. The uniforms should have no pockets, so that items such as pens and pencils cannot be carried in the pockets and fall into the product. Maintenance workers must necessarily use many tools in the production area. These must be clean and used with care so that they cannot be left in the production equipment.

Pest control

A great deal of effort must be expended to keep insects, rodents and birds out of food plants. Should they get into the plant, the food products can be contaminated with animal parts, insects, faecal droppings and feathers. In our experience a large bird (pigeon) got into a flour bin and was beaten through a rotary sifter as the flour was metered into a horizontal dough mixer. Not noticed at the time, a large product recall was later necessary to retrieve the products that contained pigeon feathers (and other parts, quite likely). Similar stories are told in many sectors of the food industry.

Insect activity can be minimised inside facilities by the use of insect light traps that attract flying insects with an ultraviolet light. The insects are killed by electrocution (electric insect killers) or by entrapment on glue boards. The light traps should be mounted on interior walls near entry doors, facing inward, so that insects are not attracted from outside the facility. They are to be used in areas peripheral to the facility's food handling and production areas, thereby preventing flying insects from entering these areas.

Many facilities ban the use of rodent bait stations inside the facility. Rather, mechanical traps and glue boards can be used to monitor and limit internal rodent activity. Poison bait stations are permitted outside the plant, the intent being to control rodents before they get inside the plant. Because of the many technical and regulatory difficulties in the application of pesticides, many facilities employ professional pest control operators for this purpose. Pesticides are used in maintenance and other areas peripheral to the food production areas.

Bird control programmes include keeping the plant exterior of the plant free of food, especially grains and other seeds, and using irregularly timed air guns to frighten the birds away. To the maximum extent possible, all plant openings to the exterior should be kept closed or screened.

7.4.2   Removal techniques

Control of metal contamination

Routine equipment maintenance and inspection are essential to prevent contamination with 'tramp metal' or 'swarf', e.g. pieces of machinery or its fasteners that can become loose, break off or be ground off into the food product stream. Many types of in-line magnets are used by food processors on incoming ingredients, processing equipment and packaging operations, both to protect the equipment from damage by tramp metal and to avoid product contamination. Used primarily with dry powder or liquid materials, magnets are usually installed in gravity flow systems that permit all the material in the food stream to pass over the magnets. Magnets are typically composed of alnico (aluminium–nickel–cobalt), ceramic material (pressed barium ferric oxide) or rare earth metals (neodymium–iron–boron). The last type of magnetic material has substantially greater pulling power than the other types. Magnets can be constructed in many shapes that can be used in routine or specialised applications. Typical shapes include plate, hump, bar, grate, ring, pulley, drum and cartridge (Imholte and Imholte-Tauscher, 1999). It is important that the food processor use an expert resource or vendor to select the magnet type and establish performance specifications. Generally, installed inside pipes, magnets must be accessible for regular inspection and cleaning. To function effectively, magnets must be cleaned regularly to maintain pull strength. The types of metals found at each inspection and cleaning should be monitored and recorded. The type, size and shape of metal contaminants are noted; the metal pieces can be kept in the record book for further reference. This information is often necessary to determine the need for further investigations and equipment repairs when unusual findings are observed.

Control of foreign material in product streams

In addition to magnets and metal detectors, a wide variety of filters, screens and sifters can be used to detect or remove physical contaminants in ingredient and product streams. Often used for product quality considerations, filters can be used in bottling or loading operations for oils, syrups and other clear liquids, thereby simultaneously reducing the hazard of foreign material contamination in the product. Sifters and sieves are often used to separate foreign materials from dry food materials. The screens for these devices should be made of non-metallic ('Nytex') materials. If metal must be used to construct a screen, it should be made of magnetic (ferrous) metal that could be removed or detected by magnets and metal detectors. Aspirators use pressurised air in a falling product stream to separate dense materials, such as cereal grains, from lighter contaminants, such as insects, weed seeds and dirt that may have entered during growing, harvesting and transportation.

7.4.3 Detection techniques

Several technologies are available to detect the presence of foreign materials in a food. The most common of these are metal detectors that can be used online or for packaged products (Imholte and Imholte-Tauscher, 1999). Materials in which metal is detected are diverted for further inspection to determine the need for corrective actions. X-ray devices can be used to inspect containers before packaging operations. With image-enhancing capabilities, x-ray devices can be used to detect dense foreign material inside food products, such as bone chips in meat products (Graves *et al.*, 1998).

Optical technologies, using visible or ultraviolet light, are used with fruits, vegetables and nuts to detect surface defects and the presence of extraneous vegetable matter, stones, etc. Peanut product producers use optical scanning to remove dark-coloured nuts, thereby reducing the risk of aflatoxin contamination in finished products. Additional detection technologies are being used on a research basis but are not widely used in commercial product testing. These include electrostatic techniques with parallel plate capacitors, microwave, nuclear magnetic resonance and ultrasonics.

7.5 CONCLUSION

The design of process controls by the product development team is a most important first step in the production of a safe food product. If appropriate extrinsic measures to control the potential hazards are not researched and validated at this step, significant time may be lost to correct the early mistakes. Worse, a product that is potentially unsafe or of reduced quality and shelf life might be produced and distributed.

Part Three
Systematic Food Safety Management

8 Overview of a world-class food safety programme

8.1 INTRODUCTION

A world-class food safety programme is a multifaceted approach to the management of food safety and the protection of consumer health. Traditionally, food safety programmes would have been thought of as the systems put in place by manufacturers and larger food service operators to ensure the safety of their individual products and processes; however, in reality, the picture is now a lot more complicated and needs to reflect control of food safety throughout the global food supply chain, as we have seen in the chapters of Part 1. In practice, the systems are still managed at the company level; however, we now see increasing communication and sharing within the supply chain, as companies seek to understand and manage the hazards and risks inherent in their own materials supply and product distribution chains.

When applying food safety management in the 21st century, we must learn from the experiences of the past, both the positive experiences of what works well and the more traumatic experiences when things have gone wrong (see Chapter 2). This has led the authors to be firm believers in a 'back-to-basics' approach, particularly in the need to revisit HACCP and prerequisite programmes and ask challenging questions about the ability of existing systems to protect the consumer. It is our experience that a shadow of complacency has crept over food safety systems, resulting in beliefs that HACCP has been 'done' and something that manufacturing companies have had in place for years so does not need worrying about. The incidents shown in Table 2.1 (Chapter 2), and many others like them, show us that this is simply not the case. We need to pay careful attention to how we manage food safety and make sure that systems are effectively planned, developed, implemented, verified and reviewed/updated. These requirements tie in with the principles of continuous improvement of management systems and the Deming 'plan-do-check-act' cycle, discussed further in Chapter 9.

This short chapter provides an introduction to the following six chapters on specific elements of systematic food safety management. Taken together, the seven chapters of this part will tell you how to develop, implement and maintain a world-class food safety programme in practice. From the point of view of both companies who are starting out on the journey towards food safety management and companies who have recognised the need to review and strengthen their existing systems, this will provide knowledge of best practice approaches backed up by experience and research in real food manufacturing environments.

Food Safety for the 21st Century, First Edition By Carol A. Wallace, William H. Sperber and Sara E. Mortimore
© 2011 Carol A. Wallace, William H. Sperber and Sara E. Mortimore

8.2 PRELIMINARY CONCEPTS AND DEFINITIONS

8.2.1 What is a world-class food safety programme?

The fundamental elements of a world-class food safety programme are safe product/process design, prerequisite programmes and HACCP supported by management practices which ensure consistent application of system elements both within the business and outwards through the links of the global supply chain. Our definition, therefore, includes these essential elements but also indicates the need for effective management practices to provide the necessary foundations and ongoing support for continuous delivery of safe food products (Figure 8.1).

World-class food safety programme: a programme based on the principles of safe product/process design, prerequisite programmes and HACCP that is supported by essential management practices, thus controlling the operational, environmental and process conditions necessary for consumer health protection through the consistent production of safe food.

It is clear that to meet the objective of consistent safe food production, the programme must cover all operations at each facility, must be fully implemented in practice and regularly challenged with stringent verification procedures to demonstrate ongoing effectiveness and currency.

In addition to 'food safety', 'food protection', 'food defence' and 'food security' are three further terms that are used in socio-political circles and, because these may lead to confusion within the global supply chain, it is useful to consider the different meanings here.

Although 'food protection' is a commonly used term, few definitions exist in the literature. However, this term is generally used to describe all measures and programmes in place to protect the safety of the food supply. As such, food protection includes both 'food safety' programmes that are in place throughout the supply chain to ensure safety of food products through the prevention and control of significant hazards, and also 'food defence' measures.

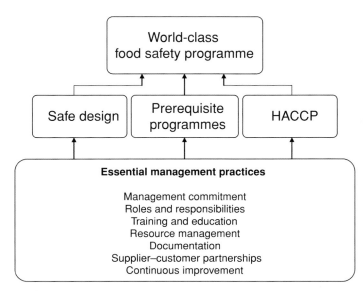

Fig. 8.1 Fundamental elements of the world-class food safety programme.

'Food defence' (*food defense*) is the collective term used in the United States by the Food and Drug Administration (FDA), the US Department of Agriculture (USDA), the Department of Homeland Security (DHS) and others to encompass activities associated with protecting the nation's food supply from deliberate or intentional acts of contamination or tampering (FDA, 2009). Recent history includes a number of well-publicised events associated with economic adulteration. With the predicted increase in population, this may be a trend set to continue where food supplies are scarce. Other groups have also produced definitions of 'food defence', and a useful interpretation was produced by an expert working group at the USDA sponsored 2007 workshop on food defence at the Maritime Institute of Technology and Graduate Studies:

> Food defense means having a system in place to prevent, protect, respond to and recover from the intentional introduction of contaminants into our nation's food supply designed specifically to cause negative public health, psychological, and/or economic consequences. (Yoe *et al.*, 2008)

Thus, food protection can be considered to include both food safety to control the naturally occurring hazards in the food chain and food defence to prevent and manage deliberate acts of hazardous contamination.

Food security is a completely different concept relating to access to food. The World Food Summit of 1996 defined food security as existing 'when all people at all times have access to sufficient, safe, nutritious food to maintain a healthy and active life' (WHO, 2010). Although safety is mentioned within this definition, it is relating to availability of safe food rather than food safety management.

From the above paragraphs, it is clear that the world-class food safety programme will form a cornerstone of food protection. Although prevention of deliberate contamination is normally considered to be outside the scope of food safety programmes, the work that goes into developing, implementing and maintaining a world-class food safety programme will also make it easier to implement food defence. This relates to the detailed understanding of food processes and supply chains that comes from food safety management, therefore making it possible to identify areas that are vulnerable to tampering and establish preventive measures.

We now consider the elements of the world-class food safety programme in more detail.

8.2.2 World-class food safety programmes – fundamental elements

As shown in Figure 8.1, the fundamental elements of a world-class food safety programme are safe product/process design, prerequisite programmes and HACCP systems, and these are supported by the essential management practices (Section 8.2.3).

Safe product/process design

Safe product/process design relies on both the understanding of hazards and formulation/process control capabilities at the development stage and the application of formal procedures to evaluate and sign off the safety of each new development prior to its implementation. The chapters of Part 2 provide an in-depth discussion of food safety hazards (Chapter 5) and give guidance on the design of safe products (Chapter 6) and processes (Chapter 7). In this part, we outline how to evaluate the safety of proposed products and processes using product safety assessment (Chapter 11).

Prerequisite programmes

Prerequisite programmes (PRPs) are the practices and conditions needed prior to and during the implementation of HACCP and which are essential to food safety (WHO, 1999). PRPs provide a hygienic foundation for the HACCP system (NACMCF, 1997) by enabling environmental conditions that are favourable for the production of safe food (CFIA, 1998). Like the HACCP system, there is international agreement on the general principles required (Codex, 2009a), and these essential characteristics of PRPs are discussed in Chapter 10, against the following headings for application to food chain establishments:

- Design and facilities
- Control of operation
- Maintenance and sanitation
- Personal hygiene
- Transportation
- Product information and consumer awareness
- Training

HACCP

The HACCP system has already been introduced right at the start of this book (Chapter 1) and, since the system has been in the public record for nearly 40 years, it is likely that many readers of this book will have been exposed to HACCP systems previously. However weaknesses in HACCP systems within a sample of multinational manufacturing facilities (Wallace, 2009) lead us to believe that deeper focus on the application of HACCP principles (Table 8.1) is essential. Chapter 12, therefore, provides a detailed discussion on the application of HACCP principles for effective HACCP plan development. This is followed by a thorough examination of the important considerations in implementing a HACCP system so that it really works in practice (Chapter 13). Maintenance of these fundamental elements of the world-class food safety programme is considered in Chapter 14.

Table 8.1 The HACCP principles.

Principle 1	Conduct a hazard analysis
Principle 2	Determine the critical control points (CCPs)
Principle 3	Establish critical limit(s)
Principle 4	Establish a system to monitor control of the CCP
Principle 5	Establish the corrective action to be taken when monitoring indicates that a particular CCP is not under control
Principle 6	Establish procedures for verification to confirm that the HACCP system is working effectively
Principle 7	Establish documentation concerning all procedures and records appropriate to these principles and their application

Source: Codex, 2009b.

8.2.3   World-class food safety programmes – essential management practices

Whilst it is commonly understood that there will be management practices that are essential to the effective running of any food safety or quality management programme or any food business, when considering systems for the consistent realisation of safe products on a continuous basis, there is less agreement on exactly what these practices entail. Our definition of essential management practices is as follows:

Essential management practices (**for food safety**): are management practices and procedures that support effective application of safe product/process design, prerequisite programmes and HACCP systems, and assure their ongoing capability to protect the consumer.

These essential management practices are all about making sure that there is ownership and responsibility throughout the company structure and that resources supporting the fundamental elements (safe product/process design, prerequisite programmes and HACCP) are appropriately administered, supervised and controlled.

This will include, as a minimum, the following:

- Management commitment
- Roles and responsibilities
- Training and education
- Resource management
- Documentation
- Supplier–customer partnerships
- Continuous improvement

It is likely that many of these practices will be under the framework of a structured (quality) management system that may be externally certified. We discuss these essential management practices in further detail, with reference to food safety requirements, in Chapter 9.

8.2.4   World-class food safety programmes – further supporting elements

In addition to the fundamental world-class food safety programme elements and essential management practices, it is likely that there will be further supporting elements in many businesses and in parts of the global food supply chain. These may include elements such as:

- *Quality management systems* – as the framework to manage the food safety programme, quality systems based on Total Quality Management (TQM) principles are useful in ensuring ongoing effectiveness and continuous improvement of food safety programmes. Companies may wish to consider externally certified systems such as ISO 9001 (ISO, 2008), ISO 22000 (ISO, 2005a) and/or schemes meeting the requirements of the GFSI Guidance Document (GFSI, 2007).
- *Best practice programmes* – for example:
  - Good laboratory practice – to provide confidence in laboratory results used in monitoring and verification of the food safety programmes.
  - Good distribution practice – to maintain food safety in transit; this may also be considered as part of the prerequisite programmes.

- *Sustainability programmes* – both to ensure supply chain sustainability through prevention of supply problems and to promote corporate responsibility in the way that the supply chain is managed.
- *Continuous improvement programmes* – such as the 'Lean' approach where tools such as process, activity and value stream mapping can help us to understand our processes in much more detail, seeing which activities are adding value in consumer terms, in this case to ensure safe food production. This might also include specific programmes such as Total Productive Maintenance (TPM) to maximise the reliability and effectiveness of production equipment, focused improvement to assist with failure, root cause analysis and Six Sigma to improve business and product process levels to as close to perfection as possible.

Although some of the above might first seem more about the functioning of business rather than food safety, the complexities of operating in the global food supply chain mean that all business systems need to be looked on as an integrated whole rather than as a collection of approaches that address different issues. This helps to ensure that we can progress away from the 'flavour of the month' approach where new concepts and systems come in periodically to replace the old ways of operating, and the importance of food safety can become diluted where the new concept might initially seem targeted at a different area of business, e.g. TPM. Instead of working against each other, food safety management programmes need to be seen as core to all business operations and all approaches being used within each company should work together supporting this and other business objectives.

8.3   WORLD-CLASS FOOD SAFETY PROGRAMMES IN THE GLOBAL FOOD SUPPLY CHAIN

World-class food safety programmes need to apply to the entire food supply chain (Figure 8.2). This means that all types of business, be they small or large and based in any link of the chain, i.e. agricultural, manufacturing/processing and food service/retail, need to apply and manage effective food safety programmes. It also requires that food products produced for any consumer, i.e. human or animal, should be subjected to the same stringent control mechanisms,

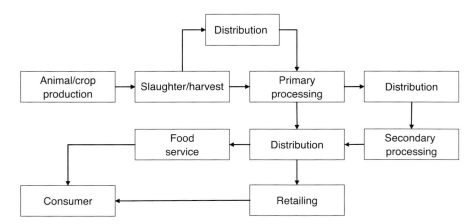

Fig. 8.2   Supply chain model. (Adapted from Sperber (2005).)

and that those handling and preparing food for any consumer, including in the home, should be aware of safe food-handling procedures.

It is no accident that HACCP evolved in the middle of the food supply chain and for this reason many of the procedures are based on activities that can be performed in manufacturing. Although this gives much attention to the middle of the supply chain, particularly with historical experience of HACCP application, the importance in applying food safety programmes in all links must not be forgotten. However, from a practical point of view, it is likely that there will be more emphasis on prerequisite programme elements at the ends of the chain, for example in primary agriculture and in the home. The importance of animal health and welfare also needs to be recognised as primary agriculture forms the interface between animal health and public health. Food safety issues around animal health have already been discussed (Chapters 4 and 5), for example with reference to concerns about antibiotics in the food chain and the ability for animals to become vectors for new human pathogens (zoonosis).

8.4 CONTINUOUS IMPROVEMENT OF THE WORLD-CLASS FOOD SAFETY PROGRAMME

A key aspect of the world-class food safety programme must be that it continually evolves to meet the requirements for food safety in the global food supply chain and, in doing so, continually improves and offers better protection for consumers and food businesses. This requires defined responsibilities and effective management of resources such that safe products meeting all food safety requirements result from the implementation of the world-class food safety programme. Through measurement and analysis of programme implementation, both monitoring and verification activities, areas where improvement is essential and/or possible will be identified and actions can be taken to continually strengthen the programmes (Figure 8.3).

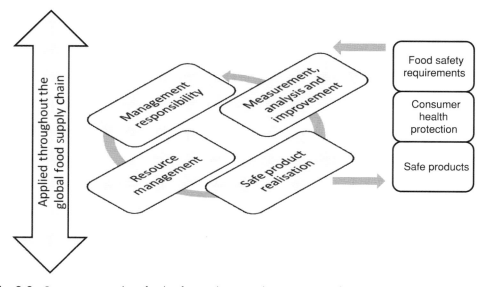

Fig. 8.3 Process approach to food safety and continual improvement of the world-class food safety programme. (Adapted from ISO (2008).)

8.5 CONCLUSIONS

World-class food safety programmes are based on effective systems for safe product design, prerequisite programmes and HACCP, supported by essential management practices and integrated into business management programmes. Applying a continuous improvement mindset towards achieving a world-class programme will enable food businesses to consistently meet both their obligations to consumers and their food safety regulatory requirements. This will result in a live and vibrant food safety culture operating 24/7 throughout the entire food supply chain.

9 Building the foundations of a world-class food safety management programme: essential steps and practices

9.1 INTRODUCTION

'Prepare to fail if you fail to prepare' is a well-known saying. There is a lot of truth in this much used phrase and yet there is very little written about the best practices in preparing to develop and implement a new or improved food safety programme. Perhaps this is because best practice is to regard it as a continuum – the cycle of continuous improvement as described in Deming's 'plan-do-check-act' (PDCA) cycle (Figure 9.1) (Cleary, 1995).

This is complementary to the four key stages (Mortimore and Wallace, 2001) of the Hazard Analysis and Critical Control Point (HACCP) implementation where preparation and planning was highlighted as a separate and important factor for success:

Stage 1 – Preparation and planning
Stage 2 – HACCP study and plan development
Stage 3 – Implementation
Stage 4 – Verification and maintenance

A food safety programme that will be sustainable, i.e. one that will withstand changes in personnel and will strengthen and develop as new information becomes available, requires firm foundations. Sufficient time needs to be spent thoughtfully considering all aspects of the business units, with the goal of having an ongoing operational food safety programme that is a way of life and is world-class.

Chapter 8 described the elements of a world-class food safety programme (Figure 9.2). This chapter starts by focusing on one of these elements – essential management practices. These are important not just for planning a programme, but need to be in place on an ongoing basis. It will then go on to discuss the planning and preparation activities that are important whether starting to develop a new programme or reviewing and strengthening an existing programme.

Prerequisite programmes (PRPs) are discussed in detail in Chapter 10, but it should be remembered that PRP review and improvements are almost always one of the most time-consuming and costly elements when strengthening a food safety programme, and this is also a key component of the preparation process.

Food Safety for the 21st Century, First Edition By Carol A. Wallace, William H. Sperber and Sara E. Mortimore
© 2011 Carol A. Wallace, William H. Sperber and Sara E. Mortimore

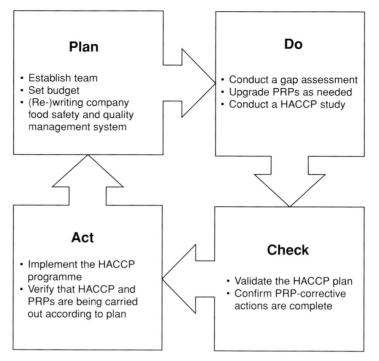

Fig. 9.1 PDCA as a template for planning.

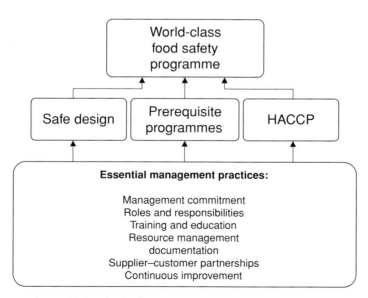

Fig. 9.2 Elements of a world-class food safety programme.

9.2 ESSENTIAL MANAGEMENT PRACTICES

9.2.1 Management commitment

There is no point in doing anything without having real commitment from company senior management, which is why we are starting with it in this section. The senior management team needs to be involved early on, and based on experience, we would contend that they need to stay involved and be visible often. For many, gaining real genuine commitment is a serious challenge and if not done well will be discussed at the water cooler, or in the break room. Here is some guidance that will be helpful.

Start with the most senior managers in the organisation – those in the middle can follow later. Ensure that they are clear with regards to their responsibilities and the likely consequences (for them personally as well as the company) if there is a food safety incident. To do this, examples of failure within the industry can be very helpful. Several are included in Chapter 2, but the media can be used to search for recent, local and similar category examples to make it relevant.

The senior team will need an indication of the status of food safety within their organisation. Many senior executives assume everything is perfect and mistakenly believe that nothing would ever go wrong in their own plants. Information shared should be fact-based and kept at a fairly high level. Often, the senior team has not really had to think about this before or has not had even this level of detail. Being able to include examples of the company's 'near misses' – will be even more effective in gaining commitment, particularly if a cost of failure can be compiled.

When briefing the team, an overview of the gaps in the organisation together with a top level plan (with timing) to close those gaps will be needed to ensure alignment and full support to proceed. Whilst senior management commitment has long been believed to be important in HACCP and food safety success, recent research (Wallace, 2009) has provided more information on what this means in practice. In particular, this has highlighted the importance to the workforce that they have 'promotional managers', i.e. managers who demonstrate their commitment both through ongoing promotion of the food safety requirements and by being instrumental in providing necessary resources and support for its implementation. The senior team needs to understand the importance of this role as they commit to taking the project forward.

The role and responsibility of the senior team would include the following:

- Showing visible signs of support and alignment, for example, through vocal support at staff meetings or plant visits, i.e. acting as 'promotional managers'.
- Involvement in regular updates and reviews on progress.
- Commitment to engaging other functional leaders to show uniformity in their support (Sales, Marketing, R&D, Operations).
- Ensuring that required resources are made available.

Finally, some basic education can be provided on what a world-class food safety programme looks like. This would include the following:

- Design for food safety
- HACCP
- Prerequisite programmes
- Culture and behaviour of people
- Benefits, including the effect on the bottom line
- Driving external forces (customers, regulatory)

Senior managers need to understand that success or failure rests on their shoulders. Without their commitment, there is not a food safety programme, and without cross-functional alignment on what the gaps are and how they will be addressed, there will be no lasting progress. Gaps and gap assessments are presented in detail in Section 9.3.3.

9.2.2 Assignment of roles and responsibilities

The company needs to assign roles and responsibilities, but this may change once more detailed gap assessments have been done and the workload is clarified. That said, there are some obvious roles to assign at the beginning, and they may differ slightly depending on whether the company is large or small, has a corporate office or stands alone, whether this is an upgrade to the existing food safety programme or a brand new activity. We anticipate that 50 years on from the birth of HACCP and more than 15 years after Codex (1993) principles, for most readers this will be an upgrade.

Whatever the circumstances, a management steering group will be needed to provide oversight. In a larger organisation, this should be a senior-level team but it may be a few key members of staff in a smaller company. The team should scope out its objectives but may need some additional education and training in order to be able to do this thoroughly.

A food safety team (could also be called the HACCP team) will also need to be appointed. This team will be cross-functional and be able to provide the more detailed technical knowledge that the company will need in order to be able to develop and maintain the programme. This team must be capable of using judgement and experience to aid risk-based decision-making. They will also conduct training and awareness sessions within the rest of the organisation.

9.2.3 Training and education

Training and education is increasingly being recognised as critical in food safety assurance. This is not really new thinking; a decade ago the question was asked, 'When HACCP appears to fail, is it the fault of the HACCP system itself or does the real failure lie with the people who are trying to implement it?' (Mitchell, 1998). Unfortunately, it has taken a decade to shift thinking and recognise more widely that training and education has been inadequate as evidenced by the number of foodborne illnesses that still occur each year. And, as can hardly be pointed out often enough, we have also learned that many food safety failures are caused by inadequate attention to prerequisite programmes, possibly also due to lack of understanding of their importance through insufficient training and education. A few important questions need to be considered early on.

What are the desired outcomes of the training?

This question needs to be answered not simply in terms of the knowledge content of the training, i.e. what the trainee will *know how* to do, but what the actual business objectives are, for example, to implement a HACCP programme that will:

1. reduce consumer complaints by X,
2. reduce microbiological product testing by Y and
3. reduce product on hold/dispositional by Z.

A monetary value could be placed upon these, also perhaps a productivity value. Doing this will help with the cost challenges that often come with the request for additional resources (refer to cost of quality model, Chapter 2).

Who needs to be trained?

The simple answer is everyone but not all to the same degree.

Overview sessions. Senior managers and hourly employees require an overview of HACCP, of prerequisite programmes, of the cause for change (i.e. what is wrong with what we have been doing so far), and some examples of companies who did not change and had a food safety incident. A reminder of the few essential rules that apply to everyone should be included (hand-washing, hygiene zone protocols, reporting signs of pest activity, cleaning up spillages, minimise use of water). Having both senior managers and hourly employees in the same sessions is an excellent way of emphasising the shared responsibility for food safety.

Detailed food safety training. The food safety team needs in-depth training and education, both at the start of an implementation project and on an ongoing basis to ensure that they keep up to date. A detailed session on HACCP and food safety hazards lasting 2–3 days is a good introduction for this group, but they will not be experts even with this level of training. The more effective classes include content on both HACCP and food safety hazards and emphasise the connection between the theory (microbiological hazards, chemical hazards and physical hazards) and practical application during hazard analysis. Since hazard analysis has been identified as an area of HACCP where teams have difficulty (Wallace, 2009), clear focus on how to apply this, along with the other HACCP principles, is fundamental.

The two sessions described earlier are usually sufficient at this stage. Later, and during implementation, there will be a need for more training of critical control point (CCP) monitors, their supervisors and PRP leaders.

How should they be trained?

Not through slide presentation alone is the simple answer. The learning pyramid as proposed by the National Training Laboratories (Figure 9.3) indicates a mere 5% average retention rate of information by lecture alone. Combine that with the fact that the typical attention span for adults is 20 minutes and it starts to become clear that thought needs to go into how training will be delivered. Improved learning comes from using the knowledge in practice (average 75% retention rate), but that increases to 90% when having to teach others and by using the new knowledge immediately after the training session. This is really important to remember when providing new information on any subject and is a strong argument for a train-the-trainer approach. By training production supervisors and other functional managers (outside the food safety team) to deliver food safety training, more of the knowledge will be retained and a sense of a shared responsibility is reinforced.

Research (Wick *et al.*, 2006) suggests that there should be a new start and finish line where learning is concerned. We need to think of it as a process which does not start and end with the 2- or 3-day training session (which is traditionally the way HACCP training has been delivered) but to be more effective the process begins much earlier and it continues for a period after the training session itself. This can be explained by means of six disciplines (Table 9.1). Using this training approach requires more effort on the part of the trainer and the trainees' supervisor, but experience shows that the additional time invested is very worthwhile.

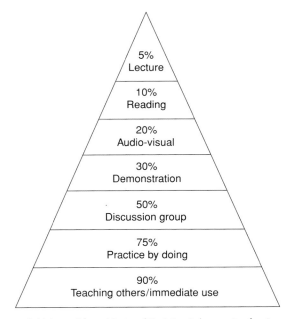

Fig. 9.3 Learning pyramid. (Adapted from National Training Laboratories for Applied Behavioral Sciences, Alexandria, VA.)

Training for the food safety team needs to cover Codex principles and guidelines and their application, plus information on hazards and their control (see 'Detailed Food Safety Training' above). When it comes to conducting PRP training, consideration of the approach outlined in Table 9.1 is also important; for example, essential practices such as hand-washing still need to be a learned behaviour for far too many people. Do not assume that this is happening 100% of the time.

Table 9.2 shows possible HACCP training subject matter and learning objectives for different groups of personnel. This can be used to cross-check in-house training materials or those of a third-party consultant.

Has the training delivered the required knowledge, skills and ability?

Although implied by disciplines 4 and 5 of Table 9.1, it is worth taking time out to consider the effectiveness of training in terms of the delivered knowledge, skills and ability. In many cases, this will be possible to determine via supervision as suggested earlier; however, for effective HACCP development it is important to establish if the HACCP team has the necessary knowledge of HACCP principles to be able to apply them in practice. Research has confirmed that certain areas of HACCP knowledge are problematic following traditional style training (Wallace *et al.*, 2005), and this suggests that it is important both to consider appropriate training methods and to understand the levels of HACCP knowledge attained following the training intervention. This can be achieved via knowledge testing, possibly offered as part of externally certificated HACCP training or via published instruments (e.g. Wallace *et al.*, 2005). Recent research (Wallace, 2009) has shown some further interesting findings with regard to HACCP training, namely:

Table 9.1   The six disciplines of breakthrough learning as applied to HACCP.

Discipline 1: Define outcomes in business terms

As discussed earlier, the team needs to think ahead of time about their objectives in doing the training. This includes deciding specifically what they really need to achieve in order to advance the business, e.g. completion of a HACCP plan, reduction in consumer complaints, reduced costs associated with internal or external company failure and reduced costs of testing. These and more can be utilised in setting improvement objectives. Each objective should also have a timeline attached to it.

Discipline 2: Design the complete experience

Involve the trainee's supervisor ahead of the session by having some discussions about training outcomes. This is where the training can be personalised to suit each individual. Poorly prepared trainees usually start any session by thinking 'what is in this for me'. Set some pre-work and assignments so that trainees come in eager to learn more and are already engaged in the subject matter, seeing it as a route to improved food safety performance.

Discipline 3: Deliver for application

This is the actual training event, so demonstrate food safety management and the HACCP concept in terms that are relevant to the trainee. Practical work and discussions should reflect processes being used in their actual workplace.

Discipline 4: Drive follow-through

By confirming food safety goals ahead of the training, it is easier to galvanise the trainees into action immediately following the training event. Regular milestones for review should be established and progress should be appropriately communicated within the organisation.

Discipline 5: Deploy active support

The more successful training programmes put a follow-through support plan in place at the start. This might involve prescheduled coaching and mentoring of the trainees as they work at applying the new knowledge. This can be achieved via support from a corporate team, a reputable consultant or by the trainers themselves. It is also important that the trainees' immediate supervisor sees the work as important and stays actively engaged and supportive. This is important if HACCP is to be seen as the way of working on an ongoing basis and not just the latest activity until the next project comes along.

Discipline 6: Document results

Not simply in terms of how many people got trained, or what they learned but in the sense of what they actually applied and how the company benefited as a result of the training. When training funds are reduced in companies, it is usually because there has been insufficient evidence of the return on the investment.

Source: Adapted from Wick *et al.* (2006).

- HACCP team knowledge of HACCP principles is not necessarily as good as the knowledge of individual team members.
- HACCP team knowledge of HACCP principles can be used to predict the effectiveness of the HACCP plans that they develop.

This suggests that understanding of the HACCP knowledge of trained HACCP teams and their individual members is important to delivering success in this area.

9.2.4   Resource management

Early in the process, the management steering group will need to review the resources required. This could also be done by the food safety team once set up. Estimates need to be made

Table 9.2   Possible training subject matter and learning objectives for different groups of personnel.

Group	Learning objective
Senior management	1. Understand the general principles of HACCP and how they relate to the food business. 2. Demonstrate an understanding of the training and knowledge requirements for food safety team members and the workforce as a whole. 3. Demonstrate an understanding of the links between HACCP and other quality management techniques and programmes and how a combined product management system can be developed. 4. Understand the need to plan the HACCP system and develop a practical timetable for HACCP application in the whole operation.
Food safety team leaders	*HACCP system and its management* 1. Demonstrate an up-to-date general knowledge of HACCP. 2. Explain how a HACCP system supports national and international standards, trade and legislative requirements. Describe the nature of prerequisite programmes and their relationship with HACCP. 3. Demonstrate the ability to plan an effective HACCP system. 4. Demonstrate a knowledge of how to lead a food safety team. 5. Demonstrate an understanding of the practical application of HACCP principles. 6. Demonstrate the ability to design, implement and manage appropriate programmes for verification and maintenance of HACCP systems. 7. Explain the methods to be used for the effective implementation of HACCP. *Additional topics* 1. Demonstrate an understanding of the nature of hazards and how they are manifested in food products/operations and give relevant examples. 2. Demonstrate an understanding of the intrinsic factors governing the safety of product formulations and methods that can be used to assess safety of new products. 3. Carry out the steps to identify significant hazards relevant to the operation and determine effective control measures, i.e. assessment of risk (likelihood of occurrence and severity). 4. Demonstrate an understanding of the training and knowledge requirements for food safety team members and the workforce as a whole. 5. Develop appropriate training programmes for CCP monitoring personnel. 6. Demonstrate an understanding of the links between HACCP and other quality management techniques and how a combined product management system can be developed.
Food safety team members	*HACCP system* 1. Justify the need for a HACCP system. 2. Show how the legal obligations on food business proprietors to analyse food hazards and identify critical steps in the business activities should be met in their appropriate industries. 3. List and explain the importance of the principles of HACCP. 4. Describe the method by which hazard analysis may be carried out and appropriate control measures ascertained to assess the practical problems. 5. Identify critical control points including critical limits to ensure their control. 6. Develop suitable monitoring procedures for critical points and explain the importance of corrective action procedures. 7. Verify the HACCP system by the use of appropriate measures. 8. Carry out the steps to introduce and manage a fully operational HACCP system.

Table 9.2 (Continued)

Group	Learning objective
	Additional topics
	1. Demonstrate an understanding of the nature of hazards and how they are manifested in food products/operations, and give relevant examples.
	2. Demonstrate an understanding of the intrinsic factors governing the safety of product formulations and methods that can be used to assess safety of new products.
	3. Carry out the steps to identify significant hazards relevant to the operation and determine effective control measures, i.e. assessment of risk (likelihood of occurrence and severity).
	4. Develop appropriate training programmes for CCP monitoring personnel.
CCP monitors	Understand the general principles of HACCP and how they relate to the food handler's role. Perform CCP monitoring tasks, record results and initiate appropriate actions.
Auditors of HACCP systems	*HACCP and regulatory auditing*
	1. Provide up-to-date general knowledge of HACCP and its relationship with national and international standards, trade requirements and legislative requirements.
	2. Examine the role of good hygiene practices as a foundation for HACCP-based food safety management systems.
	3. Provide a comprehensive revision of the application of HACCP principles for the development of HACCP-based systems for food businesses.
	4. Consider the design and management requirements associated with the application and implementation of HACCP-based food safety management systems in food businesses.
	5. Enhance the skills required for the assessment of HACCP-based food safety management systems.
	6. Consider the tools available to educate food business operators in the principles of HACCP and to provide advice and support during development and implementation of food safety management systems.
	Additional topics
	1. Understand the need for audit preparation including the development of suitable checklists.
	2. Perform HACCP audits using sampling, questioning, observation and assessment skills.
	3. Construct audit reports giving clear indication of findings and corrective action needed.
General workforce	1. Understand the general principles of HACCP and how they relate to the food handler's role.

Source: Adapted from Wallace (2001).

concerning the food safety budget and resource requirements at this stage. Examples of budget considerations include the following:

- Training, which may include the external trainer, room hire, travel time, test papers (if external certified training is used) and overtime costs for the hourly paid workforce.
- Administrative support if the team requires some short-term additional help.
- Consultant costs, where insufficient knowledge exists internally.

- Equipment and service costs where relevant. For example, analytical test equipment may be in need of an upgrade, and external testing laboratories may be needed or an upgrade on the third-party sanitation or pest control contract.
- Plant infrastructure upgrade. This will be unknown until after the more detailed gap analysis, but experience shows this to be the key element of the cost associated with implementing a HACCP programme, i.e. the PRP upgrades as opposed to HACCP itself.

It is also a good time to review what activities are truly necessary via techniques such as value stream mapping. HACCP can provide the insights in terms of food safety value, but an example of existing activities which add no value might be the receipt and filing of hundreds of unnecessary Certificates of Analysis (COAs). HACCP will help the organisation to focus on only getting in useful information on raw materials that are considered as being a significant hazard if uncontrolled. They then become part of the verification documents and have meaning.

9.2.5 Documentation

This tends to worry many smaller businesses, but whether the business is large or small there is a need for documentation as evidence of all the good work being done to protect consumer safety. It is important to think about 'measuring what matters', and this usually includes quality or regulatory compliance work as well as food safety.

Standard forms are provided in Codex (2009b) for the HACCP control chart. Those used as examples later in this book can also be used as a template. However, these are the formal elements of the HACCP plan and, the food safety team is encouraged to take additional notes or extend the forms to ensure they have a record of their thought process and the various actions that occurred. These additional working documents are unlikely to be shared with external assessors but, used internally, can be very helpful when updating a programme. Similar templates can be used to document PRPs.

Plan to control documents in terms of traceability and approval (via reference numbering, dates and approval signatures), also think about where to store them and allocate responsibility to someone on the team for this role. Weaknesses in the archiving of documentation have been highlighted (Wallace, 2009), where a study showed that some manufacturing plants had difficulty in retrieving key documents when their food safety systems were being assessed for effectiveness. If there is no documentary evidence that activities took place then for many assessors (including customers and regulatory inspectors) the activity must be assumed to not have occurred. This is also a problem with systems using only exception reporting when problems occur. If nothing at all is written down, it is not possible to demonstrate without doubt that the systems have been under control.

9.2.6 Supplier–customer partnerships

Partnerships between customers and their suppliers are an important part of any food safety programme. Confidence in raw material safety and quality can be facilitated through partnerships based on trust between the parties that requirements will be met. This will take time to build up and will likely include visits to the supplier premises as well as audits and the monitoring of specifications (see Chapter 3).

9.2.7 Continuous improvement

An essential management role is to ensure that the company is constantly on the alert for improvement opportunities. Some companies fail because they get complacent and feel that their programmes are effective, but the mindset should be one of continuous improvement. The team should be looking for new sources of information on food safety management, reviewing industry failures for root cause and assessing whether there are learnings that can be applied. Food safety management is a continuum – it is never 'done', so unlike many other projects this one does not have a real end point.

9.3 PREPARATION ACTIVITIES FOR FOOD SAFETY PROGRAMMES

Project management techniques can be used very successfully to organise all the activities that need to occur and to ensure focus.

9.3.1 Preparing a project plan

As indicated earlier, the company food safety or HACCP team, if that name is preferred along with the team leader is the 'action team', i.e. the team that will do the development work. They will be trained, have a budget and the full support of the senior management team.

At the start of the project and before starting the technical work, the team must do a few managerial things in terms of project management:

- *Confirm their objectives* (and also confirm whether third-party certification of the programme is one of them).
- *Confirm the deliverables* – what needs to get done in order to achieve the stated objectives. This can become a fairly detailed breakdown of tasks, which can be organised into categories and assigned to individuals or teams. The deliverables will be fairly detailed and will almost certainly require that the teams determine the structure of the HACCP programme, complete a gap assessment against PRP requirements, prioritise and complete any identified corrective actions.
- *Assign roles and responsibilities* – per the stated deliverables.
- *Confirm the scope of the project* – what does it include and what does it not include. For example, it might be limited to an update of an existing programme, or it might be that a new production line is being commissioned and a new HACCP plan is being prepared.
- *Determine the project milestones* and agree the project timeline.
- *Confirm assumptions and highlight any dependencies*, e.g. a product safety assessment (Chapter 11) may need to occur ahead of doing a HACCP study.
- *Confirm the budget*.
- *Set regular review dates* with senior management.
- *Agree a communication plan* so that everyone knows the status.
- *Agree the measures of success*.

Once the project plan is agreed, the team can get started on completing the assigned activities in terms of the deliverables. A few of the major ones are discussed in the remainder of this chapter.

9.3.2 Structure the HACCP programme

The structure of the HACCP programme is important not just at the development stage but also for maintenance of the system. It can be modular with multiple smaller plans or one large all-encompassing plan. Usually, this depends on the product and plant, and the company will need to evaluate options. As a guideline, a company making a single product line with few raw materials (e.g. canned fruit) may find that one process flow chart and HACCP plan is simple enough. A company with hundreds of ingredients and product lines (e.g. manufactured processed foods such as baked goods, confectionery and prepared meals) will typically use a modular approach. This involves breaking down the process into units or 'modules' which could be generically categorised as defined in the example given in Figure 9.4.

9.3.3 Carry out a gap assessment

There are two main areas to review in terms of assessing the gap between where the organisation is at the start and where it needs to get to. The areas are:

World-class food safety programme
- Safe product design
- Prerequisite programmes
  - ○ Including risk evaluation at plants
- HACCP
- Essential management practices

Human resources
- Organisation structure
- Knowledge and skills
- Headcount

We look at each of these in turn.

World-class food safety programme gap assessment

Safe product design. The gap assessment consists of establishing whether any formal evaluations are being done and whether validation data exist, for example, through challenge testing, and the compilation of literature references (see Chapters 6, 7 and 11). The organisation must know what is making its products safe.

Prerequisite programme gap assessment. Prerequisite programmes are described in Chapter 10, so detail of the specific PRP elements is not provided here. An on-site self-assessment should be conducted by the food safety team against best practice standards. Alternatively, it could be helpful to hire a reputable third-party expert consultant or audit company with a good track record in auditing against best practice standards to carry out the gap assessment with the team. In conducting a gap assessment as opposed to formal audit, there will be more

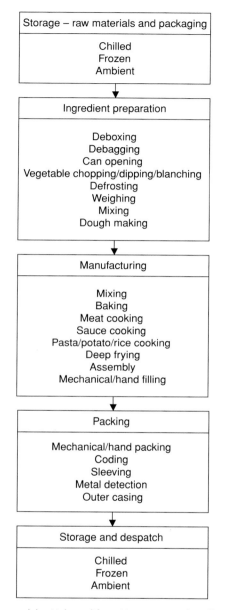

Fig. 9.4 A process operation module. (Adapted from Mortimore and Wallace (1998).)

opportunity to make this a real team effort and to ensure that as much as possible gets reviewed. Engaging the workforce in this activity can help to reinforce that this is a shared commitment for improvement as opposed to it being seen as QA team 'police' work. The gap assessment should cover a review of compliance against all elements of a best practice PRP standard, such as outlined in Chapter 10.

The team should consider other inputs whilst going through this exercise:

- Previous gap assessments
- Previous audit reports (internal/external)
- Results of regulatory inspections
- Consumer complaint data
- Any key performance indicators such as sanitation pre-operational inspections, first time quality and environmental/product microbiological test data

Whilst all PRPs are important, there are two areas, sanitation and supplier quality assurance (SQA), that we want to use as examples to illustrate the gap assessment process and which will demonstrate the in-depth investigation required.

Example 1: sanitation. There is often real value in bringing expert resources for this part of the assessment, as in-house resources rarely have sufficient expertise for the evaluation of sanitary design and sanitation programme and practice assessment. This is obviously essential for high-risk foods, but all food manufacturers need solid programmes to ensure that products are both safe and wholesome for consumption.

Here are a few key areas to consider:

- *Sanitation risk evaluation.* The team will need to evaluate how the programme was developed. Was the plant sanitation programme developed as a result of a formal determination of what needed to be cleaned, why, how, and at what frequency? Risk factors will include potential microbiological contaminants (e.g. pathogenic and spoilage organisms), potential chemical contaminants (e.g. allergens and additives) and physical contaminants (e.g. pest infestation and foreign material).
- *Efficacy of the sanitation programme.* Many companies use a combination of measures on a routine basis including ATP bioluminescence and environmental *Salmonella* and *Listeria* species monitoring plans. During an on-site assessment, there should be an opportunity to really take a closer look at the programme during active periods (Behling, 2006):
  - Assessment of the manufacturing process – the equipment, the environment, the people – during and at the end of a production shift. This can be achieved via observation and by taking targeted environmental swabs from the equipment and environment.
  - Observation of the clean-up process. This will involve talking with the operatives, observing what equipment is dismantled for cleaning and what is not. The assessment team will want to question people regarding the rationale. 'Too difficult to break apart' as a response is an excellent pointer for the team to do exactly that. They will watch whether the cleaning process itself can lead to cross-contamination, for example, the use of high-pressure hoses to clean the floors after the equipment is clean or worse, when there is an exposed product in the vicinity. The team will also be measuring flow rates, chemical concentrations, and will cross-check actual activities against documented work instructions – the aim is to be able to assess whether the company is doing what it said it would do in the procedures, and also to assess whether these are adequate.
  - Post-cleaning sampling and pre-operational inspection. This is part visual inspection and part environmental sampling. Taking samples of first production off a newly cleaned line can also be used as an indicator through carrying out microbiological testing. However, if

there are 'niches' such as cracks or crevices, poor welds or dead ends somewhere in the system then contamination may not show up until later. In the case of pathogen testing, all production will need to be held until the test results of any product sampling are known. Allergen tests will also be important where the purpose of cleaning includes the requirement to remove allergens.
○ Analyse pre-clean and post-clean data, draw conclusions from observations and make recommendations. The pest control programme can be reviewed at the same time given that sanitation is one of the preventive controls for this programme.

Example 2: supplier quality assurance. Raw material quality is fundamentally important in ensuring that a safe, wholesome product gets to the consumer. There would be very few companies where the quality of their raw materials had little impact on finished goods. For a number of companies, this can actually be a CCP, i.e. where a high-risk raw material is being purchased but receives no further kill step as it gets incorporated into the finished product. In this case, the reputation of the company relies on ability of their raw material supplier to consistently manufacture a safe product. It is therefore essential to have an effective SQA programme.

In doing a gap assessment of the SQA programme, considerations include the following:

- Raw material specifications
  - Do they exist?
  - Are they agreed and signed by the supplier?
  - Do they include the hazards of concern (microbiological, chemical, physical)?
  - Is there a written guarantee of continuing supply to the agreed specification?
- Risk evaluation
  - Do you know which raw materials are a high risk?
  - Are the hazards associated with raw materials included in the HACCP plan (via the hazard analysis)? This will be a gap in companies just starting to implement HACCP.
  - Have you confirmed which ingredients are a CCP (safety controlled by the supplier)?
- Product safety questionnaires (some form of document which questions the suppliers about their programmes)
  - Are they used?
  - Have they been reviewed?
  - Are there instances where potential hazards were missed (e.g. multiple allergens on the site but no action taken)?
  - Have third-party audits been done and are the reports on file?
  - How is the validity of test results assured?
- On-site supplier audits
  - Were these carried out for the high-risk raw materials?
  - Is there evidence that the auditor was trained and calibrated to do this audit?
  - Were any corrective actions required and is evidence on file that these were completed?
- Supplier maintenance
  - Are COAs needed for verification of the food safety programme?
  - Are any microbiological tests ever done on samples of incoming deliveries and would this be useful?
  - Does the supplier inform you of issues?
  - Are all supplies formally approved?

In plant PRP risk evaluations. Experience has shown the benefit of doing this alongside a PRP gap assessment. It is an area that requires a certain amount of expertise and is often buried amongst all the other information gathered during generalised quality management system audits. Yet it is *essential* to have this knowledge in order to upgrade or implement an effective food safety programme. The focus of this type of plant assessment is usually microbiological, but consideration should be given to chemical (particularly allergens) and physical hazards. This activity needs to be completed at the plant as the main aims are to:

(a) identify the areas where cross-contamination can occur and
(b) identify the areas or niches where bacteria can get established and grow.

Starting with the layout map of the plant, track the flow of the product, people and air. Include drains (and flow), garbage removal activities, raw materials, laundry, pallets, anything that comes in, goes out or moves around the plant. Knowledge of the process and product is extremely important as the evaluation of risk if the product becomes contaminated cannot be done in isolation. However, bear in mind that just because the product does not support the growth of microorganisms, it does not mean that they will not survive. There are many examples of low water activity (dry) and frozen products that have caused foodborne illness. Peanut butter, chocolate, dry pet food and ice cream unfortunately being good examples (Chapter 2) of how cross-contamination within a process environment can result in major foodborne illness outbreaks.

At the plant, assumptions made during a review of the layout can be confirmed and a closer look can be taken to identify niche areas where microbial growth may be occurring.

As indicated in Figure 9.5, this comes back to basic microbiology (Chapter 4). Anywhere that moisture can get trapped, food (even minute particles transferred via dust) is available, and for a period at the right temperature can present a problem, as indicated in Figure 9.5 by the exclamation point (!).

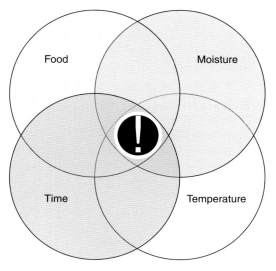

Fig. 9.5   Microbial growth requirements. (Adapted from Kornacki (2009).)

Once the risk evaluation is completed, a list can be drawn up, corrective actions confirmed and a timeline established for closing the gaps.

HACCP programme gap assessment. This is only relevant if you are updating an existing programme. Some key areas to look at are:

- Process flow diagram
  - When was it last validated on the production floor?
  - Is it complete?
  - Is it accurate?
- Was hazard analysis carried out at every process step?
  - Does the hazard analysis go into detail or is it at the 'biological hazard' level?
  - Have emerging pathogens been considered where relevant (e.g. *Cronobacter sakasakii*)?
  - Do you have sufficient control measures, e.g. do you need to do a physical hazard control device (e.g. metal detectors, sifters, sieves and magnets) audit?
- Is there sufficient data on the intrinsic product safety?
  - Does the current formula match that which was in place when the original product safety assessment was done? (This is a common weakness.)
  - Are challenge tests needed?
  - Have mathematical models been used?
- Is the required validation data on file?
  - Literature references
  - Plant data
  - Was the validation sufficient?

These are just a few key questions. With training and experience, the food safety team should be able to compile a detailed list of areas to look at.

Essential management practices. The ultimate responsibility for food safety resides with the senior management level in the company. This is why 'management responsibility' is increasingly being recognised as a critical success factor for food safety programmes and is now included within some of the Global Food Safety Initiative (GFSI) benchmarked audit schemes. There are a few things which indicate how well this is implemented, such as whether the company has a food safety and quality policy and whether the roles and responsibilities for food safety are clearly defined in job descriptions. Within the plant itself, the assessment team should be looking for signs of whether short cuts are being taken, also look for evidence when procedures are not being followed and where there is an apparent lack of accountability for enforcement.

Attitudinal assessment is difficult, but the team could consider a discussion around productivity targets versus quality targets. Also consider the following:

- How often are managers seen in the plant?
- How much training employees (supervisors and hourly) are given per annum? How is this achieved?
- Is there evidence that any new behaviours are being reinforced (health and safety as well as PRPs are a good indicator)?

Are regular management reviews taking place? The agenda should include the following:

- Audit results (internal or third party)
- Corrective action status
- Consumer/customer complaint data
- Key performance indicators such as HACCP compliance, sanitation metrics, first time quality and environmental monitoring events

Basically, does the management team take all necessary steps to ensure that product is produced under sanitary conditions, that specifications and HACCP standards are known and adhered to, and that adequate resources (trained and knowledgeable) are in place to deliver on commitments?

Human resource gap assessment

It is suggested that this is done after the world-class food safety programme gap assessment. The reason for this is that the outcomes of the food safety programme gap assessment will be a key indicator of whether there are organisational gaps that also need to be addressed. The human resource gap analysis should be fairly straightforward, but here is some guidance below.

Review organisation charts (include corporate teams where they exist as well as plant locations) and consider individuals on the basis of the following:

- Educational background
- Experience base (number of years, relevance)
- Proven leadership ability

Some of the relevant elements of leadership are:

- Being action oriented
- Acting as a change agent (can motivate and lead others)
- Holding people accountable
- Being willing to speak up
- Acting with high integrity

Obtain organisational benchmark data if available – from industry colleagues, suppliers and customers. Then, collate the data to determine whether you have both the sufficient number of people and the appropriate knowledge base in terms of education and experience. Work with human resource colleagues to make recommendations where needed. This could include additional training, staff and access to external resources.

9.4 PRIORITISATION OF CORRECTIVE ACTIONS

Having completed a gap assessment, there will almost certainly be a long list of items requiring corrective actions. Whether starting a new programme or upgrading an existing one, the lists are likely to be extensive and it can feel overwhelming unless some prioritisation and planning occur. Corrective actions should be prioritised on a risk basis. To refer to the start of this chapter, the PDCA cycle can be used to good effect or at least as a starting point (Figure 9.1).

Throughout the process, a communication plan needs to be included in order to keep the company apprised of progress and involved. Also, responsibilities must be assigned to each corrective action and timelines estimated and adhered to. Gantt charts can be helpful for tracking progress, but a simple wall calendar will also work.

Setting timelines for corrective actions will need input and oversight from senior management as well as the food safety team, and a risk evaluation of each of the issues identified can be used to assign priorities. The food safety concerns should be rated in terms of likelihood of occurrence versus severity of effect. For example, if in a cooked meat operation, one of the issues identified was 'no physical separation of raw and processed product' then the food safety hazard associated with that gap needs to be specified. If procedures exist to manage this, then they need to be taken into account to determine the likelihood of a hazard occurring. A quadrant diagram can be used such as the one shown in Figure 9.6.

Severity of effect:

Low = minor injury
Med = serious injury or short-term illness, possible hospitalisation
High = long-term illness, chronic effects or death

Likelihood of occurrence:

Low = unlikely to occur but might
Med = probably could occur (no history)
High = highly probable (known history or it will happen at some time)

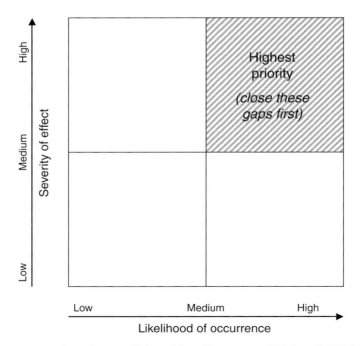

Fig. 9.6   Prioritisation quadrant diagram. (Adapted from Mortimore and Wallace (1998).)

Any gap which is rated as being a HH should result in a closing of the plant until the actions are complete. Anything judged as an HM or an MH would be the next high priority, and the team should ensure that some immediate (short-term) management of the identified issue is put in place whilst more permanent solutions are found. Anything with an MM would be next on the list followed by anything with an 'L'. Whilst the lower severity and likelihood issues can be done later, it is still important that they remain visible and on the list. If they are easy and low investment, corrections then they should just be closed out.

Other factors to consider when prioritising by risk are the external factors – customer or regulatory obligations. This however should not be the main reason for making improvements.

9.5 CONCLUSIONS

A nice schematic to end on and which is adapted from the original (Panisello and Quantick, 2001) is shown in Figure 9.7.

'A' is the sustainable model with a firm foundation of internal commitment, supported by training and education with appropriate resources and plant infrastructure and aligned to the external requirements. 'B' shows how, when the foundation for change is all external, there is instability and potentially the programme is not sustainable – when external priorities change. This knowledge can help build the action plan. External stimulus for change is healthy but should not be the sole reason for doing it. The real prize is the peace of mind that comes

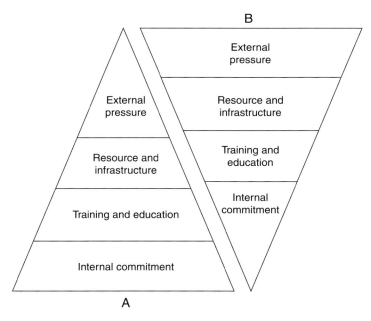

Fig. 9.7 Panisello pyramids. (Adapted from Panisello and Quantick (2001).)

with having a mindset that is continually striving to do better, and the cost benefits that come with it.

The final action plan resulting from these preparation activities should be reviewed and approved by senior management to confirm their ownership and support. Managers who have committed to the project will be keen to understand the size of the task and to monitor progress.

10 Formalised prerequisite programmes in practice

10.1 INTRODUCTION

Prerequisite programmes (PRPs) provide the hygienic foundations for any food operation. The terms 'prerequisite programmes', 'good manufacturing practice', 'good hygienic practice' and 'sanitary operating practices' are used interchangeably in different parts of the world but have the same general meaning. The term 'prerequisite programmes' has evolved to be most frequently used for systems in support of HACCP (Wallace and Williams, 2001). PRPs are not yet as standardised as HACCP principles, both in language and in practice, and this has led to different application standards and accepted practices around the world, whilst the lack of consistency in terminology helps to promulgate misunderstanding about exactly what is required.

Whilst these differences currently exist with respect to PRP programme expectations, this does enable us to look around the world for best practice. Many would agree that the US approach to allergen management and sanitation, when done well, is a centre of excellence. The UK and Japan both have very high standards with regards to personal hygiene – other centres of excellence. As we further develop global standards and elaborate on detailed requirements needed to meet expectations, we have an opportunity to discuss and share best practices and bring all countries up to the same high level, raising the bar worldwide. Multi-nationals have accelerated some of this discussion as they increasingly shop around the world for raw materials and products, and, increasingly, set up local manufacturing facilities in developing countries.

All PRPs are important for food safety assurance, and it is crucial that they are fully implemented in practice and kept up to date with best practice standards. In this chapter, the requirements for PRPs as a cornerstone of the food safety management programme are considered and examples of typical best practice expectations are provided for the formalisation and implementation of specific prerequisites.

10.2 PREREQUISITE DEFINITIONS AND STANDARDS

Several groups have suggested definitions for the term 'prerequisites' and the most commonly used are reproduced here.

Food Safety for the 21st Century, First Edition By Carol A. Wallace, William H. Sperber and Sara E. Mortimore
© 2011 Carol A. Wallace, William H. Sperber and Sara E. Mortimore

Prerequisite programmes:

> Practices and conditions needed prior to and during the implementation of HACCP and which are essential to food safety (World Health Organization, WHO, 1999).

> Universal steps or procedures that control the operating conditions within a food establishment, allowing for environmental conditions that are favourable for the production of safe food (Canadian Food Inspection Agency, CFIA, 1998).

> Procedures, including GMP, that address operational conditions, providing the foundation for the HACCP System (The National Advisory Committee for Microbiological Criteria for Foods, NACMCF, 1997).

A number of groups have published helpful material on PRPs; however, the internationally accepted requirements for prerequisites are defined in the Codex General Principles of Food Hygiene (Codex, 2009a). The PRPs listed in this document are split down into groupings as follows:

- Primary production
- Food chain
    - Establishment: design and facilities
    - Control of operation
    - Establishment: maintenance and sanitation
    - Establishment: personal hygiene
    - Transportation
    - Product information and consumer awareness
    - Training

These groupings form the essential areas where PRP elements must be developed, implemented and maintained to provide environmental conditions that are favourable to the production of safe food and thus the foundations needed for effective HACCP systems.

The intended scope of the Codex guidelines is the provision of a baseline structure for application to the entire food chain. As such, the document offers guidance to Governments on the essential elements they should encourage food businesses within their jurisdiction to apply. For the industry, it is intended that the elements of food hygiene systems described should be applied as minimum standards to provide foods that are safe and suitable for consumption, and to maintain confidence in internationally traded food commodities. Other groups have produced PRP guidance at a more detailed level, e.g. IFST (2007) and BSI (2008); however, we focus on the international requirements of Codex (2009a) in this chapter.

10.3 PREREQUISITE PROGRAMMES – THE ESSENTIALS

Focusing on the headings given in Codex (2009a), the following paragraphs describe the general requirements for PRPs in each area. It should be noted that the Codex guidelines provide only brief, 'top-level' requirements and so more detailed guidance may be needed to develop robust PRP elements for each area. Specific examples of how this might be achieved are given within the text for three PRP elements: cleaning and sanitation, pest management and allergen control.

10.3.1 Primary production

Figure 10.1 shows the four elements required of primary production PRPs. At this stage of the food chain, the intention is that food produced is safe and suitable for its intended use and primary production PRPs are, therefore, based on appropriate hygienic practices; control of contaminants, pests and diseases; and use of production areas where there are no environmental threats to the production of foods.

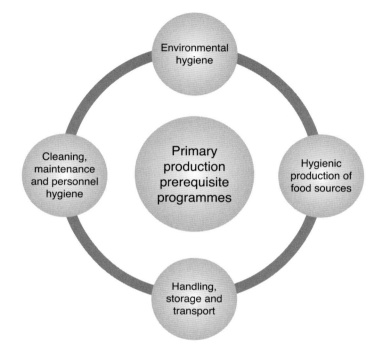

Fig. 10.1 Primary production prerequisite programmes.

Environmental hygiene requires the consideration of potential contamination sources from the environment and prevention of use of areas that could cause the presence of unacceptable levels of potentially harmful substances in the food produced.

Hygienic production of food sources relates to the need to identify likely sources of contamination and implement controls to minimise the risk of contamination, including prevention of contamination from air, soil, water, feedstuffs and other agricultural agents (e.g. fertilisers and pesticides); controlling animal and plant health to minimise threats to human health at consumption; protecting food from faecal or other contamination; and managing wastes and harmful substances appropriately.

Handling, storage and transport describes the protection, segregation and disposal requirements needed to protect foodstuffs, preventing spoilage and deterioration where possible, e.g. through control of temperature and humidity.

Cleaning, maintenance and personal hygiene at primary production refers to the necessity for appropriate procedures and facilities for the maintenance of these essential practices. Although no details of the required standards are provided, further useful information is available in the later sections of the Codex (2009b) document relating to secondary processing (see below).

10.3.2   Establishment: design and facilities

The first section on secondary processing is 'Establishment: Design and Facilities', this has four main areas of PRP elements (Codex, 2009a), as illustrated in Figure 10.2.

The **location** of food premises is important and care should be taken to identify and consider the risks of potential sources of contamination in the surrounding environment. In particular, environmentally polluted areas, areas of heavy industry which could pose contamination risks, areas prone to pest infestation, areas subject to flooding or where waste cannot be removed effectively should be avoided when planning food production facilities. Suitable controls to prevent contamination should be developed and implemented.

The design and layout of the **premises and rooms** should permit good hygiene and protect the products from cross-contamination during operation. Internal structures and equipment should be built of materials able to be easily cleaned/disinfected and maintained. Surfaces should be smooth and non-pervious and able to withstand the normal conditions of the operation, e.g. withstand the normal moisture and temperature ranges and detergents/disinfectants in use, and be inert to the food being produced. Whilst this applies particularly to the direct food contact surfaces, other surfaces within the food processing area, i.e. walls, floors, partitions, ceilings and overhead fixtures, windows and doors, should similarly be designed to minimise the build-up of dirt, condensation, etc., and the shedding of particles or contaminants that might gain

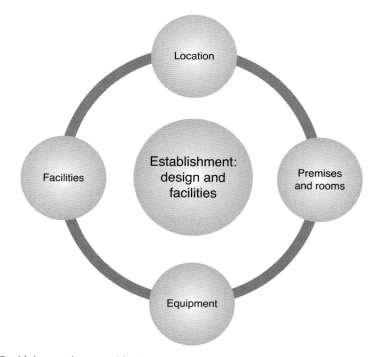

Fig. 10.2   Establishment: design and facilities – necessary prerequisite elements.

access to food products. Similar standards should be applied to the design, construction and siting of temporary/mobile premises and food vending machines.

Equipment that will come into contact with food should also be designed and constructed to facilitate cleaning and disinfection, including disassembly where necessary, and be made of materials which will have no toxic effects under the intended use. Food control and monitoring equipment should be suitable for the necessary use, e.g. able to meet rapidly and maintain the required food temperatures to eradicate microorganisms and their toxins or reduce to safe levels, or to monitor critical limits at CCPs. Containers for by-products, waste, inedible and dangerous substances must be constructed to protect food from contamination and should be specifically identifiable, including appropriate security considerations, e.g. lockable, to prevent accidental discharge or malicious contamination.

Facilities should be provided to include adequate potable water supplies, suitable drainage and waste disposal, appropriate cleaning facilities, storage areas, lighting, ventilation and temperature control. Suitable facilities should also be provided to promote personal hygiene for the workforce, including adequate changing areas, lavatories and hand-washing and drying facilities. All these facilities should be designed to minimise likelihood of product contamination, e.g. protection of light fittings to retain any breakages and design of ventilation systems to prevent airflow from contaminated to clean areas.

10.3.3 Control of operation

The rationale for operational control listed in Codex (2009a) is 'to reduce the risk of unsafe food by taking preventive measures to assure the safety and suitability of food at an appropriate stage in the operation by controlling food hazards'. This includes the need to control potential food hazards by using a system such as HACCP. The required prerequisite elements for control of the operation are outlined in Figure 10.3.

Control of food hazards (Codex, 2009a) requires the use of a system such as HACCP throughout the food chain to identify any steps in food operations which are critical to food safety; implement effective control procedures at those steps; monitor control procedures to ensure their continuing effectiveness; and review control procedures periodically, and whenever the operations change. These requirements clearly signpost the need for application of the HACCP principles to develop effective control plans (HACCP plans) for significant food safety hazards, as discussed in detail in Chapters 12–14. However, the obligation to enable control of food hygiene and safety throughout the shelf life of the product through proper product and process design is also acknowledged here.

Codex also describes **Key Aspects of Hygiene Control Systems**:

- Time and temperature control to prevent common causes of foodborne illness
- Use of specific process steps that can contribute to food hygiene, e.g. chilling or modified atmosphere packaging
- Microbiological and other specifications based on sound scientific principles
- Management techniques to control microbiological cross-contamination risks, e.g. segregation of processes/parts of processes and restricted access procedures
- Physical and chemical contamination prevention, including the use of suitable detection or screening devices where necessary

Although not mentioned specifically in Codex (2009a), the requirements to prevent cross-contamination with allergenic materials on site should be considered as key aspects of the

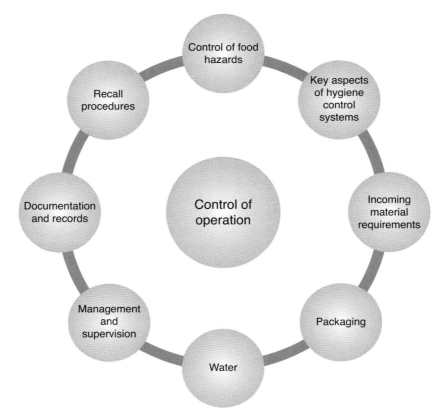

Fig. 10.3 Control of operation – necessary prerequisite elements.

control system for chemical contamination. The best way of controlling allergens as with any food safety hazard is to design them out. However, this is not always reasonable, so allergens must be controlled through:

- Avoidance of cross-contamination – where the risk comes from allergens which are inadvertently present and consequently not included on the label.
- Labelling – which can be used to manage allergens that are intentionally added to the formulation.

Effective allergen management requires an integrated approach. Each stage of the product lifecycle should be considered, i.e.:

- **During product design**
  - Is the allergen required for functionality and would a non-allergenic alternative be available?
  - Is the allergen already in use in the plant?
  - Is the allergen already in use on the same process line?
- **Hidden sources – understanding supplier control programmes**
  - What allergens are used in the supplier's plant?
  - Has the supplier programme been verified during an on-site visit?

○ Can you be sure that the supplier will notify you if and when a new allergen is introduced to the plant?
- **During Manufacture**, formalised allergen control programmes will be developed by conducting an allergen risk evaluation. This is the basis for the sequencing schedules and rework procedures which are used to prevent cross-contamination. Elements of this include:
  ○ Listing all allergenic materials which are on site. Accurate and detailed raw material specifications are needed in order to be able to identify 'hidden' allergens within compound raw materials.
  ○ Listing all product manufactured together with the allergenic raw materials that go into them. This is commonly done as a matrix of products and ingredients.
  ○ Develop an allergen process flow diagram or use the HACCP documents. This, combined with a map of the plant, is used to document where allergenic materials are situated and where they could get added into the process.
- **Transportation and labelling** include the requirements to protect food from cross-contamination during transit and the need to ensure that all allergens are clearly and accurately labelled on the package. This is governed by legislation for specific allergens in some countries. For example, at the time of writing, in the EC there is a list of 14 allergen groups that must be labelled (Table 10.1) according to Directives 2003/89/EC and 2006/142/EC.

Incoming material requirements and systems to ensure the safety of materials and ingredients at the start of processing are necessary to protect the operation and its products. This prerequisite element covers the need for appropriate specifications and acceptance procedures to prevent the acceptance of hazardous raw materials and ingredients that would not be controlled

Table 10.1 Examples of allergen labelling legal requirements: EC allergen labelling requirements.

Cereals containing gluten (i.e. wheat, rye, barley, oats, spelt, kamut or their hybridised strains) and
   products thereof
Crustaceans and products thereof
Eggs and products thereof
Fish and products thereof
Peanuts and products thereof
Soybeans and products thereof
Milk and products thereof (including lactose)
Nuts and products thereof, i.e.
     Almond (*Amygdalus communis* L.)
     Hazelnut (*Corylus avellana*)
     Walnut (*Juglans regia*)
     Cashew (*Anacardium occidentale*)
     Pecan nut (*Carya illinoinensis* (Wangenh.) K. Koch)
     Brazil nut (*Bertholletia excelsa*)
     Pistachio nut (*Pistacia vera*)
     Macadamia nut and Queensland nut (*Macadamia ternifolia*)
Celery and products thereof
Mustard and products thereof
Sesame seeds and products thereof
Sulphur dioxide and sulphites at concentrations of more than 10 mg/kg or 10 mg/L expressed as SO_2
Lupin and products thereof
Molluscs and products thereof

Source: European Parliament, Directives 2003/89/EC and 2006/142/EC.

by the process, including the use of appropriate inspection and sorting procedures along with effective stock rotation.

Although only brief details are provided by Codex (2009a), these points underline the importance of effective supplier assurance programmes to control the safety of incoming goods. In practice, this will require the development of a detailed programme, including the following:

* Raw material hazard analysis and prioritisation (tying in with HACCP; Chapter 12)
* Consideration of necessary human resources to manage supplier assurance and provision of adequate training
* Assessment of suppliers, e.g. via desktop assessment, on-site audit or use of third-party audit
* Ongoing evaluation of supplier performance and material acceptability
* Use of approved supplier databases and integration with business ordering and material acceptance procedures
* Control and maintenance of documentation, e.g. specifications, pre-supply surveys and questionnaires, conditions of supply, audit checklists and supplier performance data

Suitable packaging design to provide the necessary protection to the product during its shelf life is also highlighted by Codex (2009a). This embraces the need to ensure that the chosen packaging system is itself safe and will not pose a threat to the food product, plus the requirement to ensure hygienic conditions of reusable packaging, e.g. refillable glass bottles.

Codex also lists the importance of hygienic control of **water**, in particular the use of potable water where water is a food ingredient and for all food handling and processing operations, with the exception of specific food processes where non-potable water would not cause a contamination hazard to the food, e.g. using clean sea water for chilling in some operations. The need to treat and monitor any water being recirculated for reuse is also described, as is the requirement to make ice only from potable water, and to ensure the safety and suitability of steam for direct food contact. It is sometimes necessary to install in-plant water treatment systems such as in-plant chlorination systems to ensure a supply of potable water.

Appropriate **management and supervision**, reflecting the size of the operation and nature of its activities and processes, is also highlighted, with the need to ensure personnel have enough knowledge of food hygiene principles to be able to form a judgement on likely risks and take necessary actions. The need to keep adequate **documentation and records** and to maintain these records for a period exceeding the shelf life is also highlighted.

The development and testing of suitable **recall procedures** so that product can be effectively withdrawn and recalled in the event that a food safety problem is highlighted. This includes the need for withdrawals, public warnings and secure control of recalled products until destruction or other suitable disposition, i.e. not used for human consumption unless determined suitable or adequately reprocessed for safety.

10.3.4 Establishment: maintenance and sanitation

As shown in Figure 10.4, 'establishment: maintenance and sanitation' is a broad-ranging prerequisite grouping, which includes the elements of cleaning and disinfection/sanitation, pest management and waste management, plus the need to monitor the effectiveness of these elements in all cases.

Fig. 10.4 Establishment: maintenance and sanitation – prerequisite requirements.

Maintenance and **cleaning** are both important to keep the processing environment, facilities and equipment in a good state of repair, where it can both function as intended and prevent cross-contamination with food residues and microorganisms that might otherwise build up. Facilities should operate preventive maintenance programmes as well as attending to breakdowns and faults without delay. Cleaning can be performed by the use of separate or combined physical methods, and will normally involve the following:

- Removing gross debris
- Applying detergent solution(s)
- Rinsing with potable water
- Disinfection/sanitation as necessary

Dry cleaning or other appropriate methods may be used as appropriate to the specific situation, and all chemicals used should be handled and stored carefully, to prevent contamination of food products.

Cleaning programmes should be developed to encompass all equipment and facilities as well as general environmental cleaning, and this may require specialist expert advice. Cleaning methods need to be developed that are suitable for the item to be cleaned and should describe both the method and frequency of cleaning specific areas and items of equipment, and the responsibility for the tasks. Records of cleaning and monitoring should be kept.

A specific example of how the cleaning and disinfection/sanitation prerequisite elements might be formalised within a food business is given in Box 10.1.

Box 10.1 PRP-specific example – cleaning

A formalised cleaning and sanitation programme is one which is risk-based, is documented and is validated as being effective and routinely verified. What might each of these elements look like when properly implemented?

(A) Sanitary design

The facility and equipment will be designed and managed in a manner which enables them to:

(a) Be maintained in a clean condition and in a good state of repair such that they are not a source of contamination through build-up of soil, dislodging of foreign materials (e.g. paint), development of condensation, mould or pathogenic bacteria, and avoidance of niche areas that are difficult to clean.

(b) Avoid cross-contamination of products through the ability to control employee traffic patterns and airflow. When a food business has both raw and cooked ready-to-eat microbiologically sensitive products on site, this is critical.

(c) Provide adequate facilities to enable effective cleaning activities; ideally, a separate cleaning room with proper segregation between unclean and cleaned equipment such that cross-contamination is prevented.

(d) Manage utilities to ensure they are not a source of cross-contamination, e.g. water and air supplies.

(B) Risk evaluation

This is an evaluation of why to clean the (type of soil), what to clean, how to clean it and how often. Considerations will include potential microbiological risks, i.e. pathogenic and spoilage microorganisms, and potential chemical risks, i.e. allergens, pesticides, cleaning chemicals (e.g. cyanuric acid in chlorine-based manual cleaners) and additives; this last example being of particular concern in the feed industry, where variability in species intolerance is a major hazard.

A documented evaluation should be done for each area of the plant, and each piece of equipment in the process.

(C) Determination of appropriate cleaning methods

Determination of the cleaning method is usually done together with the external sanitation service provider. Wet cleaning using chemical cleaners and sanitizers is usual for microbiological and allergen hazards. However, dry cleaning has some significant benefits in terms of reducing the amount of water available to microorganisms and is increasingly being recognised as a better option in certain circumstances. It is generally easier to prevent environmental microbial growth in dry rather than wet conditions. If dry cleaning is used, a sanitizer (in countries where it is permitted) or alcohol wipe could follow, but it is not always necessary. This should be evaluated and appropriate data should be gathered to validate the method selected and to make adjustments as needed when setting up the programme.

(D) Sanitation schedules and cleaning procedures

A Master Sanitation Schedule (MSS) will be in place for all areas outside the regular equipment and process area cleaning. It will include, for example, overheads and light fixtures, external perimeter, coolers and freezers. This could also be incorporated into a comprehensive schedule designed to indicate tasks which are daily, weekly, monthly, quarterly or annual. The cleaning procedures will state the following:

- The equipment/area to be cleaned
- The frequency of cleaning
- How it is to be cleaned
  - Via detailed work instructions or 'one-point lesson' plans
- The time (duration) allowed for the task
- Materials to be used, for example:
  - Chemicals
  - Tools
- Chemical concentration and contact times
  - Including any routine strength testing, e.g. with titration
- Health and safety requirements, for example:
  - Safety glasses
  - Protective clothing
- Expected outcome, for example:
  - Visual standards
  - ATP bioluminescence and/or traditional hygiene swabbing
- Corrective action in the event of a problem, for example:
  - Actions required (e.g. reclean)
  - Whom to call
- Record requirements
  - Operator sign off
  - Reviewer sign off
- Verification/pre-operational inspections
  - Who is responsible
  - Method
  - Corrective actions

(E) Drain and janitorial cleaning

A separate programme will be in place with similar procedures to those described above. It will include requirements for a current schematic of drains (with an indication of flow direction), dedicated colour-coded equipment (black is usually used for this purpose). The plant will recognise that this is a major cross-contamination risk if not properly controlled.

(F) Cleaning-in-place (CIP) programmes

A plant operating a CIP system will usually have a separate documented programme. It will include the following:

- Diagrams of CIP systems and circuits
- Descriptions of each circuit
- Lists of parts that are cleaned manually together with work instructions
- Validation of hygienic design, e.g. separate circuits for raw and processed, no dead ends
- Validation records

(G) **Sanitation equipment and chemicals**

Tools and equipment:

The plant will have a programme in place to ensure the integrity of its cleaning tools such that they are not a source of contamination:

- Stored clean and dry
- Be on a regular cleaning schedule
- Have designated containers
- Never use sanitation equipment for process operations (e.g. sanitation sinks for washing, or former sanitation chemical buckets for work-in-progress food storage).

The design of sanitation tools is important. Products of an absorbent nature should be avoided as should designs that could be foreign material hazards. Avoid:

- Reusable cloths
- Reusable mops
- Wire bristle brushes (unless unavoidable and then should be controlled)
- Tools with wooden handles
- Abrasive scrub pads (can be single use and control issue)

Chemicals:

- All chemicals used will be suitable for food use and approved by the appropriate authorities.
- A material safety data sheet (MSDS) will be on file along with a supplier continuing guarantee (sometimes called a hazard data sheet).
- All chemicals will be properly labelled and *never* decanted in food containers.
- Chemicals will be stored securely and in accordance with the manufacturer's recommendations.

(H) **Validation**

Validation of PRP programmes is the same as for HACCP. It is required to establish that the programmes will be effective. In a formal programme, evidence of this will be documented. Many companies work with their sanitation provider in this area. Validation data will include the following:

(a) Evidence that the chemicals are suitable for the tasks being carried out.

(b) Evidence that the chemicals will be effective against pathogens of concern.

(c) Evidence of suitability for food industry use.

(I) **Verification (including monitoring)**

A number of monitoring and record review activities will be routinely carried out via the day-to-day measurements. This could include the following:

(a) Wet cleaning: cleaner/sanitizer concentrations:
- ATP swabs
- Visual pre-operational inspections (also post-cleaning if there is a time delay)
- Microbiological and/or allergen residue checks of rinse water

(b) Dry cleaning:
- Usually entails pre-operational visual inspections

(c) CIP Cleaning:
- Cleaner/sanitizer concentration
- Wash temperature
- Wash contact time

For COP (out-of-place cleaning), the above procedures might also be appropriate plus post-cleaning ATP and visual inspection followed by a pre-operational visual inspection.

(d)   Environmental surveillance programmes:
Microbiological surveillance programmes are an essential verification activity and all but very low-risk operations will have them. The programme will have a risk-based sampling plan, which takes account of the facility history, plant layout, product risk and includes identified sampling sites, frequencies, targeted microorganisms (often indicator organisms plus *Salmonella* and *Listeria* species), frequency and method of testing. Testing of first product off the line after cleaning is sometimes done, and this may include production being placed on HOLD status, pending results.

(e)   Audit/assessment:
Regulator hygiene audits will form part of the verification programme. In a good programme, this is supplemented by an in-depth sanitation assessment, occurring at least annually. It will include procedures and record review plus a considerable amount of time in the plant performing inspections of equipment (including tear down of, e.g. pumps and gaskets), observation of the actual cleaning activities at whatever time of day, and environmental monitoring via swabbing. A review of the efficiency of the programme will occur as part of the annual assessment or separately and will include sanitation costs (chemicals/labour) and downtime (planned or due to needed corrective actions).

(J) Training
Left until last, but training is an essential factor for an effective programme. Verification activities should provide some useful indicators for ongoing training needs. Like any programme, there will be training activities at a number of levels:

Sanitation manager:
This is sometimes the quality or production manager's responsibility. Whoever has the responsibility for the programme will need a fairly in-depth level of training. They need sufficient knowledge of the types of soil and chemicals, to be able to develop cleaning procedures. They need to be able to understand, for example, microbiology, allergen management, chemicals mode of action, the role of validation and verification, and to be very aware of potential issues if the wrong chemical or cleaning method is used.

Operators:
Hygiene and work instruction training.

CIP operators:
Require a higher level of training than other operators, particularly on operation of the specific CIP programme.

All training will be validated through testing and documented.

Pest control systems are important to prevent the access of pests that might cause contamination to the product, and good hygienic practices are necessary to prevent the creation of an environment conducive to pest infestation. Pest management is often contracted out to a professional pest control contractor. Buildings need to be made pest-proof and regularly inspected for potential ingress points. This will include sealing of holes, drains, etc., to prevent pest access and suitable screening, e.g. wire mesh on any opening windows, vents and doors.

Interior and exterior areas need to be kept clean and tidy to minimise potential food and harbourage sources. All potential food sources (including refuse) should be kept in suitable containers off the ground and away from walls. Suitable interior traps and monitoring devices should also be considered, and any pest infestations need to be dealt with promptly, without adversely affecting food safety. Regular monitoring and inspection should be performed to investigate for evidence of infestation and appropriate action taken where necessary.

Box 10.2 shows a specific example of how the pest control and management prerequisite elements might be formalised within a food business.

Box 10.2 PRP-specific example – pest management

A well-implemented pest management programme will focus around two key objectives:

1. Exclusion
2. Elimination

Emphasis should be on the preventive approach, i.e. exclusion from entering the facility. As with sanitation, the programme will be based on a *risk evaluation*. Consideration will be given to the surrounding environment, likely ingress, and exposed product zones. Having a good culture of cleaning up spills is essential – if the pests have not got a food source, they will be less attracted to the premises.

Pest control procedures

Pest control procedures will be documented in a formal pest management plan. Practices (including those aimed at prevention) will match what the plant sees as requirements, e.g. keeping external vegetation short, managing the 'graveyard' of redundant equipment and pallets, and spillages are cleaned up immediately. For most, the programme will involve the engagement of a pest control operator, and a copy of his licence should be kept on file with the procedures, along with a copy of the contract and insurance details.

The preventive measures will include the following:

- Proofing of entrances and access points
- Insect screens
- Electronic insect-killing devices
- Well-maintained dry ingredient storage (sealed, no spillage)
- Controlled use of pesticides
- Use of traps

The facility will have a schematic of the premises, with all the pest control devices clearly marked.

Pest control chemicals:

Pest control chemicals will be stored securely and clearly labelled. A MSDS will be on file, confirming suitability for food premises.

Bird control:

The plant will be designed to minimise bird activity through building construction, removal of food sources (garbage areas can be a problem if not well managed), and regular removal of roosting and nesting sites (drains and gutters will be fitted with screens and traps), doors will be fitted with air or strip curtains and kept closed, and the use of predator bird calls is often effective.

Rodent control:

- Bait stations will be tamper-resistant, secured to the location and locked. Poison bait will be solid, not granular, and only used in areas external to the plant – bringing poison into food processing areas is not recommended.
- Mechanical traps will be used in areas around entryways and regularly maintained.
- Internal traps are placed around the walls according to risk; i.e. areas of frequent catches might require a higher number of traps. Usually, they will be about 25 feet apart. All traps and bait stations will be numbered and cross-marked on the plant schematic. Rodent activity will be marked to enable identification of hot spots. For monitoring purposes, non-toxic bait stations may be used, which enables targeted and minimal use of poison.

All traps and bait stations will be regularly monitored, weekly for internal and external traps, monthly for external bait stations as a minimum.

Insect control:

- Screens will be in place if doors and windows are used for ventilation, but this is discouraged. A well-designed facility will have positive air pressure in process areas where food is exposed. Air curtains are not very effective for insects and should not be relied upon as a control device.
- Electric insect killers (EIKs) should be located outside of exposed food production areas, and if this is not possible, then well away from the exposed food areas. They should not be visible from outside the premises. The purpose of an EIK is to attract insects and that should always be remembered. Insect numbers in catch trays should be monitored for seasonal effects and infestations. Two blue bulbs plus a 'sticky tube design' are recommended, though technology advances may improve over time and new recommendations made.
- Pheromone traps where used will be set up by a trained operator. Review of catch data should be included in the programme as previous.
- All EIKs and pheromone traps should be numbered and located on the schematic diagram.
- Any pesticides used on site should be recorded by name, percentage of active ingredient, target organism, method and rate of application, area treated, licence number and name of applicator, date and signature.
- The effectiveness of the pest control programme will be routinely (at least annually) reviewed and adjustments made.

Waste management should ensure that waste materials can be removed and stored safely so that they do not provide a cross-contamination risk or become a food or harbourage source for pests. In particular, waste must not be allowed to accumulate in food handling and storage areas, and the adjoining environment and waste areas must be kept clean.

There is a need to **monitor effectiveness** of all maintenance and sanitation systems, and their effectiveness should be verified and reviewed periodically, with changes made to reflect operational changes. Audit, inspection and other tools such as microbiological environmental sampling can be used to facilitate verification of these prerequisite elements.

10.3.5 Establishment: personal hygiene

The objectives for personal hygiene stated in Codex (2009a) are:

To ensure that those who come directly or indirectly into contact with food are not likely to contaminate food by:

- Maintaining an appropriate degree of personal cleanliness
- Behaving and operating in an appropriate manner.

Food companies should, therefore, have standards and procedures in place to define the requirements for personal hygiene and staff responsibility, and staff should be appropriately trained. Figure 10.5 shows the prerequisite elements for personal hygiene.

Establishment of **health status** is important where individuals may be carrying disease that can be transmitted through food. Anyone known or suspected to be carrying such a disease should not be permitted in food-handling areas. Food-handling personnel should be trained to report illness or symptoms to management, and medical examinations should be done where necessary.

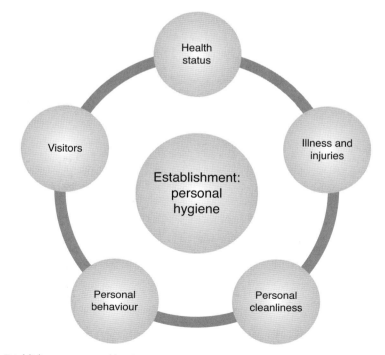

Fig. 10.5   Establishment: personal hygiene – prerequisite requirements.

Consideration of **illness and injuries** that may require affected staff members to be excluded or wear appropriate dressings should also be done. Codex (2009a) lists the following conditions that should be reported so that any need for medical examination or exclusion can be considered:

- Jaundice
- Diarrhoea
- Vomiting
- Fever or sore throat with fever
- Visibly infected skin lesions (e.g. boils and cuts)
- Discharges from the ear, eye or nose.

The need for good **personal cleanliness**, including effective hand-washing and wearing of adequate protective clothing and footwear, is also highlighted. Similarly, the prevention of inappropriate **personal behaviour** such as smoking, eating or chewing in food-handling areas should be enforced, and personal effects (e.g. jewellery and watches) should be prohibited in food-handling areas. **Visitors** to processing and product-handling areas should be adequately supervised and required to follow the same standards of personal hygiene as employees.

10.3.6   Transportation

To ensure continuation of food safety throughout transportation, transport facilities need to be designed and managed to protect food products from potential contamination and damage, and to prevent the growth of pathogens.

As illustrated in Figure 10.6, this includes **general** requirements on the need for protection of food during transit, plus further **requirements** for the design of containers and conveyances to facilitate this protection. This means construction so that the containers and/or conveyances do not contaminate the foods; permit effective segregation of different foods or foods from non-foods; protect foods from other contamination, such as from dust or fumes; and can be effectively cleaned and disinfected/sanitised.

Use and maintenance requirements for vehicles and containers include appropriate standards of cleanliness and cleaning and disinfection between loads as appropriate. Containers should be both marked for, and used for, 'food use only' where appropriate and temperature control devices should be used where necessary.

10.3.7   Product information and consumer awareness

Product information is important both for following links in the food chain and for the final food preparer and consumer. Insufficient information or inadequate knowledge can lead to products being mishandled and, ultimately, to both foodborne illness and product wastage. Figure 10.7 illustrates the four essential prerequisite elements needed in this area: lot identification, product information, labelling and consumer education.

It is important that sufficient **lot identification** information is easily identifiable on the products so that the lot or batch can be identified for recall purposes, the product can be handled correctly, e.g. stored at less than 5°C, and that stock rotation is facilitated. This will include permanent marking to identify the producer and the lot. Codex General Standard for the Labelling of Prepackaged Foods (CODEX STAN 1-1985) applies here.

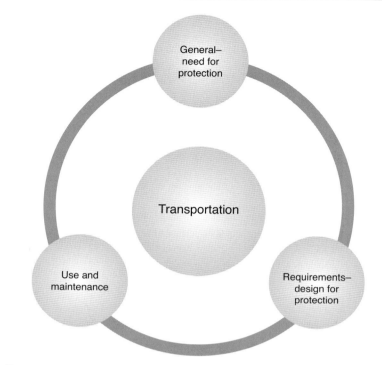

Fig. 10.6   Transportation – prerequisite requirements.

Fig. 10.7   Product information and consumer awareness.

Product information and **labelling** should be clear and sufficient such that it facilitates the correct handling, storage, preparation and use of the food by the next person in the food chain. This might include considerations both for the control of microbiological hazards, e.g. the use of temperature control, and for the provision of information about potential allergens. For example, for allergen information, considerations may include the following:

- Use of simple and straightforward language so as to provide the essential information without confusing consumers.
- Use of precautionary labelling, e.g. 'may contain' certain allergens in some cases. This is not a preferred option, but is sometimes used where there is a risk that cross-contamination may occur which cannot be adequately contained.
- Copy approval for packaging and artwork early in the development process to ensure satisfactory information is accurately listed.
- Use of bar code scanners during production to monitor whether correct packaging is being used. This is particularly useful in allergen management as a more effective alternative to periodic visual checks of packaging during production.
- Understanding of the label production process at the packaging manufacturers to assess likelihood of mix-up during printing.

Codex (2009a) also highlights the importance of **consumer education**, particularly the importance of following handling instructions and the link between time/temperature and food-borne illness.

10.3.8 Training

The final prerequisite element described by Codex (2009a) is training, which is highlighted as 'fundamentally important to any food hygiene system', since inadequate training, instruction and/or supervision can pose threats to food safety. If fact, training is an overarching requirement that impacts the success of all PRPs, the HACCP system and management requirements as well as the operation of day-to-day business procedures, and lack of or inadequate training has been implicated in food incidents (Chapter 2). Figure 10.8 shows the four essential elements of training for food hygiene, described by Codex (2009a).

Food hygiene training is essential to promote **awareness** in food-handling personnel of their roles and **responsibilities** for food control. Food handlers need the knowledge and skills to handle food hygienically, and personnel need an appreciation of the requirement to protect food from contamination, for example, when handling cleaning or pest control chemicals.

Companies should develop and implement appropriate **training programmes**, including a training needs assessment such that adequate training is developed and implemented. Training is likely to include details on the type(s) of food handled and produced and their ability to support the growth of food pathogens, plus control and monitoring procedures, such as the following:

- Process activities
- Packaging systems
- Handling and storage requirements
- Labelling and shelf life
- Specific requirements, e.g. monitoring CCPs under HACCP plans

There should be adequate **instruction and supervision** of personnel and ongoing monitoring of food hygiene behaviour. Managers and supervisors should have levels of food

Fig. 10.8   Training – prerequisite requirements.

hygiene knowledge that will allow them to judge potential food safety risks and take appropriate action.

Training should be evaluated and reviewed with **refresher training** or update training implemented as necessary.

10.4   PREREQUISITE PROGRAMMES AND OPERATIONAL PREREQUISITES

Based on the *Codex General Principles of Food Hygiene* (Codex, 2009a), the above sections describe the generally accepted hygiene requirements in any food business to provide the general environmental conditions that are favourable to the production of safe food and thus the foundations needed for effective HACCP systems. Although Codex (2009a) does not itself use the term 'prerequisite programmes', these requirements are generally accepted as key prerequisite programme elements around the world. However, a new prerequisite term has been introduced, which also needs discussion here: 'Operational Prerequisite Programmes'.

Operational Prerequisites has been introduced under ISO 22000:2005 (ISO, 2005a). It is defined as:

> Operational PRP: a PRP identified by the hazard analysis as essential in order to control the likelihood of introducing food safety hazards to and/or the contamination or proliferation of food safety hazards in the product(s) or in the processing environment.

This operational PRP definition is quite different from the definitions listed at the start of Section 10.2, and it refers specifically to the control of food safety hazards. This has been quite controversial since in HACCP (Codex, 2009b; see also Chapter 12) significant food safety hazards are expected to be controlled by CCPs. Although PRPs have always been considered important as a foundation for HACCP and therefore need to be established as formal programmes that are validated, monitored and verified, ISO 22000 now seems to be saying that operational PRPs are as critical as CCPs for the control of food safety hazards.

The idea for operational PRPs seems to have come from the fact that the manifestation of some hazards, in particular cross-contamination risks, might require elements of control that are generally thought to be part of PRPs. For example, the potential allergen contamination of a shared production line that is used to make both products containing and not containing certain allergens is likely to need specific targeted cleaning procedures as part of the management of this issue. A traditional way of dealing with this was to elevate that specific cleaning procedure to the status of a CCP, ensuring that it was validated in the same way as all other CCPs and then monitored and verified as effective. In this example, the general cleaning of all other aspects of the operation would have remained part of PRPs. Using the same example but taking the ISO 22000 definition (ISO, 2005a) into account, it is likely that the specific targeted cleaning of the line to remove the likelihood of allergen contamination would now be thought of as an operational PRP.

Some groups have proposed decision trees to help determine if the prerequisite elements required are PRPs or OPRPs, for example the prerequisite decision tree developed by Campden BRI (2009), reproduced in Figure 10.9.

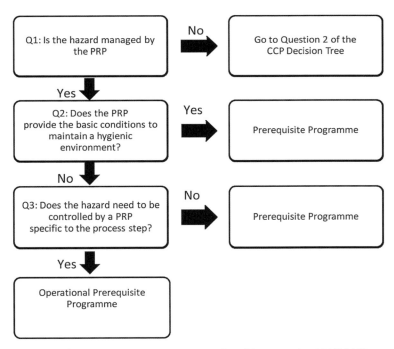

Fig. 10.9 Prerequisite decision tree (Gaze (2009). Developed by Campden BRI (2009), reproduced with permission.)

At this time, there is still debate over whether the concept of operational PRPs is useful. Some practitioners like them whilst others are worried that this is a move away from Codex HACCP. Only time will tell whether they become fully adopted into the HACCP system, and this is only likely to come as more companies get exposed to the concept through the adoption of ISO 22000 (ISO, 2005a).

10.5   VALIDATION AND VERIFICATION OF PREREQUISITE PROGRAMMES

Prerequisite programmes are the basic standards for the food facility, in which the safely designed product can be manufactured. They form the hygiene foundations on which the HACCP system is built to control food safety operation of every day. As such, it is essential that PRPs are working effectively at all times, and it is therefore necessary that each prerequisite element is validated to establish that it will be effective and that an ongoing programme of monitoring and verification is developed and implemented.

10.6   CONCLUSIONS

PRPs are necessary to provide the foundations of good hygienic practices necessary in every food operation. As such, they are essential to the production of safe food and form a cornerstone of world-class food safety programmes. The internationally accepted minimum prerequisite standards are defined in the Codex General Principles of Food Hygiene (Codex, 2009b); however, this document lists 'top-level' information and companies also need to seek further guidance to establish detailed PRPs, as illustrated in the case study examples in this chapter. All PRP elements need to be fully implemented in the facility and subjected to routine, ongoing maintenance and review procedures. Only then can they be said to be 'formalised' such that they deliver their essential role in the management of food safety.

11 Conducting a product safety assessment

11.1 INTRODUCTION

A formal and regular product safety assessment (PSA) process is essential to be sure that all critical elements for the safe production of high-quality foods have been considered, researched and adopted, as necessary, for all products produced. Of course, a safety review and assessment process should be a required element for commercial developments in essentially all types of businesses. For example, around the world we have learned that improperly designed highway bridges and buildings can collapse and that banking corporations can fail by engaging in unsound business practices. The global food industry bears greater responsibility than most types of business for adequate PSA processes, as its products are consumed daily by billions of people. PSA forms the bridge between controlling the safety of the individual product and process designs (Chapters 6 and 7) and controlling the safety of manufacturing processes, the latter being governed by the Hazard Analysis and Critical Control Point programme (HACCP) (Chapter 12).

11.1.1 Who is involved in product safety assessments?

It is in the best interest of a food company that its research and development (R&D) staff be aware of the fundamental food safety requirements that need to be considered in product development. Such awareness can be very effective in preventing food safety and quality problems after product commercialisation. Appropriate food safety training can be provided by internal or external experts.

The research staff should understand the potential hazards inherent to raw materials and finished products (Chapter 5), as well as the practical measures that can be used to control intrinsic (Chapter 6) and extrinsic (Chapter 7) hazards. The staff must also be capable to identify the needs for and to coordinate all laboratory tests and sensory evaluations during the course of product development.

Many food manufacturers have learned that it is practical to have two teams within their (R&D) organisations that are involved in the development and commercialisation of a food product: a product development team and a PSA team.

A **product development team** should be formed to manage each project. This team will include personnel who are broadly experienced in the types of products being sold by the

Food Safety for the 21st Century, First Edition By Carol A. Wallace, William H. Sperber and Sara E. Mortimore
© 2011 Carol A. Wallace, William H. Sperber and Sara E. Mortimore

company. The individuals on a product development team will vary depending on the type of project, but are likely to include food scientists, process engineers, packaging engineers, and financial and transportation personnel as necessary. Various technical functions are also part of the product development team, including chemists, microbiologists, regulatory personnel and lawyers as necessary. This latter group of individuals would likely support many product development teams. It is often useful to include a person who is expert in experimental design, analysis and mathematical modelling.

A **product safety assessment team** is used by many companies as the authoritative body to review and verify that all necessary control elements are in place to ensure the product's safety, quality and regulatory compliance. The individuals, who typically serve on the PSA team permanently, are lead scientists or managers in the company who have the expertise and responsibility to serve in this capacity. They are typically selected from the following functional areas:

- Food safety
- Quality assurance
- Packaging technology
- Regulatory affairs
- Law

In large corporations, there can be a substantial physical gap between the product development team and the several or many food manufacturing plants that can be located over large regions or even globally. Therefore, it is vital that a seamless PSA process is established and required for every product, so that information about the product and its safety is accurately documented and transferred from the R&D staff to the PSA team and, following approval, to the manufacturing staff. Large corporations typically have sufficient internal resources to ensure that this requirement is met and maintained.

In small corporations or very small companies, the physical gap between the product development team and the manufacturing locations is much smaller than it is for larger corporations. In fact, in small companies, the research, technical, product safety and manufacturing functions might be located together, and involve the same small number of people. Smaller companies, however, often do not have sufficient internal resources to conduct R&D activities, much less to ensure an effective PSA and transfer process. The management of such companies must recognise this limitation and provide the necessary external resources to complement its internal resources in order to ensure the production of safe food.

11.1.2 Timing of the product safety assessment process

The PSA process begins during product conceptualisation and is carried forward by the product development team during its development activities, including sensory and consumer testing. The process must be completed before commercial launch of the product. As elaborated in the following chapters, the PSA process is the key link in the transfer of food safety knowledge from the research and development team to the operations team that will manufacture the product. The product's HACCP team resides in plant operations. The PSA that is approved at the R&D stage contains essential information that must be considered and used in the hazard analysis that is conducted by the plant HACCP team.

11.1.3　Product safety assessment process

During the course of a project, the product development team should organise all essential information related to the formulation and production of a food, whether a new product or a reformulated commercial product is being developed. The information should include the following:

- The type of food and its intended use
- The food ingredients and their sources
- The product formulation and its intrinsic control factors
- The product process and its extrinsic control factors
- The product package and its safety features
- The product label and consumer use instructions
- The product distribution from the production plant to point of consumption
- The product handling by food service personnel or the consumer

This information is likely to be contained in a range of documentation sources, which must be collated by the product development team to provide sufficient detail for evaluation by the PSA team. Typical documentation includes the following:

- Supplier evaluations and approval forms
- Ingredient specifications
- Product specifications, including complete formulation
- Package specifications, including tamper-evident features
- Process requirements related to food safety
- Label information and consumer use instructions
- Abstract of potential hazards and control measures

The PSA team should regularly hold scheduled meetings so that it is available to consult with product development teams during product conceptualisation and development. Regular communications with the PSA team will facilitate the development process and will help ensure that unanticipated barriers are not detected during the final review and approval by the PSA team. After final review and approval, the product specifications and PSA must be signed by each functional expert member of the PSA team.

11.1.4　Previous approaches

One feature of the early approaches to PSA included the placement of food ingredients and products into particular 'hazard categories', based on the potential presence of certain hazards and the immunocompetency of the likely consumers. This approach was gradually abandoned in the first 20 years of HACCP application, but we mention it here to discourage its possible inclusion in modern food safety programmes. Through the use of this approach, for example, dried milk used in infant formulas would have been subjected to more intensive testing for *Salmonella* than would dried milk used in adult foods. As more foodborne pathogens emerged and the human population continued to grow rapidly (it has nearly doubled since 1972), it became apparent that the responsibility of food producers was to ensure that all foods were safe for consumption by all people. We needed to become more efficient in verifying the effectiveness of our food safety measures, and since specialised sampling plans based on hazard categories proved to be burdensome and ineffective for assuring food safety, thus, they were abandoned. Modern approaches to PSA involve a much more detailed review of all product elements with regard to potential presence of significant hazards and mechanisms for their control. This is a much more effective approach to protecting consumer safety.

11.2 TRAINING FOR RESEARCH AND DEVELOPMENT PERSONNEL

As the Pillsbury Company began in 1972 to apply its novel system of food safety management, HACCP, to the production of its consumer foods (Chapter 1), it conducted training of all of its R&D personnel in the critical elements to develop and document the necessary product safety control features. Remarkably, in addition to its comprehensive nature and its immediate availability, most elements of this original product safety training programme would be found in modern product safety training programmes. We have pointed out several times the 'timeless essence' of HACCP features. The same 'timeless essence' obviously can be found in effective training programmes. It is the responsibility of all food processors to operate at this high level of performance in their product development, documentation and PSA activities. Of course, even 'timeless' training programmes evolve; some features can be improved and expanded, while others might need to be minimised or eliminated. That has been our experience with the Pillsbury programmes.

Training will be necessary for personnel involved in R&D processes, including both those involved in product development teams and those involved in PSA. It is essential that these personnel be knowledgeable about the microbiological, chemical and physical hazards in food ingredients and products, and in the intrinsic and extrinsic measures that can be applied to control identified hazards. Further elements of the initial training should include the development of ingredient and product specifications that encompass relevant physico-chemical data related to water activity, moisture content, pH, storage temperature, chemical preservatives and headspace environment.

Training for PSA personnel must also include a consideration of the hazards and potential control measures that need to be in place during product processing. Consideration of these measures during product development can later assist the plant HACCP teams in their determination of critical control points. For example, time and temperature controls are often necessary in processing steps and product storage; magnets and metal detectors may be required to detect potential product contamination with metal; and sifters and screens may be required to detect and remove physical contaminants from dry ingredients and products. R&D personnel should be trained to keep accurate process flow diagrams as processes are developed. Additionally, the responsible personnel must be trained and competent in the administrative features related to product and process development, including specifications, testing procedures, and approval procedures. These features are demonstrated in the following example.

11.3 EXAMPLE OF A PRODUCT SAFETY ASSESSMENT

This fictional example has been created to illustrate the features of a PSA. Across a particular company, standardised forms and procedures should be developed to facilitate the assessment and documentation process. The first pages of the PSA include a general description of the product and its intended use and very specific descriptions of the production site, product formulation, physico-chemical features of the product, processing steps and process flow diagram.

Product safety assessment

Date	2 November 2009
PSA number	173-2009
Company	Riviera Risottos
Plant	Hendek, Turkey
Product name	Himalayan Nut Pilaf

Product description and intended use

Himalayan Nut Pilaf is a ready-to-eat, cold prepared vegetarian meal. It is manufactured from fresh and dried ingredients that are sourced globally. After manufacture, it is blast chilled, stored and distributed chilled. The package label includes a declaration of the ingredients, nutritional information, allergen information, and information for product storage, handling and shelf life. It is intended for consumption by the general population, which may include high-risk groups.

Physico-chemical features	Range
Moisture (%)	56.0–59.5
Water activity	0.975–0.99
pH	6.6–6.9
Chemical preservatives	None
Headspace environment	Air
Storage temperature (°C)	1–4
Maximum shelf life (days)	10

Ingredient	%	Ingredient Specification Number	Potential Hazards
Cooked basmati rice	69.19	27.342	*Bacillus cereus*
Sunflower oil	1.72	28.06	None identified
Fresh white onions, chopped	6.88	23.01	None identified
Fresh garlic cloves, crushed	0.86	23.24	None identified
Fresh carrots, diced	8.61	23.17	None identified
Cumin seed	0.43	29.27	Foreign material, *Salmonella*
Ground coriander	0.86	29.18	Foreign material, *Salmonella*
Sodium chloride	0.43	29.01	Foreign material
Black pepper	0.26	29.02	Foreign material
Groundnuts, unsalted	10.76	24.16	*Salmonella*, aflatoxin, allergen
	100.00		

Process description	
Rice cooking	Cook one part rice in two parts (v/v) water. The cooked rice is drained and cooled to ambient temperature. If not used in production within 4 hours, the rice is put into chilled storage at 1–4°C for a period not longer than 2 days.
Herb/spice blending	The cumin, coriander, salt and black pepper are blended, measured into batch-sized quantities, and stored at ambient temperature not longer than 1 week.
Vegetable blending	The sunflower oil, chopped onion, crushed garlic and diced carrots are blended with herb/spice mixes and used within 4 hours.
Product blending	The cooked rice is blended with the oil, vegetable/herb/spice mixture. Then, the groundnuts are blended in.
Packaging	The Himalayan Nut Pilaf is packaged into plastic trays (CPET) in 250 or 500 g portions. The containers are lidded with an oxygen-impermeable film. The product labelling information is contained on the cardboard sleeve, to which coding information is applied during manufacture. All packages are visually inspected for seal integrity and passed through a metal detector.
Storage	The packaged product is cased, palletised and put into chilled storage at 0–4°C.
Retail	The Himalayan Nut Pilaf has a maximum chilled shelf life at 4°C or less for 10 days. It is distributed to retail outlets within 3 days of its manufacturing date.

11.3.1 Process flow diagram

The product will be produced within the remit of the Riviera Risottos modular HACCP system for Risottos and Pasta Meals. A diagram indicating the specific process activities for this product from the HACCP modules is reproduced (Figure 11.1). More detail on the HACCP plan for this operation, including the other process steps that make up these modules but are not involved in this product, can be found as a case study in Appendix 1.

The last pages of the PSA include supporting references, the product label, and a summary of the PSA team's deliberations, its decisions and approval.

Supporting studies	
Riviera Risottos research notebook	Volume 169; January 1 to June 15 2009
Research report number 74-2, May 2009	Validates product shelf life
Riviera Risottos engineering notebook	Volume 37; February 1 to July 15 2009
Research report 162-1, June 2009	Validates process flow and controls, and quality of finished product

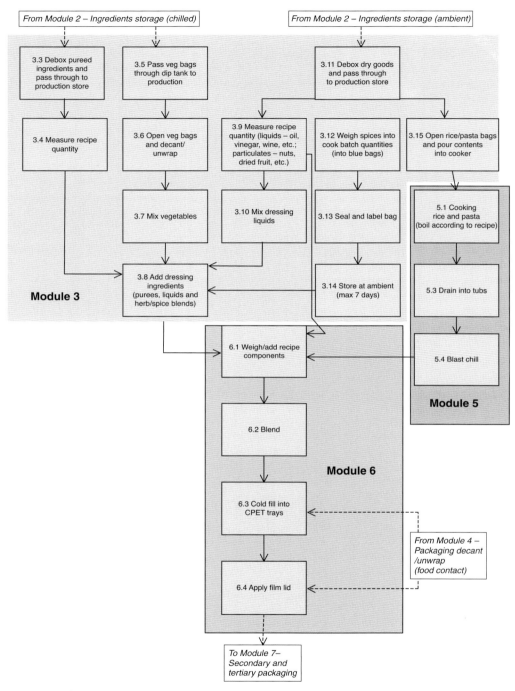

Fig. 11.1 Process flow diagram for Himalayan Nut Pilaf.

Product label

Riviera Risottos
Himalayan Nut Pilaf
Net weight 250 g

Ingredients: Cooked basmati rice, unsalted groundnuts, diced carrots, chopped onions, sunflower oil, crushed garlic, ground coriander, cumin seed, salt, black pepper

Nutrition facts		Daily values (%), based on a 2000-calorie diet
Serving size	1 cup (125 g)	
Servings per container	2	
Calories	224	11
Fat calories	72	
Total fat	8 g	11
Saturated	0	
Trans fat	0	
Protein	12 g	
Total carbohydrates	26 g	10
Fibre	<1 g	
Sugars	<1 g	
Cholesterol	0	
Sodium	200 mg	8

Contains groundnuts
Keep refrigerated: use before expiration date.
No preparation necessary. Add diced cooked
meat, poultry or seafood as desired.

Product safety assessment

The PSA team has reviewed the submitted information and supporting studies. This information is consistent with our previous experience and with published expert opinions on the control of potential foodborne hazards. When produced and handled under the specified conditions, Himalayan Nut Pilaf is highly unlikely to present a food safety hazard:

- The potential threat of *Listeria monocytogenes* infections is controlled by the washing and peeling of fresh vegetables, sanitary production conditions, temperature control and the short product shelf life.
- The potential threat of *Salmonella* in spices is controlled by irradiation by the vendor.
- The potential threat of *B. cereus* intoxication is controlled by the short-term storage of cooked rice, which is permitted only under chilled conditions.
- The potential hazard of foreign material contamination in spices is minimised by visual inspection, metal detection and vendor control requirements.
- The use of groundnuts in this product presents several potential hazards, all of which can be readily controlled by good practices. The potential hazards of *Salmonella* and aflatoxin contamination can be greatly minimised by vendor control programmes, with at least monthly verification by testing in Riviera Risottos facilities. The potential allergen hazard of groundnuts is controlled by product labelling and by thorough cleaning of all process equipment before products not containing groundnuts can be produced.

Approval signatures	2 November 2009
Responsible person or alternate	**Corporate functional responsibility**
R. Deville	Research and development
C. Basmati	Engineering and quality assurance
S. Wrappe	Packaging
B. Canu	Procurement
L. Advokat	Legal and regulatory affairs
W. Bowman	Food safety

11.4 CONCLUSIONS

Several important principles should be at the forefront of awareness and serve to guide the decisions of research and manufacturing personnel who are responsible for product safety:

1. All food prototypes must be safe for consumption. This principle pertains to R&D or pilot plant-produced samples tasted in the test kitchen, or evaluated by internal sensory panels and during external central location consumer tests, home-use tests, and market trials before product commercialisation. It is vital that the research team coordinates its development activities with the PSA team so that the safety of all analysts is assured.
2. It is essential that a thorough PSA is conducted and approved by the responsible personnel. This assessment is one of the important means to transmit important knowledge from the R&D team to the plant team, which in turn must transfer pertinent information from the PSA into the plant's HACCP plan.
3. No commercial production of new or reformulated food products can be permitted until the PSA is completed and approved by the PSA team and the product's HACCP plan has been approved and implemented by the plant HACCP team.

12 Developing a HACCP plan

12.1 INTRODUCTION

Developing a HACCP plan is a key part of the development of any food safety management programme and HACCP plans, developed by HACCP teams and unique to each production facility, are essential to the production of safe food throughout the global food supply chain. As discussed in the introduction to this book, the HACCP concept has been public knowledge for several decades, and this means that many companies already have HACCP systems in place. So, whilst some people might be starting at the beginning to develop a HACCP plan for the first time, others will be reviewing and upgrading existing systems that may have been in place within their companies for many years. It is therefore important to discuss the steps of HACCP plan development and bring together the authors' experiences of HACCP over the past 35 years, both to share best practices and to outline common pitfalls. As we have discussed in earlier chapters, there have been many limitations in the way that HACCP has been applied and very few companies have used it to best effect, particularly when considering it as part of the integrated food safety management system supported by strong prerequisite programmes. For effective food safety management in the 21st century, we need to get back to basics and use the established HACCP principles (Codex, 2009b) to best effect, such that we produce valid HACCP plans that will work to control all relevant food safety hazards when working alongside their supporting prerequisites.

In this chapter, the process of HACCP plan development is discussed in a step-by-step fashion. For those who are new to HACCP, working through the chapter will build an understanding of the necessary tasks and procedures that need to be undertaken. For those reviewing their existing programmes to ensure continued system effectiveness, this chapter could be viewed as a checklist of points to consider in strengthening the HACCP elements of food safety management practices. The process of going through these steps to develop the HACCP plan is as important, if not more important, than the document that results – 'The planning not the plan is the most important element' (Eisenhower). Only by taking the time to work through these steps methodically, using the necessary combination of expertise, knowledge and experience, can an effective HACCP plan be developed. Referring back to Chapter 2 and the misconception that 'HACCP has been done already' (Section 2.3), experience tells us that this statement is far from true. It is therefore essential that food businesses are not complacent about their control procedures but are open to the possibility that existing systems can be improved.

Food Safety for the 21st Century, First Edition By Carol A. Wallace, William H. Sperber and Sara E. Mortimore
© 2011 Carol A. Wallace, William H. Sperber and Sara E. Mortimore

12.2 PRELIMINARY CONCEPTS

12.2.1 HACCP principles

As mentioned in Chapters 1 and 8, the HACCP plan is established by applying the seven HACCP principles (Codex, 2009b). Before examining in detail how to develop a HACCP plan, it is useful to consider briefly the requirements of each principle (Table 12.1).

Table 12.1 The HACCP principles explained.

	HACCP principle	**Clarification**
Principle 1	Conduct a hazard analysis.	This requires the team to look at each process step one at a time, consider which hazards might occur, evaluate their significance and establish how best to control them.
Principle 2	Determine the critical control points (CCPs).	At this stage, the points that are critical to product safety are identified. This can be done through judgement and experience or using a structured tool – the Codex decision tree.
Principle 3	Establish critical limits.	Critical limits are the safety limits that form the boundary between safe and potentially unsafe food. These need to be established to manage all CCPs.
Principle 4	Establish a system to monitor control of the CCP.	The monitoring system needs to demonstrate that the CCP is under control on a day-to-day basis and must be capable of detecting loss of control.
Principle 5	Establish the corrective action to be taken when monitoring indicates that a particular CCP is not under control.	If the CCP is not working, action needs to be taken to protect the consumer and to put right the cause of the deviation.
Principle 6	Establish procedures for verification to confirm that the HACCP system is working effectively.	This requires checking that the system is capable of controlling relevant hazards, is working in practice and is up-to-date on an ongoing basis.
Principle 7	Establish documentation concerning all procedures and records appropriate to these principles and their application.	Documentation will include the process flow diagrams and tables created during the HACCP study (HACCP plans and development records) as well as monitoring records.

12.2.2 The HACCP plan and documentation approaches

The HACCP plan is defined as follows:

> A document prepared in accordance with the principles of HACCP to ensure control of hazards that are significant for food safety in the segment of the food chain under consideration. (Codex, 2009b)

Put simply, the HACCP plan is the documentation produced that shows how significant hazards will be controlled. The HACCP plan is a formal document, holding all details of areas critical to food safety management for a product or process. It consists of the following.

Core plan

- Valid process flow diagram
- Documented CCP management details – this is usually captured in a table known as a HACCP control chart (Mortimore and Wallace, 1998) or CCP management table

Support documentation

This comprises the preparatory documentation that has been used in developing the HACCP plan as well as details of the verification requirements, including the following:

- HACCP team details
- Product/process description (including terms of reference, consumer target group and intended use of product)
- Hazard analysis details
- CCP identification – details of approach and justification
- HACCP verification plan
- HACCP audit and review data

Documenting the HACCP study and HACCP plan development

Codex (2009b) guidance on document organisation (the HACCP worksheet) is widely used as a basis for HACCP study records, but there is no prescribed format and most companies use their own adaptations of recommended tables. A variety of templates are available in textbooks (e.g. Mortimore and Wallace, 1998) and in HACCP plan examples published on the internet. The key requirement is to understand what needs to be documented not only to help the HACCP team in their deliberations during the HACCP study but also to ensure that all food safety hazards are identified, evaluated and effectively controlled, and to provide evidence of an effective food safety system, for example, when being assessed by external auditors. Suggested documentation formats are built into this chapter to illustrate the HACCP application process and may be adapted to suit any food operation.

12.2.3 HACCP application process

The process of HACCP plan development and implementation, through the application of the Codex HACCP principles, involves a number of interlinked stages (Figure 12.1). Application of the HACCP principles is done using a logical, step-by-step approach such that each step builds on the work done in applying the previous step. As can be seen in the diagram, the HACCP principles are involved not only in development of the HACCP plan but also in HACCP implementation and maintenance, which is discussed in Chapters 13 and 14, respectively. This is a simplified schematic to illustrate the key process steps where HACCP principles are applied, and it should be noted that there will be some further overlap where individual HACCP principles apply; for example, since there are records involved in many stages of HACCP, principle 7 on documentation and records will also be involved to an extent in HACCP plan development.

12.2.4 Codex logic sequence

Although the HACCP principles provide the essential food safety management steps, they do not explain how to get started. Before applying the HACCP principles, there are a number of preparatory steps that must be completed. These are described in the Codex logic sequence for the application of HACCP, as shown in Table 12.2.

Steps 1–5 are also known as the Codex preliminary steps to HACCP application, and these are applied as part of the preparatory process before use of the HACCP principles. As previously

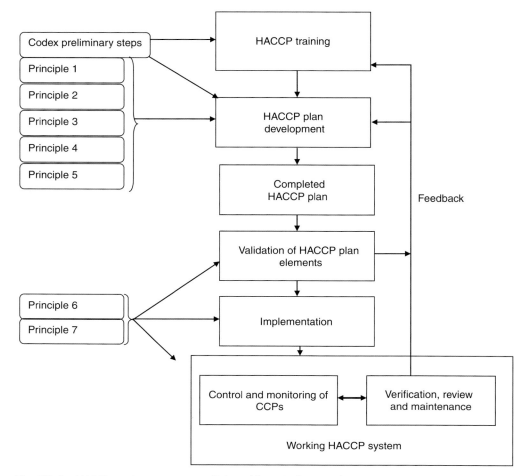

Fig. 12.1   HACCP application process. (Adapted from Wallace *et al.*, 2005.)

Table 12.2   The logic sequence for application of the Codex HACCP principles (Codex, 2009b).

Step 1	Assemble HACCP team
Step 2	Describe product
Step 3	Identify intended use
Step 4	Construct flow diagram
Step 5	On-site confirmation of flow diagram
Step 6	List all potential hazards, conduct a hazard analysis and consider control measures
Step 7	Determine CCPs
Step 8	Establish critical limits for each CCP
Step 9	Establish a monitoring system for each CCP
Step 10	Establish corrective actions
Step 11	Establish verification procedures
Step 12	Establish documentation and record-keeping

discussed, HACCP is normally applied by a multidisciplinary team so that personnel from all aspects of the operation are included. The first preliminary step includes identifying and training the team, as well as assembly of the team to carry out the study. A product (or process) description is developed as background information for the team, and the intended use for the final product is also considered. For example, is it for a specific group of consumers, e.g. infants, or for the general population? Will the product be eaten straightaway or could it be held and reheated? The team then prepares and confirms a process flow diagram that describes all the steps in the process being studied. Only when these steps have been completed is the team ready to apply the first of the HACCP principles.

12.3 APPLYING THE CODEX LOGIC SEQUENCE TO DEVELOP A HACCP PLAN

In this section, the stepwise development of a HACCP plan is discussed, according to the requirements of Codex (2009b). To illustrate this process, elements of a case study HACCP plan are included to show the application of the steps. This is for a manufacturer of chilled prepared meals known as *Riviera Risottos*, one of whose products was introduced in Chapter 11. Application of the Codex logic sequence is normally done by the HACCP team, which is discussed in detail at Step 1 (Section 12.3.2).

12.3.1 HACCP study terms of reference and scope

Although not listed as a step in the Codex logic sequence, it is standard practice to establish the scope or terms of reference at the start of any HACCP study. This should include the types of hazards to be studied, and normally this is microbiological, chemical and physical hazards. However, although all relevant hazards need to be covered in the operation, they do not need to be studied at the same time and so a particular study could focus on a specific hazard group, such as when a specialist, e.g. a microbiologist, needs to be brought into the team. Most experienced HACCP teams will cover all hazard types together, but new HACCP teams may also find it easier to focus on one group of hazards at a time.

Another important part of the terms of reference or scope is to identify exactly which part of the operation is to be covered by the HACCP study. This involves considering where the start and end points need to be and whether the HACCP study covers one product, a process involving several products or a process module, as outlined in Chapter 9 (Section 9.3.2). Modular (or process-led) systems are practical to develop and are used in most manufacturing businesses, particularly those with complex processing operations, as well as in many food service operations. This is where the operation is split into a number of process sections and HACCP is applied to each section rather than to each individual product (the individual product approach is referred to as linear or product-led HACCP). With modular HACCP, a key point is to ensure that the modules add up to the entire operation and that no processes are missed out, so it is important to identify the start and end points accurately for each HACCP study and this is defined as part of setting the terms of reference and scope.

These details may be listed as an introduction to the HACCP plan or may be included in the product/process description (see Codex logic sequence step 2). For the case study company, the terms of reference and scope for HACCP plan development were listed as follows (Table 12.3).

Table 12.3 Riviera Risottos HACCP plan terms of reference and scope.

Scope	The manufacture of chilled ready-to-eat prepared meals, which may be consumed hot or cold, including products relating to special dietary, needs requirements for the following brands: • Riviera Risottos branded products • Retailer private label products
Terms of reference	The HACCP plan will cover all relevant microbiological, physical and chemical hazards to include allergens and compounds that cause intolerance reactions. This HACCP plan covers all processes from raw material intake to chilled storage of finished products prior to dispatch.

12.3.2 Codex logic sequence step 1: HACCP teams

This step tells us to 'assemble HACCP team'; however, before a HACCP team can be assembled, the correct people need to be identified and trained. The HACCP team is a specific group of individuals who work together to apply the HACCP principles and will have different, but likely overlapping, membership from the food safety team discussed in previous chapters. The multidisciplinary HACCP team is believed to be one of the most powerful strengths of HACCP. It ensures that HACCP plans are developed by a group of people who, collectively, have the knowledge and experience to take decisions about food safety. It is important that the HACCP team includes personnel who understand not only the HACCP principles but also the food products and their ingredients, the production processes and packaging systems, the manufacturing and handling environments, and the likely hazards associated with all these aspects.

The essential expertise within the HACCP team, therefore, includes the following:

• Personnel who understand the process operations, ingredients and products on site.
• Personnel who have knowledge and experience of the equipment, how it works to achieve process conditions and the likely failure modes.
• Personnel who understand the likely hazards and appropriate control mechanisms, including how to validate process controls, including the necessary validation requirements.

This expertise is most likely to be gained by including personnel from manufacturing/operations alongside quality/technical and engineering disciplines. Additional specialists may also be required to provide knowledge and experience of specific aspects, e.g. microbiologists, toxicologists, supplier/vendor assurance personnel, storage and distribution personnel or product developers. Within the HACCP team, a leader needs to be appointed and a scribe or secretary identified. These are two crucial roles to the success of HACCP, ensuring that the HACCP development programme is coordinated and kept on track, and that accurate records of all team discussions are maintained.

The total size of a HACCP team is normally kept to a maximum of four to six personnel for ease of management, although this 'core team' may not include the additional specialists who may be called in for specific tasks. In small operations, and even in some larger ones, it may be difficult to achieve a multidisciplinary HACCP team of this nature due to the limited number of appropriate personnel on site. Although it is likely that the fewer members of staff will have wider job responsibilities and, therefore, an understanding of the whole operation allowing them to contribute the same way as a multidisciplinary team, it is also likely that personnel in small businesses will have less knowledge of food safety hazards, and this will need to be compensated for by bringing in external support.

The multidisciplinary approach to HACCP works well and ensures that the system does not rely on the knowledge and experience of one individual. However, it is important that a balance of individuals is found and a 'sharing' environment promoted where job roles are left outside the door. This helps to overcome any difficulties from existing group norms such as inability to challenge more senior/dominant staff when necessary (Wallace, 2009).

For a HACCP team to work effectively, all team members need to understand the application of HACCP principles. For best results, the whole team should be trained using a practical training intervention that covers both theory and practical application of HACCP. Whilst the multidisciplinary aspect of the team is essential to cover all necessary aspects of the operation, it is unlikely that all team members will have the same level of knowledge of HACCP principles, even after the same training intervention. Wallace (2009) found that there is a 'levelling out' of HACCP knowledge within HACCP teams such that the team knowledge is not necessarily better than and, in fact, is sometimes worse than that of the individual team members. It is therefore important to understand the balance of HACCP knowledge within the team such that the HACCP study process is guided by the team members with the best knowledge of HACCP principles. This might mean that one or two people with good HACCP knowledge are given the task of ensuring that HACCP plan development proceeds effectively whilst the remaining team members focus on their functional input to the team deliberations (Wallace, 2009). Published knowledge testing formats are available to help determine HACCP knowledge levels (e.g. Wallace *et al.*, 2005).

Table 12.4 shows details of the Riviera Risottos HACCP team.

Table 12.4   Riviera Risottos HACCP team.

R. Arborio – Quality Manager (HACCP team leader)
L. Grain – Production Manager
C. Basmati – Engineering Supervisor
M. Wild – Production Supervisor
T. Jasmine – Technical Consultant

12.3.3   Codex logic sequence step 2: product/process descriptions

This Codex preliminary step tells us to 'describe product'. In practice, this step considers both the product and the process. The reason for the product/process description is that it is important for all members of the HACCP team to understand the background to the operations that they are about to study. This is achieved by discussing the operation and noting key information. Although some HACCP teams prefer just to have the familiarisation discussion, it is most useful if the information is recorded formally as a 'product description' or 'process description'. This document then becomes a historical point of reference to the situation when the HACCP plan was developed. It will be useful at later stages as a training tool for new personnel and briefing aid for internal or external auditors or regulatory personnel who need to gain an understanding of the food safety management approach.

Topics normally included in the team's discussions are:

- Main ingredient groups to be used or 'work-in-progress' inputs to process modules
- Main processes and how materials are prepared/handled
- Production environment and equipment layout

- Hazard types to be considered, if known
- Key control measures available through processes and prerequisites
- Packaging/wrapping if appropriate to scope of study

Much of this information will have been collected at the product safety assessment stage (Chapter 11) and should be transferred to the HACCP team by the product development and/or product safety assessment team. As an example, the case study HACCP team constructed a process description (Table 12.5).

Table 12.5 Description of product.

Ready-to-eat prepared meals are manufactured from fresh, frozen and dried raw materials. Raw materials contained in the recipes include dairy products, fish and prawns, chicken, turkey, beef, lamb, bacon and pork. Allergens are used on site but are strictly controlled. Ingredients are sourced through approved suppliers globally.

All prepared meals are cooked to pasteurisation temperatures, then blast chilled, stored and distributed chilled. The shelf life of the products is determined by prescribed storage and usage conditions and is verified during production trials and confirmed microbiologically.

Fit for purpose food grade packaging is used. All packaging carries full ingredient listing, nutritional information, allergen information, heating and storage instructions and shelf life information.

When applying HACCP in food service operations, in addition to the general product description information, such as the example in Table 12.5, it is normal practice to group all the different menu/food items into like process groups. This can also be included as part of the product/process description step, as per the example in Table 12.6.

12.3.4 Codex logic sequence step 3: identify intended use

It is important to identify the intended use of the product, including the intended consumer target group. Different consumer groups may have varying susceptibilities to the potential hazards, e.g. the elderly, young children or immunocompromised individuals. However, it must be emphasised that all products should be safe for all consumers.

Intended use considerations need to be examined throughout the product supply chain, including further manufacturers/processors, food service, retailers, and through to handling and use by the final food preparer and consumer. Different uses of the food items may also need

Table 12.6 Catering process groupings – example foods.

Process group 1 (food preparation with no cooking step)	Process group 2 (food preparation for same day service)	Process group 3 (complex food preparation)
Salad greens Fresh vegetables Coleslaw/dressed salads Fish for raw consumption (sushi) Sliced sandwich meats Sliced/grated cheese Meat salads (made with pre-cooked meats)	Fried chicken Grilled or fried fish Hamburgers, sausages, etc. Roasted, fried or grilled meats Hot vegetables Cooked eggs	Soups Gravies Sauces Rice dishes Prepared meals, e.g. chilli, rice and pasta dishes Meat salads (made with meats that require pre-cooking)

Source: Adapted from FDA (2005a).

Table 12.7   Intended consumer use and potential misuse – Riviera Risottos.

Intended consumer use	The products are intended for the general population which may include high-risk groups
	Some products may contain allergens so are not suitable for the whole population
	All allergens are stated on pack and all packs carry the relevant warnings
	Products may be consumed cold or reheated as per instructions
	All products may be held under refrigerated or frozen storage prior to use
Envisaged consumer misuse	Temperature abuse
	Consumed after 'use by' date

to involve different hazard considerations, e.g. food items that may be cooked or used without any further heat process. The HACCP team needs to think about any ways that the product could be abused or used other than that intended. An example of this is low-calorie powdered drink mixes (e.g. hot chocolate) that are intended to be made up with boiling water. In some cases, diet/weight loss groups have recommended to their members the use of these products to flavour desserts such as low-calorie yogurts. If the drink mix was relying on boiling water to make it safe then this secondary use as a dessert additive would be potentially unsafe. Other good examples include refrigerated cookie dough products that are intended to be baked but are frequently consumed raw, particularly in the United States, and soup mixes being blended with sour cream to make salad dressings.

Intended use and consumer group information is usually included as part of the process description record (from Step 2). In many cases, it will be important to provide information to the consumer about how to handle, store and prepare (including cooking as appropriate) the food item safely, and this can be derived once the intended use and potential misuse of the product are established. Table 12.7 shows the intended use and misuse considerations from the case study company.

12.3.5 Codex logic sequence step 4: construct process flow diagram

A process flow diagram, outlining all the process activities in the operation being studied, needs to be constructed. This should list all the individual activities in a stepwise manner and should show the interactions of the different activities. The purpose of the process flow diagram is to document the process and provide a foundation for the hazard analysis (Step 5).

To produce a flow diagram, it is necessary to separate the process into a series of steps. In the context of HACCP, the word 'step' refers not only to obvious processing operations but also to all stages that the product goes through, e.g. incoming raw materials and storage. The diagram should progress logically and relate to how the product is actually produced, and should contain enough detail to allow an understanding of the process. The steps should be listed as 'activities', i.e. what is happening at this step, and time and temperature information should be included where relevant. Equipment design features such as mesh sizes on sieves and filters used in sieving/filtration steps can be included to provide additional information to the HACCP team. A common error in HACCP is to list the names of the process equipment rather than the process activity and to miss out transfer steps, as in the example of milk processing shown in Figure 12.2. This often results in an incomplete process flow diagram, which makes the process difficult to follow.

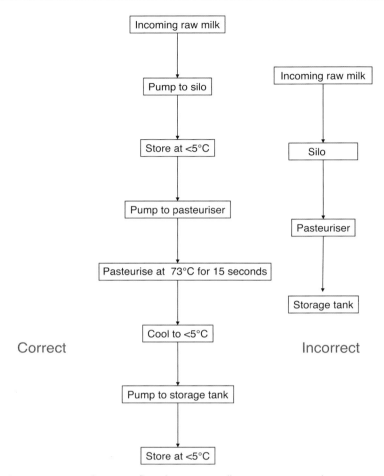

Fig. 12.2 Common errors with process flow diagrams – milk processing example.

The reason that incorrect labelling of process steps in this way can be a problem is that there are some cases where more than one process step takes place in one piece of equipment, and these may have different hazards. If only the equipment name is recorded then there is a chance that a process step and its associated hazards may be missed. Figure 12.2 shows a classic example of this: the pasteuriser used in milk processing. In this case, both the pasteurisation heat process and the cooling process take place in the pasteuriser, and the raw milk heating up and pasteurised milk cooling down are separated only by a thin metal plate. Hazards associated with the heat process would include survival of vegetative pathogens, while in the cooling process, a different hazard of cross-contamination with pathogens due to leak in the pasteuriser plate pack may occur.

The most commonly used type of flow diagram for use in HACCP studies shows ingredients or groups of ingredients and how they are stored and handled until they are combined. This gives a realistic interpretation of what actually happens from the starting point of listing ingredients along the top of the page through to the end point with the finished products at the bottom.

The style of process flow diagram will also depend on how the HACCP system is structured for the operation and the terms of reference/scope of the HACCP study. In most manufacturing operations, unless the process is very simple, the modular approach to HACCP will be used (as discussed in Section 12.3.1). This means that there will be a series of process flow diagrams comprising the different process modules and they should fit together to cover the whole operation. Only the initial modules will show the handling of ingredients, but later modules should show the incoming inputs from the previous module, e.g. work-in-progress or part-produced items. In food service operations, current thinking is to group all recipe items into a set of common processes, as shown in Table 12.6. This means that, similar to modular HACCP in manufacturing, food service process flow diagrams will be generalised and will cover the processes, but will not show individual ingredients or specific menu items. It will also be important to identify and analyse hazards that are associated with the individual ingredients.

Full details of all process activities, storage and transfer steps are needed in the HACCP study to allow a thorough hazard analysis to take place. This requires a very detailed layout to be prepared; however, many companies also use briefer outline process flow diagrams, e.g. to gain an oversight of the operations or for sharing with customers. These may be at the level of showing how the HACCP modules fit together or somewhere in between, thus there may be three levels of process flow diagrams in operation:

- Level 1 – Top level; normally shows how HACCP modules fit together.
- Level 2 – Intermediate level; overview of main operations suitable for discussion with customers, general familiarisation, etc.
- Level 3 – Detailed level; full detail on all process activities and steps, allowing hazard analysis to be performed.

Tips for constructing process flow diagrams

There are a number of conventions for flow diagrams:

- All ingredients or inputs to process modules are listed along the top of the page.
- The final product or output from the process module is placed at the bottom.
- The diagram is made up of a series of text boxes connected by arrows that denote the direction of process flow.
- Process steps are described as activities and not confused with equipment names.
- All variations on processing activities are included, e.g. when things are done differently on different shifts.
- If water, air, steam, etc., are used in processes, e.g. washing, cooking or drying, they are not listed as ingredients but the place they enter the process is indicated. However, water as an ingredient should be included.
- Layout should be designed such that lines do not have to cross wherever possible.
- Reworking or backward flow must also be identified on the flow diagram. This common practice is easily forgotten, but it is essential to include here for the hazard analysis.
- Normally, all steps are numbered to allow ease of transfer of data onto the HACCP plan.

The case study process flow diagrams developed by the HACCP team at Riviera Risottos are shown in Figures 12.3a–c. Note that Figure 12.3a shows how the modules fit together;

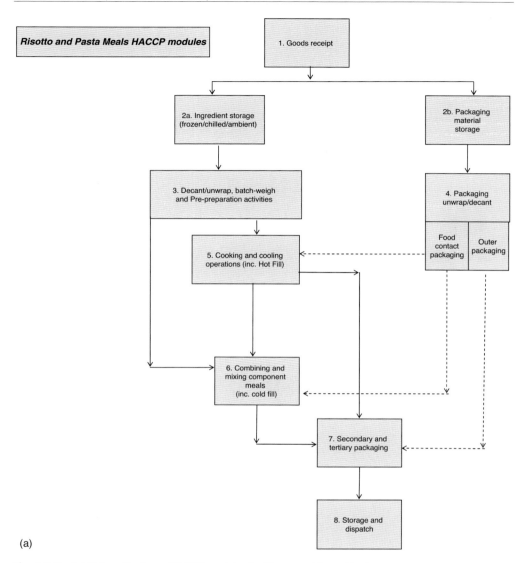

(a)

Fig. 12.3   (a) Riviera Risottos – HACCP modules for Risotto and Pasta Meals.

Figure 12.3b has been developed as an overview flow diagram to show the linkage of the different processes used to make Risottos and Pasta Meals, including both hot meals and cold rice and pasta-based salads. This diagram would need to have further detail added to understand exactly how the different subprocesses are conducted, and this has been achieved at Riviera Risottos by developing a series of flow diagrams covering the subprocesses, an example of which is shown in Figure 12.3c (Module 5 cooking and cooling activities). The complete set of process flow diagrams for all the HACCP modules at this company can be found in the Appendix 1 case studies.

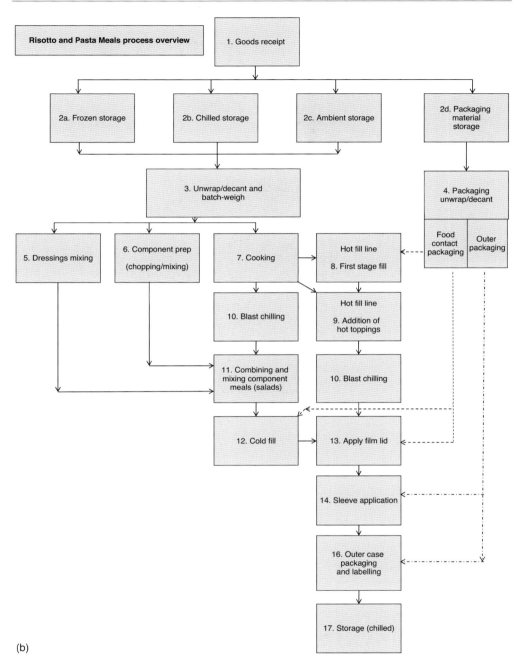

(b)

Fig. 12.3 (*Continued*) (b) Case study overview process flow diagram.

Module 5 Cooking and Cooling Activities (inc. Hot Fill)

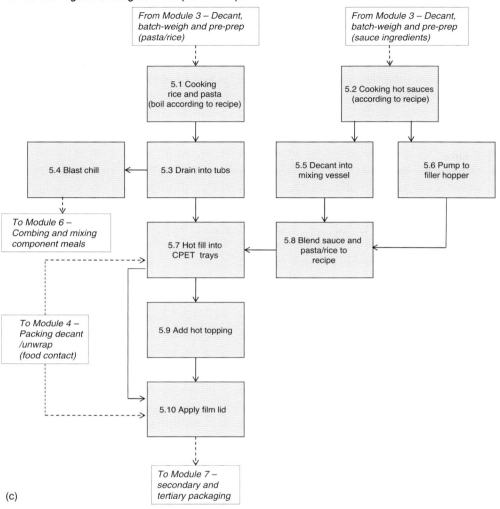

Fig. 12.3 (*Continued*) (c) Riviera Risottos – detailed process flow diagram, HACCP Module 5.

12.3.6 Codex logic sequence step 5: on-site confirmation of flow diagram

Since the process flow diagram will usually be developed in the office away from the processing activities and it will be used as a tool to structure the hazard analysis, it is important to check and confirm that it is correct. This is done simply by going into the process area and comparing the documented diagram with the actual process activities, noting any changes necessary, and making sure that all variations, for example on different shifts, are covered. This exercise is normally done by members of the HACCP team or production personnel, but it is good to have someone independent to confirm the process flow as the on-site HACCP/production team may be too close to the processes and either miss points out or make assumptions. The completed process flow diagram should then be signed off and dated as valid, and it is important to make sure that this is done before the hazard analysis commences.

Common problems with flow diagrams

- Insufficient detail is recorded for the process to be understood.
- Grouping together of process steps or ingredients results in omission of necessary steps/ingredients.
- Diagram is too complex and not easy to understand. This often occurs where engineering detail is included, e.g. technical outline of the process equipment.
- Inclusion of too much non-product/process information, e.g. quality checks such as weight control.
- Diagram does not include all possible permutations of product flow, e.g. additional holding stages and rework.
- Diagram is not representative of what really happens. This usually means that:
  - it has not been verified and
  - the HACCP team members do not have realistic site-specific knowledge or have made assumptions since they are too close to the process.

12.3.7 Codex logic sequence step 6: list all potential hazards, conduct a hazard analysis and consider control measures (apply HACCP principle 1)

Using the process flow diagram(s), the HACCP team now needs to consider each process activity in turn and list any potential hazards that might occur. They should then carry out an analysis to identify the significant hazards and identify suitable control measures. These terms are defined by Codex (2009b) as follows:

> **Hazard:** A biological, chemical or physical agent in, or condition of, food with the potential to cause an adverse health effect.
> **Hazard analysis:** The process of collecting and evaluating information on hazards and conditions leading to their presence to decide which are significant for food safety and therefore should be addressed in the HACCP plan.
> **Control measure:** An action or activity that can be used to prevent, eliminate or reduce a hazard to an acceptable level.

The NACMCF HACCP guidelines (1997) provide very similar wordings to the Codex (2009b) definitions given, illustrating the agreement that exists at international levels. However, the NACMCF (1997) definition of a hazard differs slightly:

> **Hazard:** A biological, chemical or physical agent that is reasonably likely to cause illness or injury in the absence of its control.

This provides a useful reminder of the need to evaluate risk to consumer health, which will be done alongside consideration of likelihood of occurrence during the hazard analysis, and to ensure that effective control systems are developed.

Hazard analysis is a key element of HACCP and will determine the strength of the resulting HACCP plan. The hazard analysis needs to be accurate and specific – including detail about the type of hazard and its source or cause, as well as how the significance of specific hazards was determined and justified. If the hazard analysis is too brief or general, then the following steps in the HACCP study will be more difficult and the HACCP plan is likely to be weak.

For **microbiological hazards** it is possible to generalise to a certain extent, but consideration should be given to specific pathogens. Microbial hazards are normally listed as specific organisms or using the collective terms, 'vegetative pathogens' and 'spore-forming pathogens'. A third term, 'toxin-forming pathogens', is also sometimes used, but organisms in this group could be either vegetative organisms or spore formers. The cause or source of the hazard needs to be established along with how the hazard is manifested in the process, i.e.:

- *Presence* of the hazard in a raw material
- *Contamination* with the hazard during processing and handling
- *Growth* of microorganisms during production
- *Survival* of microorganisms through a failure in a process designed to destroy them

For **physical hazards**, it is important to consider whether the item would genuinely cause physical harm to the consumer. Physical hazards are:

- Items which are sharp and may cause injury
- Items which are hard and may cause dental damage
- Items which could block airways and cause choking

For **chemical hazards**, the hazard analysis will consider the likelihood of presence of toxic chemicals in the raw materials and contamination by chemicals during processing, which may raise the toxicity to an unacceptable level.

Further detailed information about hazards can be found in Chapter 5.

The process of hazard analysis includes the following:

- Hazard identification – identifying which hazards may occur and where
- Assessment of significance – establishing which hazards are likely to occur and cause an adverse health effect
- Identification of control measures – establishing an effective mechanism for ongoing control of the hazard

The hazard identification step is often approached by team 'brainstorming'. This is a technique used to pool together ideas from members of the group. All team ideas need to be captured and recorded for evaluation of significance. A common approach to documentation of the hazard analysis is the use of hazard analysis charts (Mortimore and Wallace, 1998). Hazard analysis charts (Table 12.8) are used to help structure the hazard analysis, allowing HACCP teams to document the important aspects with respect to potential hazard identification, reasoning and decision-making regarding significance and determination of appropriate control actions. This level of detail is important to the production of the HACCP plan.

The process of hazard analysis requires the team to transcribe each process activity from the process flow diagram to the hazard analysis chart, consider any potential hazards along with their sources or causes and then evaluate their significance. The source, cause and manifestation information is useful as this helps in identifying an appropriate control measure, and the inclusion of justification columns provides useful information for any future challenges of the HACCP plan, for example through external audit.

Table 12.8 Hazard analysis chart headings.

Process step	Hazard: source, cause and manifestation	Likelihood of occurrence (high/low)	Severity of outcome (high/low)	Significant? (yes/no)	Justification of significance decision	Control measure(s)	Justification of control measures

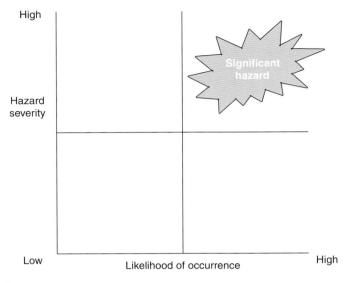

Fig. 12.4   Significance assessment.

Determination of hazard significance

Codex (2009b) requires 'control of hazards that are of such a nature that their elimination or reduction to acceptable levels is essential to the production of a safe food' and states that the process of hazard analysis is intended to 'identify those hazards that are significant for food safety and therefore should be addressed in the HACCP Plan'. Although the term 'significant hazard' is not defined by Codex, the International Life Sciences Institute (ILSI, 1999) has put these two phrases together to form a useful definition:

> **Significant hazard:** Hazards that are of such a nature that their elimination or reduction to an acceptable level is essential to the production of safe foods.

To identify the significant hazards, it is necessary to consider the likelihood of occurrence of the hazard in the type of operation being studied as well as the severity of the potential adverse effect. A significant hazard, therefore, is one that is **both** likely to occur **and** cause harm to the consumer (Figure 12.4).

Most companies will assess significance of hazards using judgement and experience, but structured 'risk evaluation' methods, where different degrees of likelihood and severity are weighted, are sometimes used to help with the significance decision. Risk is the probability or likelihood that an adverse effect will be realised. Risk evaluation decisions should be taken from a sensible viewpoint, based on knowledge and experience. Structured risk evaluation methods often involve significance assessment tables which aim to consider the degree of likelihood and the severity of effect by rating these as 'high', 'medium' or 'low' (Table 12.9). This is similar to the concept shown in Figure 12.4 but aims to put individual hazards in boxes to assist with the significance decision.

Although these tools are generally believed to make significance assessment more straight-forward by the companies using them, they do still require training in their application and use of judgement to position the identified hazards in the correct subcategories, i.e. a tool such

Table 12.9   An example of a significance assessment table.

Severity of effect	Likelihood of occurrence		
	High	**Medium**	**Low**
High			
Medium			
Low			

as Table 12.9 is only useful if there is also guidance on which boxes are equivalent to significant hazards. Currently, there are no definitive rules on this within HACCP, although a previous approach to HACCP (NACMCF, 1992) did include this kind of structured system. There has been a resurgence in the use of these risk evaluation tools in recent years, which is believed may be linked to uptake of the ISO 22000 audit standard, *Food Safety Management Systems – Requirements for Any Organization in the Food Chain* (ISO, 2005a). This requires formal records of hazard assessment to be maintained, although there is no specific requirement in the standard for any particular tool to be used.

Nowadays most experienced HACCP practitioners would weigh up the likelihood and severity to make a judgement on significance while working through the hazard analysis. The most important thing is that the decision on hazard significance is based on appropriate knowledge and experience, and is not clouded by the inappropriate use of unproven decision-making tools. Risk evaluation tools need expert knowledge in their application but can offer a structured framework for significance assessment in the right hands. However, in the wrong hands, errors in application of risk evaluation tools, including fundamental misunderstandings either in likelihood of occurrence or in severity evaluation can lead to both inappropriate identification of extra hazards as significant or, more dangerously, non-identification of significant hazards (Wallace, 2009).

Further assistance to consider when carrying out the hazard analysis is provided by Codex (2009b), which lists some brief points (Table 12.10), and NACMCF (1997), which lists a series of questions to help the HACCP team discuss different hazard issues (Table 12.11).

Table 12.10   Codex guidance on application of HACCP principle 1.

List all potential hazards associated with each step, conduct a hazard analysis and consider any measures to control identified hazards

The HACCP team should list all the hazards that may be reasonably expected to occur at each step according to the scope from primary production, processing, manufacture and distribution until the point of consumption. The HACCP team should next conduct a hazard analysis for the HACCP plan to identify which hazards are of such a nature that their elimination or reduction to acceptable levels is essential to the production of a safe food.

In conducting the hazard analysis, wherever possible the following should be included:

- The likely occurrence of hazards and severity of their adverse health effects
- The qualitative and/or quantitative evaluation of the presence of hazards
- Survival or multiplication of microorganisms of concern
- Production or persistence of toxins, chemicals or physical agents in foods
- Conditions leading to the above

Consideration should be given to what control measures, if any exist, can be applied for each hazard.

More than one control measure may be required to control a specific hazard, and more than one hazard may be controlled by a specified control measure.

Source: Codex (2009b).

Table 12.11 Examples of questions to be considered when conducting a hazard analysis.

1. Ingredients
    *   Does the food contain any sensitive ingredients that may present microbiological hazards (e.g. *Salmonella*, and *Staphylococcus aureus*), chemical hazards (e.g. aflatoxin, antibiotic or pesticide residues) or physical hazards (e.g. stones, glass and metal)?
    *   Are potable water, ice and steam used in formulating or handling the food?
    *   What are the sources (e.g. geographical region and specific supplier)?
2. Intrinsic Factors – physical characteristics and composition (e.g. pH, type of acidulants, fermentable carbohydrate, water activity and preservatives) of the food during and after processing
    *   What hazards may result if the food composition is not controlled?
    *   Does the food permit survival or multiplication of pathogens and/or toxin formation in the food during processing?
    *   Will the food permit survival or multiplication of pathogens and/or toxin formation during subsequent steps in the food chain?
    *   Are there other similar products in the marketplace? What has been the safety record for these products? What hazards have been associated with the products?
3. Procedures used for processing
    *   Does the process include a controllable processing step that destroys pathogens? If so, which pathogens? Consider both vegetative cells and spores.
    *   If the product is subject to recontamination between processing (e.g. cooking and pasteurising) and packaging, which biological, chemical or physical hazards are likely to occur?
4. Microbial content of the food
    *   What is the normal microbial content of the food?
    *   Does the microbial population change during the normal time the food is stored prior to consumption?
    *   Does the subsequent change in microbial population alter the safety of the food?
    *   Do the answers to the above questions indicate a high likelihood of certain biological hazards?
5. Facility design
    *   Does the layout of the facility provide an adequate separation of raw materials from ready-to-eat foods if this is important to food safety? If not, what hazards should be considered as possible contaminants of the ready-to-eat products?
    *   Is positive air pressure maintained in product packaging areas? Is this essential for product safety?
    *   Is the traffic pattern for people and moving equipment a significant source of contamination?
6. Equipment design and use
    *   Will the equipment provide the time–temperature control that is necessary for safe food?
    *   Is the equipment properly sized for the volume of food that will be processed?
    *   Can the equipment be sufficiently controlled so that the variation in performance will be within the tolerances required to produce a safe food?
    *   Is the equipment reliable or is it prone to frequent breakdowns?
    *   Is the equipment designed so that it can be easily cleaned and sanitised?
    *   Is there a chance for product contamination with hazardous substances, e.g. glass?
    *   What product safety devices are used to enhance consumer safety?
        o   Metal detectors
        o   Magnets
        o   Sifters
        o   Filters
        o   Screens
        o   Thermometers
        o   Bone removal devices
        o   Dud detectors
    *   To what degree will normal equipment wear affect the likely occurrence of a physical hazard (e.g. metal) in the product?
    *   Are allergen protocols needed in using equipment for different products?

(Continued)

Table 12.11 *(Continued)*.

7. Packaging
   - Does the method of packaging affect the multiplication of microbial pathogens and/or the formation of toxins?
   - Is the package clearly labelled 'keep refrigerated' if this is required for safety?
   - Does the package include instructions for the safe handling and preparation of the food by the end user?
   - Is the packaging material resistant to damage thereby preventing the entrance of microbial contamination?
   - Are tamper-evident packaging features used?
   - Is each package and case legibly and accurately coded?
   - Does each package contain the proper label?
   - Are potential allergens in the ingredients included in the list of ingredients on the label?
8. Sanitation
   - Can sanitation have an impact upon the safety of the food that is being processed?
   - Can the facility and equipment be easily cleaned and sanitised to permit the safe handling of food?
   - Is it possible to provide sanitary conditions consistently and adequately to ensure safe foods?
9. Employee health, hygiene and education
   - Can employee health or personal hygiene practices impact upon the safety of the food being processed?
   - Do the employees understand the process and the factors they must control to ensure the preparation of safe foods?
   - Will the employees inform management of a problem which could impact upon safety of food?
10. Conditions of storage between packaging and the end user
    - What is the likelihood that the food will be improperly stored at the wrong temperature?
    - Would an error in improper storage lead to a microbiologically unsafe food?
11. Intended use
    - Will the food be heated by the consumer?
    - Will there likely be leftovers?
12. Intended consumer
    - Is the food intended for the general public?
    - Is the food intended for consumption by a population with increased susceptibility to illness (e.g. infants, the aged, the infirmed, immunocompromised individuals)?
    - Is the food to be used for institutional feeding or the home?

Source: NACMCF (1997).

The preceding series of questions (Table 12.11) is designed to be used as part of hazard analysis, as appropriate to the process under consideration (NACMCF, 1997). The purpose of the questions is to assist in identifying potential hazards. Whilst some of these questions (NACMCF, 1997) may help HACCP teams during the hazard analysis, many are more general questions about hygiene conditions at the facility and might prove more useful when evaluating the prerequisite programme requirements (see Chapters 9 and 10). In addition, there may be further questions to ask with the experience gained from more recent food incidents, such as when products are used for different means than intended (e.g. consumed raw rather than baked/cooked).

Control measures

Once the significant hazards have been established, effective control measures need to be identified for each significant hazard. As defined previously, control measures are the actions that can be used to prevent, eliminate or reduce a hazard to an acceptable level. Control for each significant hazard is essential, but there may be more than one control measure for any hazard. There will also be control measures operating for prerequisite programmes.

Control measure options include the following:

- Process steps, e.g. cooking, sieving and metal detection
- Product intrinsic factors
- Use of approved suppliers
- Temperature-controlled storage or holding
- Handling procedures
- Controlled segregation

An important point about control measures is to make sure that they are capable of ongoing control of the hazard at all times. Often, HACCP teams mistakenly identify monitoring checks rather than controls – the measure must be control not monitoring and effective control measures must:

- relate to the hazard and source,
- be comprehensive and appropriate and
- be validated – will it really control the hazard?

When deciding on control measures, it is important to consider the different options that may be available to control the particular hazard in order to establish the best method for control. This can include an evaluation of the measures currently in place, but it is important to decide whether these are strong enough or whether additional control is necessary. Chapter 6 provides further detailed information on designing control options.

12.3.8   Codex logic sequence step 7: determine CCPs (HACCP principle 2)

Critical control points (CCPs) are the points in the process where the significant hazards must be controlled, and are defined by Codex (2009b) as follows:

> **Critical control point:** A step at which control can be applied and is essential to prevent or eliminate a food safety hazard or reduce it to an acceptable level.

CCPs can be identified using HACCP team knowledge or experience and by using tools such as the Codex CCP decision tree (Figure 12.5). The Codex CCP decision tree is a useful tool, and although it may seem daunting to use at first, it does become easier with practice. In some cases CCPs are specified by legislation, e.g. milk pasteurisation.

To use the decision tree, the questions are asked in sequence for each significant hazard that has been identified for each process activity as follows.

Q1: Do control measures exist for the identified hazard?

In most cases, when conducting the hazard analysis (Step 6), control measures will have been identified for the significant hazards. Therefore, it is most common to answer 'yes' to Q1. However, if the HACCP team could not identify a control measure, then it is necessary to answer 'no' and move on to the subquestion, Q1a.

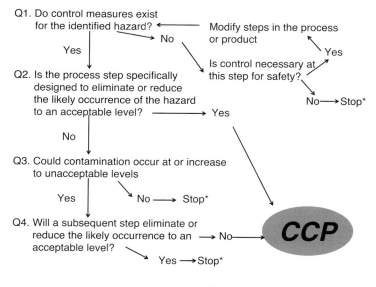

Fig. 12.5 CCP decision tree. (Adapted from Codex, 2009b.)

Q1a: Is control at this step necessary for safety?

Perhaps control is not necessary at this process step for safety; for example, there might be a control measure later in the process. In this example the answer, 'no' results in the decision that this step is not a CCP for that particular hazard and the instruction to stop/proceed (move on to the next hazard). However, if the HACCP team considers that control is necessary at this step for safety (perhaps there is no control later) then the answer 'yes' results in the decision tree instruction to 'modify step, process or product', i.e. carry out some modification to allow a control measure to be built in.

For example, if you were concerned about metal hazards entering a process but had no control measure later on that could remove them, it would be possible to carry out a modification to build in control through a suitable metal hazard removal system such as magnets or metal detection later on in the process.

Once a modification has been determined, the team needs to go back and ask Q1 again. Now, they will be able to answer 'yes' and move on to Q2. A key point to remember, though, is that **if you cannot establish a CCP for a significant hazard then you cannot make the product** as it would always be potentially unsafe for consumption.

Q2: Is the process step specifically designed to eliminate or reduce the likely occurrence of the hazard to an acceptable level?

This is a key question in the decision tree and one that people often have difficulty with. The question provides, in effect, a short cut to a CCP decision for those process steps that are designed to control hazards, e.g. most cooking processes. The important thing to remember is that the question is asking about the process step and not the control measure. This is because control measures are always designed to control hazards so would always result in the answer 'yes'. Because you are asking the question about control measures, and a CCP is identified every time you answer 'yes', the result could be many more CCPs than are actually required. To make matters slightly more confusing, some process steps are also control measures – these are the ones that this question is designed to find.

Where cooking processes are not specifically designed to control hazards, then the team should answer 'no' rather than 'yes'. Not all cooking steps are CCPs – some are at such high heat processes, designed to change the physical structure of the product rather than for safety (e.g. some baking processes), that a significant hazard such as vegetative pathogens simply could not survive. In other words, the likelihood of occurrence should have been established as 'low' during the hazard analysis, and it will not be a significant hazard. Although this should have been identified at the hazard analysis stage, and therefore not be an issue during CCP decisions, some companies find in practice that customers and/or regulators insist on such steps being CCPs, even when there is clear justification that this is not required.

If the team believes the answer to this question is 'no', they should progress onto Q3.

Q3: Could contamination with identified hazard(s) occur in excess of acceptable level(s) or could these increase to unacceptable levels?

(*Note*: Acceptable levels are safe for consumption and unacceptable levels may cause harm to the consumer.)

If a significant hazard has been identified, then this is really already saying that something unacceptable could occur. Therefore, in most cases the answer to this question will be 'yes'. However, the question does give the chance to just think again and confirm whether it is unacceptable or acceptable. Sometimes, use of the loop at Q1a, resulting in a process modification, might mean that the hazards are no longer considered unacceptable (perhaps they have been designed out of the process completely) and so the answer would be 'no' in this case. Where you have answered 'yes' to this question, move onto Q4, otherwise stop and proceed with the decision tree for the next hazard.

Q4: Will a subsequent step eliminate the identified hazard(s) or reduce their occurrence to acceptable levels?

This last question allows the presence of a hazard at one process step if it is going to be effectively controlled at a later process step. It is helpful in keeping the CCPs to a manageable number, whilst making sure that the essential ones are identified. If there is a subsequent step in the process where the hazard will be controlled ('yes' answer) then the current process step is not the CCP but the later step will be. It is important to check that the later process step is properly identified as a CCP when the team gets to the end of the study. If there is no subsequent process that will control the hazard, then the current step needs to be made a CCP and managed accordingly.

In the same way as for hazard analysis, when working with the decision tree it is useful to keep a record of the team's discussions and justification of the decisions for future reference. This is normally done using a CCP decision record sheet (Table 12.12). Even if the team is not using the decision tree, it will be important to keep a record of the decisions so that full evidence of the HACCP process is available to show regulators and auditors.

Once the HACCP team has worked through the processes for all the hazards, a list of CCPs will be available. These are the points in the processes that must be carefully managed to make

Table 12.12 CCP decision record.

Process step (hazard)	Control measure	Q1	Q1a	Q2	Q3	Q4	CCP Y/N?	HACCP team notes (justification)

sure that the food produced is safe. For each of the CCPs, it is now important to define how they will be controlled and managed on a day-to-day basis. HACCP Principles 3, 4 and 5 are applied to set these standards, and normally this information is recorded in a HACCP control chart or table (Table 12.13).

12.3.9 Codex logic sequence step 8: establish critical limits for each CCP (HACCP principle 3)

Critical limits are the safety limits that must be achieved for each CCP to ensure that the food is safe. If the process operates beyond the critical limits, then products made will be potentially unsafe. Critical limits are defined by Codex as follows:

Critical limit: A criterion that separates acceptability from unacceptability.

Critical limits are expressed as absolute values (never a range) and often involve criteria such as temperature and time, pH, acidity and moisture. Critical limits must be measurable and established for all CCPs. The choice of critical limit can be based on scientific and experimental data, industry or legislative standards and historical evidence.

Critical limits are rarely the same as existing control parameters. For example, in cooking boiled eggs, one hazard might be identified as presence of *Salmonella* spp. in the raw egg, and this should be controlled by the cooking process. The critical limit to kill *Salmonella* spp. will be the equivalent of 70°C for 2 minutes in the centre of the egg. However, it would be unusual to cook eggs using these parameters; instead, the egg would be boiled in water at 100°C for, say, 6 minutes. It is important to know that these everyday process parameters would achieve the critical limit, along with what the margin of error is, and this is done by validating the process, which is discussed in more detail later.

As mentioned previously, critical limits never involve a range as they are the absolute value that defines the barrier between 'safe' and 'potentially unsafe'. Therefore, another measure that is often used for practical purposes in food operations is the 'target level' or 'operational limit'. The difference between critical limits and operational limits is illustrated in Figure 12.6. Operational limits provide a buffer zone for process management and indicate if a CCP is going out of control.

12.3.10 Codex logic sequence step 9: establish a monitoring system for each CCP (HACCP principle 4)

Once the critical limits, and operational limits, have been established, the next step in HACCP involves developing a monitoring system for ongoing measurement, which will demonstrate

Table 12.13 An example of HACCP control chart.

CCP No.	Process step	Hazard	Control measure	Critical limit	Monitoring			Corrective action	
					Procedure	Frequency	Responsibility	Procedure	Responsibility

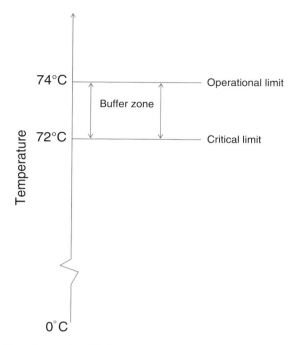

Fig. 12.6 Critical limits and operational limits.

that the CCPs are working effectively. Codex (2009b) defines monitoring as:

> **Monitoring:** The act of conducting a planned sequence of observations or measurements of control parameters to assess whether a CCP is under control.

Monitoring is necessary to demonstrate that the CCPs are being controlled within the appropriate critical limits and monitoring requirements need to be specified by the HACCP team during the HACCP study. Each monitoring activity should have a person who is allocated to carry it out (CCP monitor) and record the results and take any necessary actions. In manufacturing, monitoring is usually done by production line personnel who are involved in operating the processes where the CCPs are located. The frequency of monitoring should also be defined, and this will relate to the throughput of product in the particular process; i.e. in a fast-moving process that produces large amounts of product in a short time, the monitoring occasions may need to be closer together than in a slower process with a smaller throughput. This relates to the ability to handle the amount of product that is produced between monitoring checks if a CCP is found to be out of control; for example, if monitoring checks are done every 30 minutes then there will be less potentially unsafe product to handle than if the checks are done every 4 hours. The ideal is to have continuous monitoring systems linked to alarm and action systems. For example, metal detection is frequently used as a CCP in food manufacturing, and traditionally, metal detection effectiveness has been monitored manually via the use of test sticks at a given frequency. More recently, continuous monitoring of metal detectors by online computer systems has become possible and is used in some companies. In this case, the computer system does the monitoring, and this can be verified as effective by manual checks on the system, e.g. 4 hourly, as illustrated in Table 12.14.

Table 12.14 Comparison of regular and continuous CCP monitoring.

CCP metal detection	Process step	Hazard	Control measure	Critical limit	Monitoring			Verification
					Procedure	**Frequency**	**Responsibility**	
Example 1: Manual monitoring	Metal detection	Metal in product	Effective metal detection and removal system in place	2 mm ferrous; 3 mm non-ferrous; 4 mm stainless steel	Check operation of detector and rejection mechanism with metal test pieces	Every 30 minutes	Line operator	Check and sign-off line records by supervisor in every shift
Example 2: Online monitoring	Metal detection	Metal in product	Effective metal detection and removal system in place	2 mm ferrous; 3 mm non-ferrous; 4 mm stainless steel	Online monitoring and recording by computer system	Continuous	Line supervisor	Check operation of detector and rejection mechanism with metal test pieces every 4 hours

12.3.11 Codex logic sequence step 10: establish corrective actions (HACCP principle 5)

Corrective action needs to be taken where monitoring shows that there is a deviation from a defined critical limit. Corrective actions must deal **both** with the product produced while the process is out of control (it may need to be destroyed or reprocessed) **and** with the process fault that has caused the CCP deviation to bring the process back under control. Codex (2009b) defines corrective action as follows:

> **Corrective action:** Any action to be taken when the results of monitoring at the CCP indicate a loss of control.

As for monitoring, the corrective action procedures and responsibility need to be identified by the HACCP team during the HACCP study, but will be implemented by the appropriate operations personnel if deviation occurs. Corrective action is not 'contact the quality manager' for every event. The multidisciplinary team should use their collective knowledge to brainstorm the likely failure modes and identify appropriate specific corrective actions for each eventuality. The effectiveness of the proposed corrective action plan needs to be verified and challenged because this is the last defence mechanism protecting the consumer from receiving potentially unsafe product should a CCP fail.

12.3.12 Codex logic sequence step 11: establish verification procedures (HACCP principle 6)

Verification requires that procedures are developed to confirm that the HACCP system can work and is working effectively. There are two different types of confirmation required:

1. It is important to check that the HACCP plans developed by the HACCP team will work effectively to control all the relevant hazards. This is done once the HACCP control charts have been completed, before implementation of the HACCP plan in the operation, and is known as **validation**. Validation is also done periodically after the HACCP plan has been implemented to check that the HACCP plan is still appropriate for the control of all relevant hazards, taking into account any changes that have occurred in the operations, processes, products and ingredients, as well as any updates to knowledge on hazards.
2. The HACCP team needs to consider how to determine if the HACCP system is working effectively over time, once it has been implemented. This is known as **verification**, and involves various procedures and methods that will be used to demonstrate compliance with food safety requirements.

Many people find the terms 'validation' and 'verification' confusing, partly because the words sound similar and they are both part of the verification principle. They are actually two separate and different activities and would benefit from being separated out into two HACCP principles (see Chapter 4). However, although they are both part of principle 6, it is helpful to consider the definitions in more detail to help understand the difference (Table 12.15).

Table 12.15 Defining validation and verification.

Term	Codex definition	Clarification
Validation	Obtaining evidence that the elements of the HACCP plan are effective	• Is the HACCP plan capable of controlling all relevant hazards if correctly implemented? **or** • Will it work?
Verification	The application of methods, procedures, tests and other evaluations, in addition to monitoring, to determine compliance with the HACCP plan	• Is there compliance with food safety requirements defined in the HACCP plan? **or** • Is it working in practice?

Validation will include the following:

- Cross-checking through the HACCP plan to make sure that all the principles have been correctly applied.
- Checking that the hazards will be controlled, i.e.:
  - The control measures are suitable.
  - Correct CCPs have been identified.
  - Critical limits are set correctly for the hazard, e.g. using literature values, and challenge testing.
  - Process will achieve the critical limit(s); e.g. the process is capable of always achieving this limit within normal process variation.
  - Monitoring will detect loss of control if it happens.
  - Corrective action will prevent the potentially unsafe food being consumed.

Validation can be done by HACCP team members working with other managers within the business. As with preparing the HACCP plan, it will be better to involve more than one person if possible, and, like hazard analysis, this will be an area where many companies will need to use expert resource from outside the company to assist in validation.

Commonly used **verification** procedures include the following:

- HACCP audits
- Review of CCP monitoring records
- Product testing – microbiological and chemical tests
- Review of deviations, including product disposition and customer complaints

Verification can also be done by HACCP team members or other personnel within the business, e.g. supervisory staff. It is important to have independence from the system to audit effectively, so consideration can be given to using external resource or other personnel who were not involved in developing or in the day-to-day running of HACCP. Auditors should be competent as studies have shown (Wallace, 2009) that weaknesses exist in HACCP systems, even when audited by professional third-party auditors (see also Chapter 2). In the aforementioned research (Wallace, 2009), a wide range of weaknesses were found in HACCP systems at production sites of multinational manufacturers. Although all sites had a range of verification procedures in place, including internal and external audits, the procedures had failed to pick up the weaknesses identified, and this underlines the need for agreed standard audit approaches and effective training of HACCP auditors. It is, therefore, recommended that food companies question both the competency and the experience of external HACCP auditors before their engagement.

12.3.13 Codex logic sequence step 12: establish documentation and record-keeping (HACCP principle 7)

It is important to document the HACCP system and to keep adequate records. The HACCP plan will form a key part of the documentation, outlining the CCPs and their management procedures (critical limits, monitoring and corrective action). It is also good practice to keep documentation showing how the HACCP plan was developed, i.e. the hazard analysis, CCP determination and critical limit identification processes.

When the HACCP plan is implemented in the operation, records will be kept on an ongoing basis. Essential records include the following:

- CCP monitoring records
- Records of corrective actions associated with critical limit deviation
- Records of verification activities
- Records of modifications to processes and the HACCP plans

In Europe, the legal requirements for food safety management include documentation commensurate with the size of the business, so smaller businesses would not be expected to keep the same level of documentation as larger ones. However, the key consideration for all businesses should be to have sufficient documentation to demonstrate the effective working of the HACCP system. Maintenance and archiving of HACCP records is therefore an important element of effective HACCP, and will be discussed further in Chapter 14. Records may be kept as paper archives; however, companies are increasingly turning towards computerised record-keeping systems. This may involve simple scanning of paperwork for storage purposes or integrated systems where monitoring is done through hand-held computer terminals, perhaps accessed by the CCP monitor using a swipe card, and the data are archived within the site's computer systems.

12.4 CONCLUSIONS

Application of HACCP principles is achieved by following a straightforward stepwise procedure outlined by the Codex (2009b) logic sequence. This will only result in an effective HACCP system if performed by HACCP teams made up of personnel with the correct blend of training, skills and experience. The outcome of this HACCP study process should be a HACCP plan that clearly defines how all significant hazards relevant to the operation will be controlled.

13 Implementing a HACCP system

13.1 INTRODUCTION

Implementation of the HACCP plan is a crucial part of food safety system effectiveness. It is very important to get implementation right so that the critical control points (CCPs) identified in the HACCP plan development stage (Chapter 12) will work every day to protect the consumer. However, there is limited advice on HACCP implementation in the literature, and it is often assumed that if companies can develop HACCP plans, then effective implementation will follow. This is not true as implementation is an easy place to go wrong and weaknesses in HACCP plan implementation feature strongly on the list of reasons for HACCP failure (Chapter 2). It is important to note that many people see having a HACCP plan as the end point of their HACCP endeavours – it is actually only the beginning. Understanding these issues is fundamental to success in 21st century HACCP, i.e. the full application of HACCP principles along with implementation and maintenance of the resulting management systems.

As discussed in Chapter 12, Codex (2009b) describes how to develop HACCP plans and their associated verification and documentation requirements; however, it does not discuss how to implement the HACCP plans into everyday practice, and this has already been outlined in Chapter 4 as an area where more advice would be helpful. Implementing HACCP requires careful preparation and training of the workforce and is best managed as a change management process. Depending on the maturity of the operation, this may be a straightforward implementation of the HACCP requirements or may require a culture change, i.e. to include more fundamental prerequisite requirements.

The implementation of a HACCP plan requires the input of a range of different personnel within the operation. The HACCP team members who have developed the plan will have key roles, as will the wider food safety team and, in particular, the line operators, supervisors and managers who will be involved in the day-to-day running of HACCP in practice. The process of HACCP implementation involves the handover of responsibility and ownership of HACCP from its developers to its operators, and thus the HACCP plan documents are activated as a working system.

In this chapter, we focus on how to make the HACCP plan documents work in the production environment and discuss best practice approaches both to implementation of new HACCP plans and to implementing changes and updates to existing HACCP systems. Covering implementation theory in a stepwise fashion, this chapter aims to build on experiences and learnings about

Food Safety for the 21st Century, First Edition By Carol A. Wallace, William H. Sperber and Sara E. Mortimore
© 2011 Carol A. Wallace, William H. Sperber and Sara E. Mortimore

HACCP plan implementation to deliver tips for success and help avoid the common pitfalls that may result in HACCP system weaknesses.

13.2 ACTIVITIES FOR IMPLEMENTATION OF A HACCP PLAN

The implementation stage is where the HACCP plans are handed over from the HACCP team that has worked on the development process to the operations personnel who will manage the CCPs on a day-to-day basis. Training is therefore a key requirement, and this will include training for the personnel who will monitor CCPs and take corrective action, along with HACCP awareness training for the wider operations workforce. Implementation needs to be carefully planned, with responsibility for the various actions given to the appropriate people. It is not simply a case of handing the HACCP plan documentation over to the operations personnel; rather, there is a need for detailed and careful planning such that all the required activities for successful implementation can be identified and progressed.

When implementing a new HACCP plan, this is best achieved by breaking the necessary activities down into steps as follows (Figure 13.1). When implementing updates to existing HACCP plans, this will be more straightforward and may be done via a less formal approach; however, the importance of reviewing the completeness of each amendment and verifying that effective HACCP plan implementation has been achieved cannot be understated in either case.

Most companies choose a phased approach to implementation of HACCP plans, implementing a bit at a time to make sure each part works smoothly before moving on to the next. Implementation planning is, therefore, crucial to make sure that each piece of the 'puzzle' is completed and links with the others. Implementation always starts with the validated HACCP plan.

13.2.1 The validated HACCP plan

The requirements for validation of HACCP plan elements were discussed in Chapter 12. It is important to make sure that new HACCP plans have been validated and signed off as correct and suitable for control of all likely hazards in the operation before the implementation process commences. This is equally true for any amendments to existing HACCP plans, and in this case, validation will make sure that the amendments enhance and strengthen the existing food safety systems. It would be a great shame to implement a poorly thought-out HACCP plan amendment that actually weakens food safety control. Ensuring that the HACCP plan (or amendment) has been validated prior to implementation will prevent the need to go back to this step, should deficiencies be identified during implementation, as this will result in loss of focus and credibility.

13.2.2 Implementation action planning

The main HACCP plan implementation activities will involve setting up CCP management systems, including monitoring and corrective action, alongside training of personnel and completion of any other initiatives required to support HACCP, e.g. prerequisite programme improvements from the gap analysis (Chapter 9), modifications to the processes or equipment – often, issues which came through the decision tree loop at Q1a (Chapter 12). It

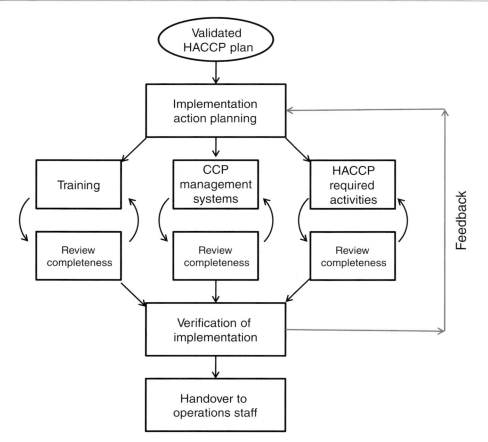

Fig. 13.1   Steps to HACCP implementation. Note that these steps of HACCP plan implementation do not have to occur in a set order as several activities could be going on at the same time involving different personnel. (Adapted from Mortimore and Wallace (2001).)

is important to construct a detailed activity list for all implementation activities, which can be checked off as each task is completed. This should include details of who is responsible and also the deadlines for action, i.e. an implementation of timetable. Project planning techniques, as described in Chapter 9, including the use of Gantt charts (Figure 13.2), are useful to keep the implementation plan on track. These form a pictorial representation of the timetable and capture a list of all the main and subactivities plus their priorities; however, a simple listing of what needs to be done, when, and by whom will suffice.

Key considerations at the action planning stage would include the following:

- Review of CCP management requirements from the HACCP plan against current control systems, identification of gaps and development of necessary procedures, e.g. availability and calibration requirements of monitoring equipment.
- Identification of additional support initiatives for HACCP, comparison with existing support systems and development of procedures to fill gaps in current systems.
- Listing of any modification requirements identified during the HACCP study, communication with appropriate personnel and planning the necessary work.

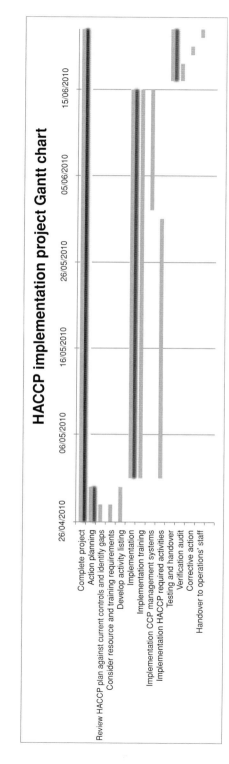

Fig. 13.2   HACCP implementation project Gantt chart.

- Consideration of personnel resources available and necessary to manage the HACCP plan on a day-to-day basis. Identification of knowledge and skill gaps and planning of appropriate training.
- Consideration of existing documentation, identification of gaps and planning the redesign.
- Consideration of the general HACCP awareness levels of management and operations personnel, identification of implementation champions and planning of management support strategy.

The action plans produced should be seen as a 'work-in-progress' and should be regularly updated as actions progress and/or new action requirements are identified.

13.2.3 Training

As discussed in Chapter 9, all staff in the operation will need some awareness training about the HACCP system and food safety requirements. Personnel who carry out monitoring and corrective action will need more detailed training about their roles. Specifically, they will need to understand the monitoring procedure (i.e. exactly what they have to do), the frequency, where they should record results and what to do if the results show that the CCP is out of control, i.e. the corrective action that must be taken. Management personnel, particularly those overseeing operations areas where the HACCP plan is being implemented, will also need both HACCP awareness training and an understanding of their key role to support the implementation. 'Promotional management' (Wallace, 2009) is seen as a key factor in the success of HACCP, i.e. the managers who are committed to the HACCP project and show this commitment both through ongoing promotion of HACCP requirements to their staff and by being instrumental in providing necessary resources and support for its implementation. Foremost in gaining this level of commitment is appropriate training about HACCP, the current status and necessary steps to get the HACCP plan working.

HACCP training at the implementation stage is often done by HACCP team members or managers who have been involved in the HACCP process, and who now have experience of applying HACCP to supplement their own knowledge. It is important that the personnel involved in this key step have appropriate training skills and that transfer of the key messages via training can be verified. In addition to the HACCP training, training in, and/or communication of, the application of the HACCP implementation action plan will be necessary, in particular how the implementation affects and amends current working practices. External trainers could be used to support this training, but will need to be fully briefed on how HACCP fits the specific applications on site.

CCP monitor training will be a key aspect of the implementation training requirements. This may be done by HACCP team members but could equally be achieved by personnel from the production or engineering teams, or the wider food safety team, including human resource personnel. Training of supervisory personnel who will verify and countersign the monitoring records on a day-to-day basis is also crucial to ensure that they also understand the importance of monitoring and corrective action, and they know what to expect to see on the specific monitoring records. This will help to overcome the problem of defective monitoring and corrective action records, e.g. where monitoring has been missed or where CCP deviation is recorded but no corrective action is apparent, being signed off by supervisors without challenging the specific defects.

13.2.4   CCP management systems

CCP management systems include the monitoring and corrective action requirements for CCPs, plus the appropriate process control verification, including calibration. The required monitoring and corrective action procedures should have been defined in the HACCP plan. This is the point where the content of the HACCP plan needs to be converted into work instructions or standard operating procedures (SOPs) that will be followed by the trained personnel with monitoring and corrective action roles. Now is the time to make sure all the required facilities, equipment and documentation are available and that the systems will work, i.e. will the monitoring detect deviation from the CCPs when it occurs? Will the corrective action procedure correctly identify and deal with, e.g. quarantine/destroy the potentially unsafe product and bring the process back under control? This is also an opportunity to reconfirm the process capability to control hazards at the identified CCPs – this should have been established during validation of the HACCP plan elements, but at implementation it is important to ensure that these systems will work in practice.

There is also a clear link between the CCP management systems aspect of HACCP plan implementation and training to make sure that the personnel involved in day-to-day management of CCPs understand their roles and responsibilities, and the consequences of CCP failure.

Documentation for ongoing CCP management will include written/pictorial work instructions for CCP monitoring, corrective action and, where appropriate, calibration. Record-keeping systems will also be needed, and these should be carefully designed such that it is exactly clear what the records mean after the event. Experience shows that this is an area where weaknesses can be found when CCP records are examined during audits. For example, if a dash (−) is recorded on the monitoring sheet where a 'yes' or 'no' answer is expected, does this mean that there is a 'no' answer or that the CCP was not checked or that production was not running? Careful consideration of what information is needed from monitoring to demonstrate that the CCP is operating within its critical limits will help to ensure good design of monitoring record sheets. This will need to be supported by effective training to ensure that all CCP monitors are completing the records accurately and that this is being verified by supervisors on an ongoing basis.

Although it is not necessary to always develop completely new record-keeping formats if suitable record sheets are already in place, it is important to make sure that the existing formats are capable of capturing the required information in a structured way. Consideration should also be given as to how records will be archived so that they can be readily identified at a later stage. Where electronic monitoring systems are to be used, it is important that the system can be proved as effective, including challenging to show that only trained CCP monitors can input data, that this can only be done in real time and that the data can be archived securely and cannot be changed/overridden after the event.

Where HACCP is regulated, some companies keep HACCP records completely separate from other production records for ease of regulatory inspection. It is up to each individual company to decide the best approach for record-keeping and retention for its business; however, the important point is to be able to demonstrate that food safety has been maintained during production, so the chosen approach must achieve this at a minimum.

13.2.5   HACCP-required activities

HACCP-required activities are the items that need to be actioned to support implementation of the HACCP plan. This will include prerequisite programme elements that need to be developed and implemented to strengthen or fill gaps in existing prerequisite programmes (Chapter 9), plus any process, equipment and/or process area modifications that have been identified as essential to support the HACCP plan.

As discussed in Chapters 9 and 10, it is important to perform a prerequisite programme gap analysis and to implement appropriate, formally managed prerequisites as a cornerstone for food safety. Any gaps identified at that stage must now be put in place plus further necessary amendments identified during the HACCP team's deliberations must be rectified. Since the HACCP plan can only be effective if these prerequisite programmes are also working in practice, it is normal to perform a completion check on prerequisites at the time of HACCP plan implementation.

Any other modifications identified as necessary for food safety by the HACCP team must also be implemented at this stage. This could include steps to tighten protection of product from the processing environment, such as additional/modified covers for production equipment, or removal of redundant plant and equipment, e.g. pipe work.

13.2.6 Verification of implementation

Verification of HACCP plan implementation is a vital step in achieving effective control of food safety. This forms the initial verification of the HACCP plan as a baseline for continued HACCP system verification, which will progress as part of the food safety system maintenance procedures (Chapter 14). Verification of implementation is normally done by auditing the system immediately after implementation, i.e. in the first few days. The audit should be performed by trained auditors who are independent from the system, both in terms of its development and its day-to-day management. This is best managed by trained systems auditors, who can challenge all aspects of the documented HACCP plan, the related work instructions and SOPs, and the items on the detailed implementation activity list. All findings from the audit must be documented, and necessary corrective action must be taken immediately such that the HACCP plan implementation status is confirmed.

13.2.7 Handover to operations staff

Although operations staff will clearly be involved in the steps of implementation (Figure 13.1), formal handover for the day-to-day running of the HACCP plan is necessary so that the requisite staff take ownership of the system. Like all other aspects of HACCP implementation, this needs to be carefully planned, such that sufficient personnel resources are available to take on HACCP management and that all personnel involved receive appropriate training for their roles.

When planning the handover, it will be necessary to review job descriptions for staff involved and to add their new CCP management responsibilities where appropriate. This should tie in with the work instructions/SOPs and training given to staff. It is important that operations staff understand the importance of the HACCP system and the crucial role that they play in its ongoing effectiveness. It is equally crucial that they are confident in their abilities to manage the CCPs and take necessary corrective action, and this should be verified following training. It is useful for members of the HACCP team and/or food safety team to work closely with operations staff for a short period after implementation to help build up confidence levels. This will help both to support personnel in taking the necessary actions at CCPs and to reinforce the food safety requirements.

13.3 CONSIDERATIONS FOR IMPLEMENTING UPDATES AND CHANGES TO AN EXISTING HACCP SYSTEM

A common failure in HACCP systems is the failure to keep the HACCP plan up-to-date through controlled amendment. This is often clear from HACCP audit, where process and ingredient changes may become immediately obvious when following through the process flow diagram, and changes in control procedures are detected when assessing the records of the working HACCP system. The need to review and update existing HACCP plans is discussed further in Chapter 14; however, it is important to consider the implementation of these amended HACCP plans at this stage in the HACCP application process.

The approach for implementing updates and changes to existing HACCP systems may be less formal than the stepwise approach discussed earlier; however, it is essential to make sure that these changes are working effectively before passing the responsibility for their day-to-day management over to operations personnel. Any HACCP plan amendments that affect monitoring, record-keeping and/or corrective action will require retraining of the appropriate CCP management personnel, and may necessitate further changes to job descriptions, work instructions/SOPs, HACCP record-keeping sheets, etc. It is easy for items to be overlooked, so a checklist of specific points, similar to the detailed activity list used for implementation of new HACCP plans, will be beneficial.

13.4 CONCLUSIONS

The implementation of new HACCP plans or HACCP plan amendment is an essential aspect of food safety management, which requires careful planning and attention to detail. HACCP plan implementation needs to be built on strong prerequisite foundations, therefore verification that necessary prerequisites are in place and working plays an important role. Training of operations personnel and building confidence in their ability to manage CCPs are essential parts of the handover from the HACCP team's development stage to the day-to-day operation of HACCP in practice. This requires appropriate support from the HACCP team and company management, i.e. the 'promotional managers', in addition to constant reinforcing of the importance of food safety requirements.

14 Maintaining a food safety programme

14.1 INTRODUCTION

In any food business, it is essential that the facility, ingredients, processes and products are managed effectively on an ongoing basis to ensure that only safe food is produced and served for consumption. The food safety programme elements – safe design, prerequisite programmes and HACCP – supported by effective management practices, will allow the requirement for safe food to be achieved. These essential requirements for ongoing food safety need to continue in the operations as implemented; they are 'living' systems and must not be allowed to become out of date. This is very important as any weak links in food safety programmes can have catastrophic results. Referring back to Chapter 2, we can see that the lack of, or weaknesses in, maintenance of food safety programmes is a major reason for failure. Several high-profile food safety incidents have involved large companies who had food safety programmes, and this demonstrates the importance of challenging the controls and keeping systems up to date. Going forward, it is important to assess regularly whether the systems are working and therefore demonstrate that there is continuous control of food safety.

In this chapter, the essential requirements for food safety programme maintenance are discussed. This will include consideration of prerequisite programmes, HACCP and management requirements along with the tools and approaches that can be used to maintain an effective food safety programme.

14.2 WHAT IS FOOD SAFETY PROGRAMME MAINTENANCE?

Maintenance of a food safety programme requires several key fundamentals:

- Challenging the effectiveness of the programme elements
- Ensuring that the programme remains up-to-date, both with the ingredients, processes and operations on site, and with changing knowledge on food hygiene and food safety hazards
- Making certain that the programme remains suitable, both for the provision of adequate hygiene foundations and for the effective control of all relevant food safety hazards

Food Safety for the 21st Century, First Edition By Carol A. Wallace, William H. Sperber and Sara E. Mortimore
© 2011 Carol A. Wallace, William H. Sperber and Sara E. Mortimore

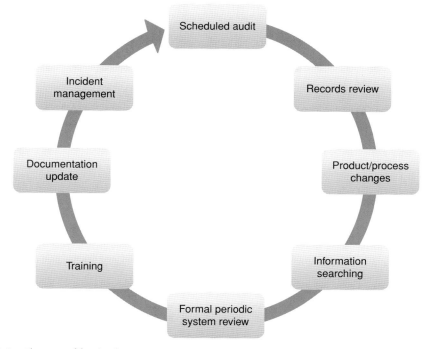

Fig. 14.1   Elements of food safety programme maintenance.

Effective food safety programme maintenance, therefore, requires the application of a range of different techniques and approaches and the involvement of personnel from different roles and areas of the operation. Tools will include audit and management review alongside a variety of specific test procedures, and it will be important for personnel to have appropriate skills, e.g. in auditing and information searching/update. The key elements of food safety programme maintenance are illustrated in Figure 14.1. These elements provide a similar ongoing cycle to the 'plan-do-check-act' process, discussed in Chapter 9 (Figure 9.1).

14.3   RESPONSIBILITY FOR FOOD SAFETY PROGRAMME MAINTENANCE

The responsibility for food safety programme maintenance lies at the food business level. Senior managers need to be involved, both as part of management review and to provide commitment and confirmation of resources for all required maintenance elements. It is important to promote an open environment and have regular discussions about food safety programme operation. Maintenance will be less effective if personnel feel that they are working in a 'blame culture' where results of maintenance elements such as audit are used to criticise staff and work practices rather than as an opportunity for continuous improvement and strengthening of the systems.

Effective maintenance requires a multidisciplinary approach, involving personnel from all different levels of the organisation. Members of the food safety and HACCP teams will be involved in certain aspects of maintenance, particularly keeping the systems up-to-date through a proactive approach to ensuring continued suitability of food safety programme elements. Managers, supervisors and line workers also have important roles to play in challenging the systems and keeping them up to date with the operations. Personnel with audit skills and responsibilities will play a key role in programme verification, and the knowledge and skills of engineering and R&D personnel will be vital, both in highlighting potential changes to products and processes and in ensuring the ongoing capability of processes to control food safety hazards.

14.4 MAINTENANCE OF PREREQUISITE PROGRAMME ELEMENTS

Part of the requirement to formalise prerequisite programmes is the need to ensure that these are maintained and are therefore working effectively over time. This is essential if prerequisite programmes are to provide the safe environmental foundations that are needed for safe food manufacture. In order to challenge the effectiveness of prerequisite programmes, audit and inspection are widely used tools that can determine if the prerequisites are working as intended and identify weaknesses in their operation. Regular hygiene inspections or audits are used by many food companies to assess their prerequisite programmes. These should include visual inspections of the condition of facilities, plant and equipment as well as examination of prerequisite programme records and documentation, and interviewing members of staff with prerequisite responsibilities.

Like all other parts of the food safety programme, it is important that the prerequisite programme elements are kept up to date with best practice, and so food companies need to keep abreast of new information on prerequisites and potential improvements that could be made. Where prerequisite programme elements are amended, there will be a need to retrain staff in the new procedures and it will also be important to include prerequisites in refresher training sessions for all staff. Updates to prerequisite documentation and records will also be needed.

14.5 MAINTENANCE OF HACCP SYSTEM ELEMENTS

After implementation (Chapter 13), the HACCP verification procedures identified in Step 11 of the HACCP logic sequence (Codex, 2009b) need to commence. A HACCP system will only achieve its purpose in managing food safety if its effectiveness is regularly **challenged** and it is **kept up to date** through continuous maintenance. This will be achieved by a variety of different activities, including audit, records review, testing and information searching.

14.5.1 HACCP verification activities

Key HACCP verification activities will include:

- *Audit of the system to check that it is working correctly:* This should be planned so that the whole system is audited at least annually, however more frequent assessment is recommended for ongoing control and this can be achieved by looking at smaller sections of the system on a regular basis. Several published tools are available to assist in HACCP audit (see Section 14.6).

- *Review of records:*
  - *HACCP records.* It is a requirement of HACCP that monitoring records are reviewed by a responsible reviewing 'official'. This is normally a supervisor or manager who can check that monitoring has been done, that corrective action has been taken where necessary, and can look for trends over time.
  - *Customer complaint records.* As for prerequisite programmes, review of external data, such as any customer complaint records, is also very useful in HACCP verification. Trending of customer complaint data can give information on the success of control measures and also give indications of where the system may be drifting towards loss of control.
  - *Product and materials test records.* Although not useful in monitoring due to the time required to obtain results and the variability of microbial contamination within foodstuffs, routine microbiological tests on products and ingredients can provide useful information for HACCP verification. This can be particularly informative if data are reviewed and trended over time as it can give confidence in the effectiveness of CCPs for microbiological hazards.
  - *Calibration activities and records.* Equipment used for CCP and prerequisite programmes (PRP) monitoring must be regularly calibrated for accuracy; therefore, reviewing calibration records will also be important.

The above verification activities are often managed within the framework of a structured quality management system, such that they are performed regularly according to a defined schedule.

14.5.2 HACCP maintenance activities

The maintenance activities for HACCP are largely based on keeping the HACCP plans current and suitable for control of all relevant significant food safety hazards. This will include the following:

- **Keeping up to date with product and process developments.** Where there are new products, new ingredients or new processes, it is important to keep the working HACCP plan up to date. This involves checking to make sure the existing HACCP plan is still valid and making amendments where necessary for continued control. If the development is a new variant of an existing item with an identical process, then it is likely that no changes are required unless there are new raw material hazards to address or the product intrinsic factors have changed as a result of a reformulation. However, if it is a completely new process or involves amendments to existing process activities, then a review of current controls will be required and it is likely that a new HACCP study will be needed. With such changes it can be expected that the process flow diagrams will need to be updated at the very least, with accompanying hazard analysis for any new/amended activities. Depending on whether any new significant hazards are present, and assuming that they can be controlled, then new CCPs or CCP amendments may also be needed, and this will obviously have resulting requirements for monitoring and corrective action procedures. Changes to HACCP plans will themselves need to be validated prior to implementation (Chapter 13) and will then pass into the ongoing maintenance cycle.

    It will be easier to capture changes to products and processes if there is regular communication with personnel involved in new initiatives such as product developers and engineers. Meetings between HACCP team leaders or senior technical personnel and development personnel are therefore vital, and this can be done within the framework of regular management meetings, perhaps via a food safety team. A formal mechanism for approval is also recommended such that the proposed changes cannot be made until food safety has been reviewed and necessary amendments to HACCP plans have been built into the operation. This will likely require a documented sign-off procedure involving senior staff as signatories and could tie in with the product safety assessment process (Chapter 11).

- **Periodic review of HACCP system elements.** In addition to reviewing suitability in response to changes to the products and processes, it is also important to perform periodic reviews of the HACCP system. This allows the system to be challenged for suitability to control all relevant significant food safety hazards, and new information gained from the information searching activities (see below) should be built into this review. Even where the HACCP plans are being updated more regularly due to product and process changes, it is recommended that a periodic formal review should be timetabled as an independent check that the defined systems are valid. This should be done at least annually in all food operations.

- **Information searching.** It is vital to keep up to date with information that might suggest the need to strengthen or amend the HACCP plans. Information may come from a variety of sources and personnel need to be given responsibility for horizon scanning to ensure that all appropriate information is identified and actioned. Specific searching requirements will be needed for the following:

  ○ *Identifying and assessing new hazard information.* As discussed in Chapter 5, a number of microbiological hazards (e.g. *Escherichia coli* O157:H7) have emerged in recent years that were previously unknown within the food industry. Similarly, information changes on what is known about potential chemical hazards and their toxicity to human populations. This necessitates keeping up to date on all potential hazards, particularly with relevance to the segments of the food chain that are affected.

  ○ *Learning from other's failures for impact on own system.* Unfortunately, foodborne illness incidents and outbreaks do happen and are generally well reported, at least in the media. It can be more difficult to find trustworthy information on the underlying causes of these incidents and outbreaks; however, many investigating authorities do publish findings either in peer-reviewed journals or on public authority websites, e.g. UK Health Protection Agency (http://www.hpa.org.uk/) and US Centre for Disease Control and Prevention (http://www.cdc.gov/).

  ○ *Regulatory changes.* The food business also needs to keep up to date with any new legislation or guidance from regulatory authorities on best practices for dealing with new food safety concerns. This should include review of information available from authorities in the countries of materials sourcing, manufacturing and sale, as requirements may differ.

- **Training.** Ongoing training is needed for the successful operation of any HACCP system. This will include training of any new staff in HACCP at a level appropriate to their position in the business. For example, new line operators may need training in CCP monitoring and corrective action, whilst new HACCP team members will need detailed training in the application of HACCP principles. Regular refresher training sessions are also important to keep focus on HACCP as an essential element of the food safety management programme.

14.6   USE OF AUDIT FOR SUCCESSFUL FOOD SAFETY SYSTEM MAINTENANCE

The verification and maintenance requirements discussed above, both for prerequisite programmes and HACCP, necessitate the application of a range of management tools and approaches. Audit is an important technique that can be used both in validation and verification of HACCP and in the assessment of prerequisite programme effectiveness. In HACCP validation, audit can be used to check for deficiencies in the HACCP plan. In verification, audit is used to check that the working system is in compliance with the requirements of the HACCP plan. Similarly, audit can be used to check suitability of newly developed prerequisite programme elements and also to establish ongoing compliance with specified programmes in practice.

The *Codex HACCP principles and guidelines for their application* (Codex, 2009b) are often used as criteria to audit against. Because the HACCP principles are very succinct, it is difficult to audit solely against their requirements, and expert judgement and experience are needed so that the auditor can identify what is acceptable or deficient practice. The Codex HACCP guidelines give more detail; however, as guidelines, these are not mandatory statements, so it would be difficult for an auditor to insist that compliance was required – this is why the development of HACCP-based ISO 22000 was an important development for the food industry (see later). In addition, the scope of the Codex HACCP documents is limited to HACCP rather than the broader requirements of a food safety management programme. The *Codex General Principles of Food Hygiene* (Codex, 2009a) would also need to be used to assess prerequisite programmes, and these documents together would be the minimum criteria used to determine if a food safety programme had been developed in line with international principles and guidelines. In addition to evaluation of an entire food safety programme, audit will often be performed at a much more specific level, e.g. to measure the effectiveness of a particular HACCP plan or prerequisite programme element in practice.

14.6.1 Audit definitions

It is useful to consider some key audit definitions. The following come from BS EN ISO 19011: 2002 *Guidelines for Quality and/or Environmental Management Systems Auditing* (ISO, 2002):

Audit: systematic, independent and documented process for obtaining audit evidence and evaluating it objectively to determine the extent to which the audit criteria are fulfilled.
[Key points here are the independence of audit, such that it is a view from 'outside' the system, and that it is performed systematically so that no important aspects are missed.]
Audit criteria: sets of policies, procedures or requirements. Audit criteria are used as a reference against which the actual situation is compared.
[For example, an audit could be done using the HACCP principles, a specific HACCP plan or sets of prerequisite requirements as the audit criteria. The term 'Audit Standard' (or just Standard) is also widely used in quality and food safety management. A Standard is a document that pulls together the audit criteria that are necessary for a particular purpose. Standards are widely used by industry and governments throughout the world and range from documents specifying technical criteria for products to requirements for management systems.]
Audit evidence: records, statements of fact or other information, which are relevant to the audit criteria and verifiable.
[Audit evidence is the information that needs to be collected to back up the auditor's conclusions about the suitability and effectiveness of the system being audited.]
Audit findings: results of the evaluation of the collected audit evidence against audit criteria.
[This normally requires a judgement from the auditor, using the audit evidence found to evaluate whether the audit criteria have been met.]

Auditee: organisation being audited.

[Depending of the nature of the audit, this could be a company, manufacturing site, department, process area or team responsible for application of a specific HACCP plan or prerequisite programme element.]

Auditor: a person with the competence to conduct an audit.

[Competence is a crucial point here: all auditors of food safety programmes must have the required training, skills and experience to determine effectiveness of the programme elements that they are required to audit.]

14.6.2   The auditor and audit skills

Food safety audits are carried out by people and, as such, rely on the skills, experience and judgement of the individual concerned. Whilst audit skills can be taught, some people have more of a natural flair for this type of activity than others. The competence of food safety auditors is related to a range of aspects, including education, training and experience; personal attributes; general knowledge and skills; and food safety (including HACCP) specific skills. Figure 14.2 outlines the essential elements of auditor competence.

Audit skills training will be required to provide personnel (who should already have relevant education, knowledge and experience) with the ability to challenge the food safety programme (see Chapter 9). Skills required for successful audits in any field include the following:

- Organisational skills
- Information handling and sampling skills
- Interview techniques

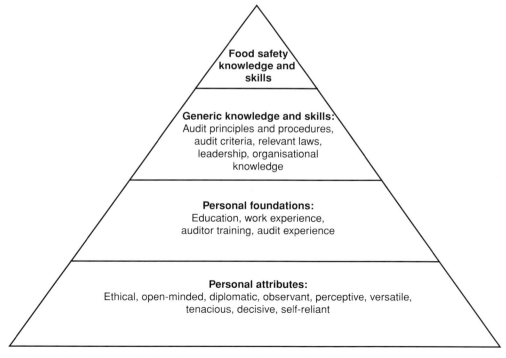

Fig. 14.2   Essential elements of auditor competence. (Adapted from ISO, 2002.)

Audit skills are required to collect information and evidence to:

- understand the operation of the systems,
- establish whether the requirements of the standard are met and
- determine how effectively the system operates.

This is achieved through interviewing, sampling and testing of information. Interviews are used as a two-way information exchange process, and questioning needs to be efficient to establish what is happening in practice. Types of questions used in audit would include the following:

- Open questions – these are most useful to allow the auditee to describe the situation in his/her own words.
- Closed questions – useful for confirming understanding.
- Hypothetical questions – useful for collecting information about how personnel would handle unusual events, such as CCP deviation.
- Leading questions – these are to be avoided in the audit situation as they give the auditee information about the expected response and may, therefore, influence the credibility of the audit findings.

Also important to the auditor are 'observational and listening' skills, including the following:

- Observational skills, e.g. reading the non-verbal response
- Understanding body language and facial expressions
- Keeping eye contact with the interviewee
- Being aware of special positioning and cultural issues, e.g. respecting personal space
- Timing of questions, ensuring that the auditee's response is heard and understood before moving on

These types of skills are often gained through auditor skills training. Food safety auditors who want to attend formal audit skills training normally attend courses on quality or food safety systems auditing to ISO standards, e.g. ISO 9001 (ISO, 2008) or ISO 22000 (ISO, 2005a) or equivalent.

14.6.3   Audit checklists

Auditors normally use a checklist or aide memoir to assist in structuring the audit. Checklists are valuable audit tools that can be used both to ensure that the important points are covered and to record the findings. In their simplest form, checklists may be a table (Table 14.1).

In preparation for the audit, the auditor lists the main points to be covered in the left-hand column, identifies the approach they will take to address these points in the central

Table 14.1   An example of an audit checklist.

Points to be assessed, e.g. process area	Considerations, questions and points to raise	Auditor's findings

Source: Adapted from Mortimore and Wallace (1998).

column – for example, whom will I talk to, what data will I look at, what questions will I ask, etc. – and uses the final column to record the findings on the day. When recording findings, auditors often develop their own coding systems to make it easier to identify deficiencies or non-conformances from their notes.

It is important that the checklist should not limit flexibility for the auditor to follow audit trails using their judgement. For this reason, very structured checklists are not favoured by some auditors. However, in professional auditing it is important to ensure consistency of approach and application of the standard by individual auditors. In these cases, highly structured checklists detailing each clause of the standard may be used, to ensure that every point is covered by every auditor.

A range of audit checklist examples can be found in the literature, particularly checklist tools for the assessment of HACCP (e.g. Sperber, 1998; Wilkinson and Wheelock, 2004; Wallace *et al.*, 2005).

14.6.4 Use of external audit and certification schemes as part of food safety programme maintenance

In HACCP and food safety, standards from a variety of sources are used. These may be legislative standards, national/international standards such as BS/EN/ISO, customer-driven, expert group or food industry sector standards. When combined with professional third-party audit and certification, these standards can provide a useful independent measure of food safety programme effectiveness. Examples of external audit and certification schemes for food safety management programmes include the following:

- *Global Food Safety Initiative (GFSI).* The GFSI aims to ensure equivalence between the various standards and auditing schemes (see Chapter 3). This is, therefore, a framework scheme, which evaluates the effectiveness and equivalence of individual schemes against the GFSI requirements. The latest copy of the GFSI Guidance Document is the fifth edition from September 2007, and can be downloaded at http://www.ciesnet.com/2-wwedo/2.2-programmes/2.2.foodsafety.gfsi.asp
- *ISO 22000:2005.* ISO 22000:2005 *Food Safety Management Systems – Requirements for any organisation in the food chain* (ISO, 2005a) is a purpose-designed audit standard based on Codex HACCP principles and also the management requirements for an effective system. This standard includes the requirement to develop PRP elements (ISO, 2005a; Clause 7.2); however, it does not include detail on what is required in each PRP element.
- *PAS 220:2008.* Because of the omission of sufficient detail on PRPs in ISO 22000, additional standards have been developed by industry, e.g. the 'Publicly Available Specification' PAS 220:2008 *Prerequisite Programmes on food Safety for Manufacturing* (BSI, 2008). PAS 220:2008 and ISO 22000:2005 have been taken together by an organisation known as the 'Foundation for Food Safety Certification' to provide an auditable framework covering HACCP, PRPs and management requirements. This has been developed into a new third-party audit scheme called FSSC 22000, which has also been accepted through benchmarking to the GFSI standard.
- *BRC.* The *BRC Global Standard – Food* is a retail-driven standard that was developed in the UK and is used throughout the world. The standard was developed by a group of retailers to help them meet their responsibilities under UK legislation (at the time, the Food Safety Act 1990, United Kingdom Statutory Instruments, 1990) to ensure safety of the supply chain.

This is a detailed standard containing requirements for HACCP, prerequisites, legal control and quality management systems. The latest copy of the standard was published in 2008 and is entitled *BRC Global Standard for Food Safety – Issue 5*. This has been benchmarked and approved against GFSI.

- *Dutch HACCP Code.* The Dutch HACCP Code was put together by a group of HACCP experts working in the Netherlands who recognised the difficulties of auditing against Codex. This standard, therefore, includes Codex HACCP principles and prerequisite programmes but also management requirements and, as such, shows parallels to BS EN ISO 22000:2005. The Dutch HACCP Code can be downloaded free of charge at http://www.foodsafetymanagement.info. It has also been benchmarked and approved against GFSI.

14.7 INCIDENT MANAGEMENT

Whilst food safety management programmes are designed to prevent food safety issues and incidents, it is generally accepted that elements of systems and procedures do fail from time to time and thus it is necessary to develop management programmes to effectively deal with food safety failures. This is usually achieved by the establishment of formal incident management programmes.

In food safety failures, the primary concern must be public health protection, so incident management programmes must include methods to trace, recall and quarantine suspect product, as well as appropriate communication methods and channels to provide essential information and instructions to customers and consumers. The incident management programme's ability to manage incidents should be tested on a regular basis to ensure that consumers will be protected in the event of food safety system failure.

When food safety management failures occur, it is important to understand the cause of the failure, and management tools such as root cause analysis can be used to assist with this activity. This should result in action to strengthen the food safety systems and to prevent the recurrence of the issue in the future. Preventive action to predict the likely failure modes within the food safety management programme should also be considered, e.g. using tools such as 'failure mode and effect analysis' to predict potential weaknesses or 'why, why' analysis to get at root cause and identify means to strengthen the programme elements.

14.8 CONCLUSIONS

To ensure ongoing control of food safety, the prerequisite programmes and HACCP systems need to work together as a cohesive system. The key points to achieve this are:

- Verification of the effectiveness of food safety system elements, using tools such as audit and results review
- Review of system elements and their suitability for food safety, with particular reference to changes in knowledge about food safety hazards and their control and changes to ingredients, products, processes and operating practices at the manufacturing site
- Change control procedures that require formal safety assessment and approval for all proposed changes to ingredients, process activities and products

- Ongoing management and update of system elements
- Training and retraining of staff, including new recruits and temporary personnel
- Incident management programmes, including testing of their ability to protect the consumer

Together with ongoing assessments of safe recipe/process design for all new products (see Chapter 11), and perhaps within the framework of an external, professionally audited certification scheme, the above maintenance procedures will ensure effective functioning of the food safety programmes on an ongoing basis. Thus, the world-class food safety management programme can be achieved and continually improved, providing ongoing assurance of consumer health protection.

References

Agricultural Research Service (ARS), US Department of Agriculture (2007) Pathogen modeling program v. 7.0. http://www.ars.usda.gov/services/software/software.htm (accessed 11 January 2008).

American Public Health Association (APHA) (1972) *Proceedings of the 1971 Conference on Food Protection*. Stock number 1712-0134. US Government Printing Office, Washington, DC.

Anonymous (2004) 47 detained for selling baby-killer milk. http://www.chinadaily.com.cn/english/doc/2004-05/10/content_329449.htm (accessed 19 October 2009).

Azanza, P.P.V. (2006) Philippine foodborne-disease outbreaks (1995–2004). J. Food Saf. **26**, 92–102.

Baird-Parker, A.C. and B. Freame (1967) Combined effect of water activity, pH and temperature on the growth of *Clostridium botulinum* from spores and vegetative cell inocula. J. Appl. Bacteriol. **30**, 420–429.

Behling, R. (2006) 4 Steps to strategic sanitation risk assessment. Food Safety Magazine, The Target Group Inc., Glendale, CA, August/September 2006.

Block, S.S. (1991) Historical review. In: S.S. Block (ed.) *Disinfection, Sterilization and Preservation*, Fourth Edition. Lea and Febiger, Philadelphia, pp. 3–17.

Bouyer, M. (1970) Process for packaging and sterilization of bread. US Patent No. 3,542.568.

Breuer, T. (1999) CDC investigations: the May 1998 outbreak of *Salmonella agona* linked to cereal. Cereal Food World. **44**(4), 185–186.

British Retail Consortium (BRC) (2008) *Global Standard for Food Safety*. Issue 5. TSO, London.

Bruhn, C. M. (2009) Understanding 'green' consumers. Food Technol. **63**, 28–30, 32, 35.

BSI (2008) *PAS 220: 2008, Prerequisite Programmes on Food Safety for Food Manufacturing*. BSI, London.

Buzby, J.C., L.J. Unnevehr and D. Roberts (2008) *Food Safety and Imports: An Analysis of FDA Food Related Import Refusal Reports*. Economic Information Bulletin No. 39 USDA September 2008, http://www.ers.usda.gov/Publications/EIB39 (accessed 23 August 2009).

Byrne, D. (2001) *Consumers Expect Safe Food*. Round table on food quality, safety and production. SPEECH/01/498 London, 29 October 2001.

Campden BRI (2009) *HACCP – A Practical Guide*, Fourth Edition. Guideline No. 42. Campden BRI, Chipping Campden.

Canadian Food Inspection Agency (CFIA) (1998) Prerequisite programs. http://www.inspection.gc.ca/english/fssa/polstrat/haccp/manue/ch2sec1-2e.shtml#2.2 (accessed February 2010).

CAST (1994) *Foodborne Pathogens: Risks and Consequences*. Task Force Report No. 122. Council for Agricultural Science and Technology, Ames, IA, USA.

Centers for Disease Control and Prevention (CDC) (2007a) Preliminary FoodNet data on the incidence of infection with pathogens transmitted commonly through food – 10 states, 2006. MMWR. Morb. Mortal. Wkly. Rep. **56**, 336–339.

Centers for Disease Control and Prevention (CDC) (2007b) Provisional cases of infrequently reported notifiable diseases. MMWR. Morb. Mortal. Wkly. Rep. **55**, 1143.

Centers for Disease Control and Prevention (CDC) (2007c) Provisional cases of selected notifiable diseases. MMWR. Morb. Mortal. Wkly. Rep. **55**, 1398–1403.

Chapman, P. (2006) Maintaining world leadership in food and drink manufacturing. Food Sci. Technol. **20**(2).

Christian, J.H.B. (2000) Drying and reduction of water activity. In: B.M. Lund, T.C. Baird-Parker and G.W. Gould (eds) *The Microbiological Safety and Quality of Food*. Aspen Publishers, Gaithersburg, pp. 146–174.

Chu, F.S. and G.Y. Li (1994) Simultaneous occurrence of fumonisin B$_1$ and other mycotoxins in moldy corn collected from the People's Republic of China in regions with high incidences of esophageal cancer. Appl. Environ. Microbiol. **60**, 847–852.

Chung, K.C. and J.M. Goepfert (1970) Growth of *Salmonella* at low pH. J. Food Sci. **35**, 326–328.

Cleary, B.A. (1995) Supporting empowerment with Deming's PDSA cycle. Empowerment. Organ. **3**(2), 34–39.

Cockey, R.R. and M.C. Tatro (1974) Survival studies with spores of *Clostridium botulinum* Type E in pasteurized meat of the blue crab *Calllinectes sapidus*. Appl. Microbiol. **27**, 629–633.

Code of Federal Regulations (CFR) (1969) Title 21, Part 110. Current good manufacturing practices in manufacturing, packing, or holding food. US Government Printing Office, Washington, DC.

Code of Federal Regulations (CFR) (1996) Title 9, Part 304. Pathogen reduction: Hazard analysis and critical control point (HACCP) systems; Final Rule. US Government Printing Office, Washington, DC.

Code of Federal Regulations (CFR) (1997) Title 21, Part 123. Fish and fishery products HACCP regulation. US Government Printing Office, Washington, DC.

Code of Federal Regulations (CFR) (2001) Title 21, Part 120. Juice HACCP regulation. US Government Printing Office, Washington, DC.

Code of Federal Regulations (CFR) (2002) Part 113. Thermally processed low-acid foods packed in hermetically sealed containers. Part 114. Acidified foods. US Government Printing Office, Washington, DC.

Code of Federal Regulations (CFR) (2003) Title 9, Part 430.4. Control of *Listeria monocytogenes* in post-lethality exposed ready-to-eat products; Final Rule. US Government Printing Office, Washington, DC.

Code of Federal Regulations (CFR) (2008a) Title 21, Part 113. Thermally processed low-acid foods packaged in hermetically sealed containers. Government Printing Office, Washington, DC. http://www.access.gpo.gov/nara/cfr/waisidx_08/21cfr113_08.html (accessed 8 December 2008).

Code of Federal Regulations (CFR) (2008b) Title 21, Part 114. Acidified foods. Government Printing Office, Washington, DC. http://www.access.gpo.gov/nara/cfr/waisidx_08/21cfr114_08.html (accessed 8 December 2008).

Code of Federal Regulations (CFR) (2008c) Title 9, Part 590.570. Pasteurization of liquid eggs. Government Printing Office, Washington, DC. http://edocket.access.gpo.gov/cfr_2008/janqtr/pdf/9cfr590.570.pdf (accessed 8 December 2008).

Code of Federal Regulations (CFR) (2008d) Title 9, part 590.575. Heat treatment of dried whites. Government Printing Office, Washington, DC. http://edocket.access.gpo.gov/cfr_2008/janqtr/pdf/9cfr590.575.pdf (accessed 8 December 2008).

Codex Alimentarius Commission (1993) Recommended international code of practice. General principles of food hygiene. Annex to CAC/RCP 1-1969. Rome.

Codex Alimentarius Commission (2009) Code of hygienic practice for milk and milk products. CAC/RCP-2004. http://www.codexalimentarius.net/web/more_info.jsp?id_sta=10087 (accessed 23 October 2009).

Codex Alimentarius Committee (1999) *Principles and Guidelines for the Conduct of Microbiological Risk Analysis* CAC/GL-30. FAO/WHO, Rome.

Codex (Joint FAO/WHO Food Standards Programme, Codex Alimentarius Commission), Committee on Food Hygiene (1993) *HACCP System and Guidelines for its Application*. Annex to CAC/RCP 1-1969, Rev. 4 in Codex Alimentarius Commission Food Hygiene Basic Texts. Food and Agriculture Organization of the United Nations, World Health Organization, Rome.

Codex (Joint FAO/WHO Food Standards Programme, Codex Alimentarius Commission) (1997) Recommended international code of practice. General principles of food hygiene. Annex to CAC/RCP 1-1969. Revision 3. Rome.

Codex (Joint FAO/WHO Food Standards Programme, Codex Alimentarius Commission), Committee on Food Hygiene (2009a) *Recommended International Code of Practice, General Principles of Food Hygiene*. CAC/RCP 1-1969, Rev 4 (2003) in Codex Alimentarius Commission Food Hygiene Basic Texts, 4th Edition. Food and Agriculture Organization of the United Nations, World Health Organization, Rome.

Codex (Joint FAO/WHO Food Standards Programme, Codex Alimentarius Commission) (2009b) Hazard analysis and critical control point (HACCP) system and guidelines for its application. *Food Hygiene Basic Texts*, Fourth Edition. Joint FAO/WHO Food Standards Programme, Food and Agriculture Organization of the United Nations, Rome. http://www.fao.org/docrep/012/a1552e/a1552e00.htm (accessed July 2010).

Commission of the European Communities (Commission) (2002) Recommendation on the reduction of the presence of dioxins, furans, and PCBs in feeding stuffs and foodstuffs. Official. J. Eur. Community. **L 67**, 69–73.

Congressional Research Service (2008) CRS Report for Congress: *Food and Agriculture Imports from China* (Becker), CRS Report Order Code RL34080.

Cook, F.K. and B.L. Johnson (2009) Microbiological spoilage of cereal products. In: W.H. Sperber and M.P. Doyle (eds) *Compendium of the Microbiological Spoilage of Foods and Beverages*. Springer, New York, pp. 223–244.

Cormier, R.J., M. Mallet, S. Chiasson, H. Magnusson and G. Valdimarsson (2007) Effectiveness and performance of HACCP-based programs. Food Control **18**, 665–671.

Csonka, L.N. (1989) Physiological and genetic responses of bacteria to osmotic stress. Microbiol. Rev. **53**, 121–147.

Davies, W.P., R.N. Baines and J.X. Turner (2006) Red alert: food safety lessons from dye contaminants in spice supply. Acta Hort. **699**, 143–150.

Deak, T. and L.R. Beuchat (1996) *Handbook of Food Spoilage Yeasts*. CRC Press, Boca Raton.

Department of Energy (DOE) (2009) United States renewable energy consumption in the nation's energy supply, 2008. http://www.eia.doe.gov/fuelrenewable.html (accessed 21 August 2009).

DeVries, J.W. (2006) Chasing 'Zero' in chemical contaminant analysis. Food Saf. Mag. **2**(4) The Target Group CA 91201.

de Waal, C.S., G. Hicks, K. Barlow, L. Alderton and L. Vegosen (2006) Foods associated with foodborne illness outbreaks from 1990 through 2003. Food Prot. Trends. **26**, 466–473.

Dominguez, S.A and D.W. Schaffner (2007) Development and validation of a mathematical model to describe the growth of *Pseudomonas* spp. in raw poultry stored under aerobic conditions. Int. J. Food Microbiol. **120**, 287–295.

Downes, F.P. and K. Ito (eds) (2001) *Compendium of Methods for the Microbiological Examination of Foods*, Fourth Edition. American Public Health Association, Washington, DC.

Doyle, M.P. (ed.) (1989) *Foodborne Bacterial Pathogens*. Marcel Dekker, Inc., New York.

Doyle, M.P. (1991) Evaluating the potential risk from extended-shelf-life refrigerated foods by *Clostridium botulinum* inoculation studies. Food Technol. **45**, 154–156.

Doyle, M.P., L.R. Beuchat and T.J. Montville (2001) *Food Microbiology: Fundamentals and Frontiers*, Third Edition. ASM Press, Washington, D.C.

Economic Research Service, USDA (ERS) (2009) U.S. Processed food trade. http://www.ers.usda.gov/ (accessed 23 August 2009).

EC (2004) European Commission (EC) regulation No. 852/2004 of the European Parliament and of the Council of 29 April 2004 on the hygiene of foodstuffs. http://www.eur-lex.europa.eu/LexUriServ/LexUriServ.do?uri=OJ:L:2004:139:0001:0054:EN:PDF (accessed 1 October 2009).

Falkenmark, M. and J. Rockström (2005) *Balancing Water for Humans and Nature*. Earthscan, London.

Farkas, J. (2001) Physical methods of food preservation. In: M.P. Doyle, L.R. Beuchat and T.J. Montville (eds) *Food Microbiology: Fundamentals and Frontiers*, Second Edition. ASM Press, Washington, DC, pp. 567–591.

Foegeding, P.M. and F.F. Busta (1991) Chemical food preservatives. In: S.S. Block (ed.) *Disinfection, Sterilization and Preservation*, Fourth edition. Lea & Febiger, Philadelphia, pp. 802–832.

Food and Agriculture Organization (FAO) and World Health Organization (WHO) (2002) Risk assessment of *Listeria monocytogenes* in ready-to-eat foods. http://www.who.int/foodsafety/publications/micro/mra_listeria/en/print.html (accessed 15 September 2009).

Food and Drug Administration (FDA, USA) (1997) *Grade 'A' Pasteurized Milk Ordinance*. Publication no. 229. US Department of Health and Human Services, Washington, D.C.

Food and Drug Administration (FDA, USA) (2000) Action levels for poisonous or deleterious substances in human foods and animal feed. http://www.fda.gov/Food/GuidanceComplianceRegulatoryInformation/GuidanceDocuments/ChemicalContaminantsandPesticides/ucm077969.htm (accessed 24 June 2010).

Food and Drug Administration (FDA, USA) (2005a) Managing food safety: a manual for the voluntary use of HACCP principles for operators of foodservice and retail establishments. http://www.fda.gov/Food/FoodSafety/RetailFoodProtection/ManagingFoodSafetyHACCPPrinciples/Operators/default.htm (accessed February 2010).

Food and Drug Administration (FDA, USA) (2005b) Food code. Temperature and time control, part 3–501 (pp. 79–88). http://www.fda.gov/food/foodsafety/retailfoodprotection/foodcode/foodcode2005/default.htm (accessed 23 October 2009).

Food and Drug Administration (FDA, USA) (2006) Grade 'A' pasteurized milk ordinance. Publication No. 229. US Department of Health and Human Services. http://www.cfsan.fda.gov/~ear/pmo05toc.html (accessed 8 December 2008).

Food and Drug Administration (FDA, USA) (2007) Interim melamine and analogues safety/risk assessment. http://www.fda.gov/Food/FoodSafety/FoodContaminantsAdulteration/ChemicalContaminants/Melamine/ucm164658.htm (accessed 24 June 2010).

Food and Drug Administration (FDA, USA) (2009) Food defense acronyms, abbreviations and definitions. http://www.fda.gov/Food/FoodDefense/Training/ucm111382.htm (accessed January 2010).

Food Protection Committee (1964) *An Evaluation of Public Health Hazards from Microbiological Contamination of Foods*. Pub.1195. National Academy of Sciences, Washington, DC.

Food Safety and Inspection Service (FSIS) (2001) Risk assessment of the public health impact of *Escherichia coli* O157:H7 in ground beef. http://www.fsis.usda.gov/OPPDE/rdad/FRPubs/00-023N/00-023NReport.pdf (accessed 15 September 2009).

Food Standards Agency (2003) Reporting limits for nitrofuran and chloramphenicol residues harmonized. http://www.food.gov.uk/news/newsarchive/2003/jun/nitrofuranschloramphenicol (accessed 21 October 2009).

Foster, E.M. (1971) The control of *Salmonella* in processed foods. A classification system and sampling plan. J. Assoc. Off. Anal. Chem. **54**, 259–266.

Frazao, E., B. Meade, A. Regmi (2006) Converging patterns in global food consumption and food delivery systems. www.ers.usda.gov/amberwaves (accessed February 2008).

Frazier, W.C. (1958) *Food Microbiology*. McGraw-Hill Book Company, Inc., New York.

Friedman, M. (2003) Chemistry, biochemistry, and safety of acrylamide. A review. J. Agric. Food Chem. **51**, 4504–4526.

Friedman, M. (2007) *Enterobacter sakazakii* in food and beverages (other than infant formula and milk powder). Int. J. Food Microbiol. **116**, 1–10.

Gaze, R. (ed.) (2009) *HACCP: A Practical Guide*, Fourth Edition. Guideline G42. Campden BRI, UK

Gerba, C.P., J.B. Rose and C.N. Haas (1996) Sensitive populations: who is at greatest risk? Int. J. Food Microbiol. **30**, 113–123.

Global Food Safety Initiative (2007) *Guidance Document*, 5th Edition. CIES – the Food Business Forum, Paris, France, September 2007.

Global Trade Information Services (GTIS) (2009) Global trade atlas. http://www.gtis.com (accessed 17 July 2009).

Gorris, L. and Y. Motarjemi (2009) Training, education and capacity building in food and water safety and in nutrition. New Food **1**, 47–50.

Gould, G., P. Franken, P. Hammer, B. Mackey and F. Shanahan (2005) *Mycobacterium avium* subsp. *paratuberculosis* (AP) and the food chain. Food Prot. Trends. **25**, 268–297.

Graves, M., A. Smith and B. Batchelor (1998) Approaches to foreign body detection in foods. Trends. Food Sci. Technol. **9**, 21–27.

Greig, J.D., E.C.C. Todd, C.A. Bartelson and B.S. Michaels (2007) Outbreaks where food workers have been implicated in the spread of foodborne disease. Part 1. Description of the problem, methods, and agents involved. J. Food Prot. **66**, 130–161.

Griffiths, C. (1994) Application of HACCP to food preparation practices in domestic kitchens. Food Control **5**, 200–204.

Griffiths, C. (2009) Guest Editorial. *Is it a case of 'plus ca change, plus c'est la meme chose'?* Perspectives in Public Health. J.R. Soc Public Health **129**(2).

Gurtler, J.B., J.L. Kornacki and L.R. Beuchat (2005) *Enterobacter sakazakii*: a coliform of increased concern in infant health. Int. J. Food Microbiol. **104**, 1–35.

Guynot, M.E., A.J. Ramos, V. Sanchis and S. Marin (2005) Study of benzoate, propionate, and sorbate salts as mould spoilage inhibitors in intermediate moisture bakery products of low pH (4.5–5.5). Int. J. Food Microbiol. **101**, 161–168.

Haley, M.M. (2001) Potential U.S. implications of food-and-mouth disease. Food Technol. **55**, 228.

Hathaway, S. (1999) Management of food safety in international trade. Food Control. **10**, 247–253.

Hefle, S.L. and S.L. Taylor (1999) Allergenicity of edible oils. Food Technol. **53**, 62–70.

Herbert, R.A. and J.P. Sutherland (2000) Chill storage. In: B.M. Lund, T.C. Baird-Parker and G.W. Gould (eds) *The Microbiological Safety and Quality of Food*. Aspen Publishers, Gaithersburg, Maryland, pp. 101–121.

Hennessy, T.W., C.W. Hedber, L. Slutsker, et al. (1996) A National outbreak of *Salmonella* Enteritidis infections from ice cream. N. Engl. J. Med. **334**(20).

Higgs, W. and R. Fielding (2007) Food allergen management. Food Sci. Technol. **21**(3), 35–36.

Howard, J. (1971) The canned menace called botulism. *Life*, September issue.

Hueston, W.D. (2003) Bovine spongiform encephalopathy. In: M.E. Torrence and R.E. Isaacson (eds) *Microbial Food Safety in Animal Agriculture*. Current topics. Iowa State Press, Ames, IA.

Hueston, W.D. and C.M. Bryant (2005) Transmissable spongiform encephalopathies. J. Food Sci. **70**, R77–R87.

Hyman, F.N., K.C. Klontz and L. Tollefson (1993) Food and drug administration surveillance of the role of foreign objects in foodborne injuries. Pub. Health Rep. **108**, 54–59.

Imholte, T.J. and T.K. Imholte-Tauscher (1999) *Engineering for Food Safety and Sanitation*. Technical Institute of Food Safety, Woodinville, Washington. ISBN 0-9671264-0-1.

Institute of Food Science and Technology (IFST) (2007) *Food & Drink – Good Manufacturing Practice: A Guide to its Responsible Management*. Institute of Food Science & Technology, London.

International Commission on Microbiological Specifications for Foods (ICMSF) (1974) *Microorganisms in Foods. Book 2. Sampling for Microbiological Analysis: Principles and Specific Application*. University of Toronto Press, Toronto.

International Commission on Microbiological Specifications for Foods (ICMSF) (1988) *Microorganisms in Foods. 4. Application of the Hazard Analysis and Critical Control Point (HACCP) System to Ensure Microbiological Safety and Quality*. Blackwell Scientific Publications, London.

International Commission on Microbiological Specifications for Foods (ICMSF) (1996) *Microorganisms in Foods. 5. Characteristics of Microbial Pathogens*. Blackie Academic & Professional, London.

International Commission on Microbiological Specifications for Foods (ICMSF) (2002) *Microorganisms in Foods. 7. Microbiological Testing in Food Safety Management*. Kluwer Academic/Plenum Publishers, New York, p. 178.

International Commission on Microbiological Specifications for Foods (ICMSF) (2005) *A Simplified Guide to Understanding and Using Food Safety Objectives and Performance Objectives*. ICMSF, November 2005.

International Life Sciences Institute (ILSI) (1999) Validation and verification of HACCP. ILSI Europe Report Series. http://www.ilsi.org/Europe/Publications/R1999Val_Ver.pdf.

International Organization for Standardization (ISO) (2002) *Guidelines for Quality and/or Environmental Management Systems Auditing*. BS EN ISO 19011:2002.

International Organization for Standardization (ISO) (2005a) *Food Safety Management Systems — Requirements for Any Organization in the Food Chain*. BS EN ISO 22000:2005.

International Organization for Standardization (ISO) (2005b) *Quality Management Systems – Fundamentals and Vocabulary*. BS EN ISO 9000:2005.

International Organization for Standardization (ISO) (2008) *Quality Management Systems – Requirements*. BS EN ISO 9001:2008.

Jackson, L.S., F.M. Al-Taher, M. Moorman, J.W. DeVries, R. Tippett, K.M.J. Swanson, T-J. Fu, R. Salter, G. Dunaif, S. Estes, S. Albillos and S.M. Gendel (2008) Cleaning and other control and validation strategies to prevent allergen cross-contact in food-processing operations. J. Food Prot. **71**, 445–458.

James, S. and C. James (2006) Current food quality, safety and economic aspects of food refrigeration. Food Sci. Technol. **20**(2).

Jay, J.M. (2000) *Modern Food Microbiology*, Sixth Edition. Aspen Publishers, Gaithersburg, MD.

Jiang, J. (2008) China's rage over toxic baby milk. http://www.time.com/printout/0,8816,1842727,00.html (accessed 19 October 2009).

Jones, K.C. and P. de Voogt (1999) Persistent organic pollutants (POP): state of the science. Environ. Pollut. **100**, 209–221.

Joslyn, L.J. (1991) Sterilization by heat. In S.S. Block (ed.) *Disinfection, Sterilization and Preservation*, Fourth Edition. Lea & Febiger, Philadelphia, pp. 495–526.

Kautter, D.A., T. Lilly, Jr, R.K. Lynt and H.M. Solomon (1979) Toxin production by *Clostridium botulinum* in shelf-stable pasteurized process cheese spreads. J. Food Prot. **42**, 784–786.

Kaye, D. (2006) Botulism associated with commercial carrot juice – Georgia and Florida. Clin. Infect. Dis. **43**, III–IV.

Kornacki, J.L. (2009) The missing element in microbiological food safety inspection approaches. Part 2, *Food Safety Magazine*, The Target Group Inc., Glendale, CA, April/May 2009.

Labuza, T.P. and M.K. Schmidl (1985) Accelerated shelf-life testing of foods. Food Technol. **39**, 57–62, 64.

Larebeke, N., L. Hens, P. Schepens, A. Covaci, J. Baeyens, K. Everaert, J.L. Bernheim, R. Vlietinck and G. De Poorter (2001) The Belgian PCB and dioxin incident of January-June, 1999: exposure data and potential impact on health. Environ. Health Perspect. **109**, 265–273.

Leistner, L. and G.W. Gould (2002) *Hurdle Technologies*. Springer-Verlag, New York.

Lehmacher, A., J. Bockemuhl and S. Aleksic (1995) Nationwide outbreak of human salmonellosis in Germany due to contaminated paprika and paprika-powdered potato chips. *Epidemiology and Infection*, **115**, 501–511. Cambridge University Press, UK.

Levy, R.V. and T.J. Leahy (1991) Sterilization filtration. In: S.S. Block (ed.) *Disinfection, Sterilization and Preservation*, Fourth Edition. Lea & Febiger, Philadelphia, pp. 527–552.

Lewis, M. and N. Heppell (2000) *Continuous Thermal Processing of Foods: Pasteurization and UHT Sterilization*. Aspen Publishers, Gaithersburg, Maryland.

Lim, J.Y., J. Li, H. Sheng, T.E. Besser, K. Potter and C.J. Hovde (2007) *Escherichia coli* O157:H7 colonization at the rectoanal junction of long-duration culture-positive cattle. Appl. Environ. Microbiol. **73**, 1380–1382.

Liston, J. (2000) Fish and shellfish poisoning. In: B.M. Lund, T.C. Baird-Parker and G.W. Gould (eds) *The Microbiological Safety and Quality of Food*. Aspen Publishers, Gaithersburg, Maryland, pp. 1518–1544.

López-Malo, A., S.M. Alzamosa and E. Palou (2005) *Aspergillus flavus* growth in the presence of chemical preservatives and naturally occurring antimicrobial compounds. Int. J. Food Microbiol. **99**, 119–128.

Lund, B.M. (2000) Freezing. In: B.M Lund, T.C. Baird-Parker and G.W. Gould (eds) *The Microbiological Safety and Quality of Food*. Aspen Publishers, Gaithersburg, Maryland.

Lund, B.M. and T. Eklund (2000) Control of pH and use of organic acids. In: B.M. Lund, A.C. Baird-Parker and G.W. Gould (eds) *The Microbiological Safety and Quality of Food*. Aspen Publishers, Gaithersburg, pp. 175–199.

Lund, B.M., T.C. Baird-Parker and G.W. Gould (eds) (2000) *The Microbiological Safety and Quality of Food*. Aspen Publishers, Gaithersburg, Maryland.

Macaloon, T. (2001) HACCP implementation in the United States. In: T. Mayes and S. Mortimore (eds) *Making the Most of HACCP: Learning from Others' Experience*. CRC Press Woodhead Publishing Ltd., Cambridge, UK.

Marth, E.H., C.M. Capp, L. Hasenzahl, H.W. Jackson and R.V. Hussong (1966) Degradation of potassium sorbate by *Penicillium* species. J. Dairy. Sci. **49**, 1197–1205.

McMeekin, T.A., J. Brown, K. Krist, D. Miles, K. Neumeyer, D.S. Nichols, J. Olley, K. Presser, D.A. Ratkowsky, T. Ross, M. Salter and S. Soontranon (1997) Quantitative microbiology: a basis for food safety. Emerg. Infect. Dis. **3**, 541–549.

Mead, P.S., L. Slutsker, V. Dietz, L.F. McCaig, J.S. Bresee, C. Shapiro, P.M. Griffin and R.V. Tauxe (1999). Food-related illness and death in the United States. Emerg. Infect. Dis. **U5U**, 607–625.

Mirandé, A. (2000) Status report on the effect of SE bacteria vaccination in PEQAP flocks (1997–1999). *49th Annual New England Poultry Health Conference*, Portsmouth, New Hampshire.

Mitchell, R.T. (1998) Why HACCP fails. Food Control **9**(2–3), 101.

Miyagishima, K., G. Moy, S. Miyagawa, Y. Motarjemi and F.K. Kaferstein (1995) Food safety and public health. Food Control **6**(5), 253–259.

Morris, J.G. (2000) The effect of redox potential. In: B.M. Lund, T.C. Baird-Parker and G.W. Gould (eds) *The Microbiological Safety and Quality of Food*. Aspen Publishers, Gaithersburg, pp. 235–250.

Mortimore, S. and Y. Motarjemi (2002) Industries needs and expectations to meet food safety. *5th International Meeting: Noordwijk Food Safety and HACCP Forum, 9–10 December 2002*, Netherlands.

Mortimore, S.E. and R.A. Smith (1998) Standardized HACCP training: assurance for food authorities. Food Control **9**(2–3), 141–145.

Mortimore, S. and C. Wallace (1994) *HACCP: A Practical Approach*. Chapman and Hall, London.

Mortimore, S.E. and C.A. Wallace (1998) *HACCP: A Practical Approach*, Second Edition. Aspen Publishers Inc. (Springer), Gaithersburg.

Mortimore, S. and C. Wallace (2001) *Food Industry Briefing Series: HACCP*. Blackwell Science Ltd, Oxford, UK.

Motarjemi, Y. and L. Gorris (2009) *Safety Foremost*. IUFoST dispatch International Food Ingredients No 2 – 2009.

Motarjemi, Y. and F. Käferstein (1999) Food safety, hazard analysis and critical control point and the increase in foodborne disease: a paradox? Food Control **10**, 425–333.

Murphy, P.A., S. Hendrich, C. Landgren and C.M. Bryant (2006) Food mycotoxins: an update. J. Food Sci. **71**, R51–R65.

National Academy of Sciences (NAS) (1960) Conference report: microbiological standards for foods. Pub. Health Rep. **75**, 815–822.

National Advisory Committee on Microbiological Criteria for Foods (NACMCF) (1992) Hazard analysis and critical control point system. Int. J. Food Microbiol. **16**, 1–23.

National Advisory Committee on Microbiological Criteria for Foods (NACMCF) (1997) Hazard analysis and critical control point principles and application guidelines. J. Food Prot. **61**(6), 762–775; http://www.fda.gov/Food/FoodSafety/HazardAnalysisCriticalControlPointsHACCP/ucm114868.htm (accessed February 2010)

National Advisory Committee on Microbiological Criteria for Foods (NACMCF) (1998) Hazard analysis and critical control point principles and application guidelines. J. Food Prot. **61**, 1246–1259.

National Advisory Committee on Microbiological Criteria for Foods (NACMCF) (2006) Requisite scientific parameters for establishing the equivalence of alternative methods of pasteurization. J. Food Prot. **69**(5), 1190–1216.

National Research Council Food Protection Committee (NRC) Subcommittee on Microbiological Criteria (1985) *An Evaluation of the Role of Microbiological Criteria for Foods and Food Ingredients.* National Academy Press, Washington, DC.

National Training Laboratories (no date) *The Learning Pyramid*, NTL, Bethel, Maine, USA. http://www.ntl.org/

Nevas, M., M. Lindström, A. Virtanen, S. Hielm, M. Kuusi, S.S. Arnon, E. Vuori and H. Korkeala (2005) Infant botulism acquired from household dust presenting as sudden infant death syndrome. J. Clin. Microbiol. **43**, 511–513.

The New York Times. (1971) Pillsbury recalls cereal: boxes may contain glass. March 24, p. 27.

Notermans, S. and M. Borgdorff (1997) A global perspective of foodborne disease. J. Food Prot. **60**, 1395–1399.

Olsen, A.R. (1998) Regulatory action criteria for filth ad other extraneous materials. I. Review of hard or sharp foreign objects as physical hazards in foods. Regul. Toxicol. Pharmacol. **28**, 181–189.

Osterholm, M.J. (2006) A modern world and infectious diseases: a collision course. A lecture at Nobel Conference 42, Gustavus Adolphus College. http://www.gustavus.edu/events/nobelconference/2006/osterholm-lecture.cfm (accessed 21 August, 2009).

Panisello, P.J. and P.C. Quantick (2001) Technical barriers to Hazard Analysis Critical Control Point (HACCP). Food Control **12**, 165–173.

Parisi, A.N. and W.E. Young (1991) Sterilization with ethylene oxide and other gases. In: S.S. Block (ed.) *Disinfection, Sterilization and Preservation*, Fourth Edition. Lea & Febiger, Philadelphia, pp. 580–595.

Peariso, D. (2007) A year in foreign material contamination. Food Saf. Mag. **13**(6), 24, 26–27, 52.

Pennington, H. (2009) The public inquiry into the September 2005 outbreak of *E. coli* O157 in South Wales. HMSO. http://wales.gov.uk/ecolidocs/3008707/reporten.pdf?skip=1&lang=en (accessed 9 March 2009).

Peshin, S.S., S.B. Lall and S.K. Gupta (2002) Potential food contaminants and associated health risks. Acta. Pharmacol. Sin. **23**, 193–202.

Pitt, J.I. (1974) Resistance of some spoilage yeasts to preservatives. Food Technol. Austral. **26**, 238–241.

Raczek, N.N. (2005) Food and beverage preservation. In: W. Paulus (ed.) *Directory of Microbicides for the Protection of Materials. A Handbook*. Springer, Dordrecht, pp. 287–304.

Richardson, D.G. and R.G. Hans (1978) Process for preparing food in the package. U.S. Patent No. 4,120,984.

Rocourt, J., G. Moy, K.Vierk and J. Schlundt (2003) *The Present State of Foodborne Disease in OECD Countries*. Food Safety Department, World Health Organization, Geneva. http://www.who.int/foodsafety/publications/foodborne_disease/oecd_fbd.pdf (accessed 22 October 2009).

Ross, A.I.V., M.W. Griffiths, G.S. Mittal and H.C. Deeth (2003) Combining non-thermal technologies to control foodborne microorganisms. Int. J. Food Microbiol. **89**, 125–138.

Ross-Nazzal, J. (2007) 'From farm to fork:' How space food standards impacted the food industry and changed food safety standards. In: S.J. Dick and R.D. Launius (eds) *Societal Impact of Space Flight*. National Aeronautics and Space Administration, Washington, DC, pp. 217–236.

Ross, T. and T.A. McMeekin (1994) Predictive microbiology. Int. J. Food Microbiol. **23**, 241–264.

Route, N. (2001) HACCP and SMEs: a case study. In: T. Mayes and S. Mortimore (eds) *Making the Most of HACCP: Learning from Others' Experience*. CRC Press Woodhead Publishing Ltd, Cambridge, UK.

Reuter (1990) France: Perrier puts cost of benzene scare at $79 million. Reuter Newswire, 10 May, Western Europe.

Sakai, J. (2007) Arsenic contamination lacks one-size-fits-all remedy. http://www.news.wisc.edu/14545 (accessed 22 December 2007).

Schechmeister, I.L. (1991) Sterilization by ultraviolet radiation. In: S.S. Block (ed.) *Disinfection, Sterilization and Preservation*, Fourth Edition. Lea & Febiger, Philadelphia, pp. 553–555.

Schecter, R. and S.S. Arnon (2000) Extreme potency of botulinum toxin. Lancet **355**, 237–238.

Scott, V.N. and K.E. Stevenson (2006) *HACCP a Systemic Approach to Food Safety*. Food Products Association, Washington, DC.

Scott. W.J. (1957) Water relations of food spoilage microorganisms. Adv. Food Res. **7**, 83–127.

Shapton, N.F. (1989) *Food Safety – A Manufacturer's Perspective*. Hobsons Publishing, Cambridge, UK.

Silverman, G.J. (1991) Sterilization and preservation by ionizing irradiation. In: S.S. Block (ed.) *Disinfection, Sterilization and Preservation*, Fourth. Edition. Lea & Febiger, Philadelphia, pp. 566–579.

Sivapalasingam, S., C.R. Friedman, L. Cohen and R.V. Tauxe (2004) Fresh produce: a growing cause of outbreaks of foodborne illness in the United States, 1973 through 1997. J. Food Prot. **67**, 2342–2353.

Smith, J.P., R. Ooraikul, W.J. Koersen, E. D. Jackson and R.A. Lawrence (1986) Novel approach to oxygen control in modified atmosphere packaging of bakery products. Food Microbiol. **3**, 315–320.

Solomon, H.M. and D.A. Kautter (1986) Growth and toxin production by *Clostridium botulinum* in sautéed onions. J. Food Prot. **49**, 618–620.

Sperber, W.H. (1982) Requirements of *Clostridium botulinum* for growth and toxin production. Food Technol. **36**, 89–94.

Sperber, W.H. (1983) Influence of water activity on foodborne bacteria – a review. J. Food Prot. **46**, 142–150.

Sperber, W.H. (1998) Auditing and verification of food safety and HACCP. Food Control **9**(2–3), 157–162.

Sperber, W.H. (1999) The role of validation in HACCP plans. Dairy Food Environ. Sanit. **19**, 920, 912.

Sperber, W.H. (2001) Hazard identification: from a quantitative to a qualitative approach. Food Control **12**, 223–228.

Sperber, W.H. (2005a) HACCP does not work from farm to table. Food Control **16**, 511–514.

Sperber, W.H. (2005b) HACCP and transparency. Food Control **16**, 505–509.

Sperber, W.H. (2006) The John H. Silliker Lecture. Rising from the ocean bottom – the evolution of microbiology in the food industry. Food Prot. Trends **26**, 818–821.

Sperber, W.H. (2008) Organizing food protection on a global scale. Food Technol. **62**, 96.

Sperber, W.H. (2009a) Introduction to the microbiological spoilage of foods and beverages. In: W.H. Sperber and M.P. Doyle (eds) *Compendium of the Microbiological Spoilage of Foods and Beverages*. Springer, New York, pp. 285–299.

Sperber, W.H. (2009b) Microbiological spoilage of acidified specialty products. In: W.H. Sperber and M.P. Doyle (eds) *Compendium of the Microbiological Spoilage of Foods and Beverages*. Springer, New York, pp. 285–299.

Sperber, W.H. and the North American Millers' Association Microbiology Working Group (NAMA) (2007) Role of microbiological guidelines in the production and commercial use of milled cereal grains: a practical approach for the 21st century. J. Food Prot. **70**, 1041–1053.

Sperber, W.H., K.E. Stevenson, D.T. Bernard, K.E. Deibel, L.J. Moberg, L.R. Hontz and V.N. Scott (1998) The role of prerequisite programs in managing a HACCP system. Dairy Food Environ. Sanit. **18**, 418–423.

Statutory Instrument No. 205 (2009) The plastic materials and articles in contact with food (England) regulations 2009. http://www.opsi.gov.uk/si/si2009/uksi_20090205_en_1 (accessed 21 October 2009).

Strachan, N.J.C., G.M. Dunn, M.E. Locking, T.M.S. Reid and I.D. Ogden (2006) *Escherichia coli* O157: burger bug or environmental pathogen? Int. J. Food Microbiol. **112**, 129–137.

Sugiyama, H. and K.H. Yang (1975) Growth potential of *Clostridium botulinum* in fresh mushrooms packaged in semipermeable plastic film. Appl. Microbiol. **30**, 964–969.

Su, H.P., S.I. Chiu, J.L. Tsai, C.L. Lee and T.M. Ping (2005) Bacterial food-borne illness outbreaks in northern Taiwan, 1995–2001. J. Infect. Chemother. **11**, 146–151.

Tanaka, N. (1982) Challenge of pasteurized process cheese spreads with *Clostridium botulinum* using in-process and post-process inoculation. J. Food Prot. **45**, 1044–1050.

Taylor, J. (2002) *Perceptions of HACCP: A Narrative Interview Study*. Food Safety Express Research Information Ltd, Hemel Hempstead, Herts, UK.

Taylor, S.L. and S.L. Hefle (2005) Allergen control. Food Technol. **59**, 40–43, 75.

The Pillsbury Company (1972) *Product Safety Documentation Instructions*. Minneapolis, Minnesota.

The Pillsbury Company (1973) *Food Safety Through the Hazard Analysis and Critical Control Point System*. Contract No. FDA 72-59. Minneapolis, Minnesota.

Thiermann, A.B. (2007) The new World Organization for Animal Health standards on avian influenza and international trade. Avian Dis. **50**, 338–339.

Timbo, B., K.M. Koehler, C. Wolyniak and K.C. Klontz (2004) Sulfites – a Food and Drug Administration review of recalls and reported adverse events. J. Food Prot. **67**, 1806–1811.

Todd, E.C.D. (1997) Epidemiology of foodborne diseases: a worldwide review. World Health Stat. Q. **50**, 30–50.

Troller, J.A. (1993) *Sanitation in Food Processing*, Second Edition. Academic Press, New York.

Ulberth, F. and H. Fielder (2000) Persistent organic pollutants – a dossier. Eur. J. Lipid. Sci. Technol. **102**, 45–49.

United Kingdom Statutory Instruments (1990) The Food Safety Act 1990. http://www.opsi.gov.uk/acts/ acts1990/ukpga_19900016_en_1

United Kingdom Statutory Instrument (2008) The Transmissible Spongiform Encephalopathies (England) Regulations 2008, No. 1881. http://www.opsi.gov.uk/si/si2008/uksi_20081881_en_1 (accessed 8 February 2010).

US Centers for Disease Control and Prevention (2008) Outbreak of *Salmonella* serotype Saintpaul infections associated with multiple raw produce items — United States. Morb. Mortal. Wly Rep. **57**(34), 929–934.

US Centers for Disease Control and Prevention (2009) Outbreak of *Salmonella* serotype Saintpaul infections associated with eating alfalfa sprouts — United States. Morb. Mortal. Wly Rep. **58**, 1–3.

US Department of Agriculture (USDA), Agricultural Marketing Service (1997) Pasteurized shell eggs (pasteurized in-shell eggs). Fed. Regist. **62**, 49955–49957.

USA Today (2009) Broken links in food safety chain hid peanut plants' risks. USA Today, 26 April 2009.

Wallace, C.A. (2001) Effective HACCP training. In: T. Mayes and S. Mortimore (eds) *Making the Most of HACCP*. Woodhead Publishing Ltd, Cambridge, pp. 213–231.

Wallace, C.A. (2009) The impact of personnel, training, culture and organisational factors on the application of the HACCP system for food safety management in a multinational organisation, PhD Thesis, University of Central Lancashire, UK.

Wallace, C.A. and A. Williams (2001) Prerequisites: a help or a hindrance to HACCP? Food Control **12**, 235–240.

Wallace, C.A., S.C. Powell and L. Holyoak (2005) Post-training assessment of HACCP Knowledge: its use as a predictor of effective HACCP development, implementation and maintenance in food manufacturing. Br. Food J. **107**(10), 743–759.

Ward, D.R., M.D. Pierson and M.S. Minnick (1984) Determination of equivalent processes for the pasteurization of crab meat in cans and flexible pouches. J. Food Sci. **49**, 1003–1004.

Whiting, R.C. (1995) Microbial modeling in foods. Crit. Rev. Food Sci. Nutr. **35**, 467–494.

Wick, C.W., R.V.H. Pollock, A.McK. Jefferson and R.D. Flanigan (2006) *The Six Disciplines of Breakthrough Learning: How to Turn Training and Development into Business Results*. Wiley, CA, USA

Wilkinson, J.M. and J.V. Wheelock (2004) *Assessing the Effectiveness of HACCP Implementation and Maintenance in Food Production Plants on the Island of Ireland*. Safefood Food Safety Promotion Board.

Woolhouse, M.E.J. and S. Gowtage-Sequeria (2005) Host range and emerging and reemerging pathogens. Emerg. Infect. Dis. **11**, 1842–1847.

World Health Organization (WHO) (1998) *Guidance on Regulatory Assessment of HACCP*. Report of a joint FAO/WHO consultation on the role of government agencies in assessing HACCP. WHO/FSF/FOS/98.5, Geneva.

World Health Organization (WHO) (1999) *Strategies for Implementing HACCP in Small and/or Less Developed Businesses*. WHO/SDE/FOS/99.7, Geneva.

World Health Organization (WHO) (2007a) *Hazard Analysis Critical Control Point System (HACCP)*. http://www.who.int/foodsafety/fs_management/haccp/en/ (accessed 14 May 2009).

World Health Organization (WHO) (2007b) Food safety and foodborne illness. Fact sheet No. 237 http://www.who.int/mediacentre/factsheets/fs237/en (accessed 14 November 2007).

World Health Organization (WHO) (2007c) 20 questions on genetically modified foods. http://www.who.int/foodsafety/publications/biotech/20questions/en (accessed 14 November 2007).

World Health Organization (WHO) (2007d) Acrylamide. http://www.who.int/foodsafety/chem/chemicals/acrylamide/en (accessed 14 November 2007).

World Health Organization (2010) Food security. http://www.who.int/trade/glossary/story028/en/ (accessed January 2010).

Wrigley, B.J., S. Ota, and A. Kikucki (2006) Lightening strikes twice: lessons learned from two food poisoning incidents in Japan. Publ. Relat. Rev. **32**, 349–357.

Yoe, C., M. Parish, D. Eddy, D.K.Y. Lei, B.A. Paleg and J.G. Shwartz (2008) Risk management: the value of the food defense plan. Food Safety Magazine, http://www.foodsafetymagazine.com/article.asp?id=2394&sub=sub1 (accessed January 2010).

Part Four
Appendices

Appendix 1
HACCP case studies*

INTRODUCTION

This appendix contains four practical case studies, reflecting the application of the HACCP and food safety management to different links in the global food supply chain as follows:

Case study 1: Primary production – egg production
Case study 2: Manufacturing – prepared meals
Case study 3: Food service – event catering
Case study 4: The home – consumer responsibilities

Two of these case studies are real examples and two are fictional examples developed to illustrate the application of food safety management. The fictional examples are Case study 2: Manufacturing – Riviera Risottos, which is a continuation of the HACCP work at the fictional company used to illustrate application of principles in Chapters 11 and 12, and Case study 4: The home. Case studies 1: Primary production – egg production and 3: Food service – event catering are real examples, and we are indebted to the contributors, Jose Chipollini, Moark LLC, USA, and Erica Sheward, Global Food Standards, UK, for sharing these with us.

COMMENTARY

Case study 1 provides a detailed examination of the issues faced by egg producers at the beginning of the food supply chain. This case study provides substantial reference information, which gives a good example of some of the information that might be needed, both to give food safety and HACCP team members up-to-date knowledge on hazards and control mechanisms and to validate chosen controls and limits.

Although fictional, Case study 2 shows the level of detail needed in process flow diagrams to be able to perform a comprehensive hazard analysis. Hazard significance has been assessed using a two-step (low/high) judgement for both likelihood of occurrence and severity of the potential adverse effect.

---

*Note: Although two case studies are from real operations and two are fictional, it is important to note that all case studies are, by their nature, developed at a point in time and the findings may not be exhaustive. The scope of specific examples may be limited and the findings theoretical. Therefore, the examples given are not intended as specific recommendations for similar processes/products but as illustrations of the application of principles in different situations. The contributors are experienced in the areas covered, but the specific case study content does not necessarily reflect the views of their companies or of the book authors.

Food Safety for the 21st Century, First Edition By Carol A. Wallace, William H. Sperber and Sara E. Mortimore
© 2011 Carol A. Wallace, William H. Sperber and Sara E. Mortimore

Case study 3 includes examples of prerequisite paperwork and signage necessary to support HACCP activities, whilst Case study 4 is written in a more narrative style and identifies the key considerations for safe food handling in the home.

We hope that you find these case studies to be useful examples of HACCP and food safety management applied in practice.

Case study 1
Shell eggs – food safety case study

Jose Chipollini

This case study is an example to illustrate application of food safety management and is provided without any liability in its application and use.

INTRODUCTION

Salmonella Enteritidis (SE) in shell eggs represents a significant public health hazard in the egg industry worldwide. Every year a significant number of human salmonellosis is caused by the consumption of raw or partially cooked (*SE*) contaminated eggs.

For egg laying operations to be able to meet consumer demand and be profitable, a significant economy of scale must be achieved. To attain this economy of scale, poultry is generally reared under confinement in large populations (Li *et al.*, 2007). This large concentration of chickens within one building creates large volumes of dust and manure, contaminating every surface in the poultry house. This environment is also conducive to developing potentially large populations of flies and rodents (Kreager, 1998). Measures to control *Salmonella* in the poultry house are multiple and difficult to use effectively because there are numerous sources of *Salmonella* infection.

Prevention of hazards in the early stages of the supply chain is vital in achieving appropriate levels of public health protection. HACCP is recognised as an important management tool and the gold standard to improved food safety (Woteki *et al.*, 2004). The HACCP system has been used with great success at the processing level in a wide variety of food industries. There is fairly small information on the application of HACCP programmes in egg-producing farms. Most food safety efforts developed at the farm level have been developed as quality assurance programmes based on HACCP principles to reduce the risk of producing contaminated eggs.

FOOD SAFETY/HACCP TEAM MEMBERS

The HACCP team shall consist of individuals with specific knowledge on egg production. The team may include a veterinarian or extension poultry specialist, farm manager and egg production employee. It would also be beneficial to have a HACCP expert assisting the HACCP team and ensuring the proper application of the HACCP principles. Training and education

Food Safety for the 21st Century, First Edition By Carol A. Wallace, William H. Sperber and Sara E. Mortimore
© 2011 Carol A. Wallace, William H. Sperber and Sara E. Mortimore

of the HACCP team should involve potential biological and chemical hazards present in egg production. The farm manager should be the team coordinator responsible for the development, implementation and maintenance of the HACCP plan.

TERMS OF REFERENCE

The HACCP study reviews the biological, chemical and physical hazards throughout egg production. Biological and chemicals hazards at the poultry farm are more likely to contaminate eggs than are physical hazards. Physical hazards are unlikely to be a problem because of the protective nature of the shell. The primary hazard is the presence of SE inside eggs. The goal of the HACCP study is to reduce the number of *Salmonella* Enteritidis-contaminated eggs for the table egg market.

PRODUCT/PROCESS DESCRIPTION

Product

The product is a freshly laid shell egg. It is intended for use by shell egg-processing plants located on the same farm site where they were laid or sold to other egg-grading plants to be processed for the table egg market.

The primary function of the egg is to protect the embryo from impact and microbial challenge, permit exchange of water and gases, and provide a source of calcium for the growing embryo. The egg has natural barriers to defend the egg contents from microbial challenge. This natural defence is partially physical, the shell, its membranes, and the albuminous sac, and partially chemical, the shell membranes and albumen (Stadelmann and Cotterill, 1995).

The shell is a porous structure where gas and moisture exchange occurs. The pores on the shell are sealed by a thin protein layer, called the cuticle that is deposited on the exterior surface of the shell just before the egg is laid. This is the first physical line of defence against horizontal transmission of hazards into the egg contents. Inside the egg, next to the shell are the inner and outer shell membranes. These two layers are highly fibrous and protect the egg against microbial invasion. In addition to their function as a physical barrier, the shell and shell membranes also act as a chemical barrier. Proteins with antibacterial properties have also been associated with the shell and shell membranes (Gantois *et al.*, 2009). Following the shell membranes is the albumen that consists of four layers – two thick and two thin. The albumen is a gel-like substance of proteins with antimicrobial activity, capable of binding iron and biotin, and with the ability to break down the cell wall of some bacteria. Additionally, during storage the albumen pH rises from neutral to about 9.6, providing a hostile environment for microbial growth (Sharp and Powell, 1931; Board, 1974). Finally, the vitelline membrane (yolk membrane) contains all the nutrients in the yolk needed for the development of the embryo. The yolk provides an excellent medium for microbial growth.

Chemical hazards

Chemical hazards that could be present in eggs include agricultural chemicals (pesticides, veterinary drugs) and environmental contaminants (dioxins). These are controlled by pest control and flock health management prerequisite programmes.

Biological hazards

Salmonella Enteritidis is a major cause of foodborne illnesses that are closely associated with the consumption of contaminated shell eggs. The egg contents may become contaminated with (SE) by two routes: (1) trans-shell (horizontal transmission), penetration through the shell from the colonised gut, faeces or from contact with contaminated surfaces, And (2) trans-ovarian (vertical transmission), through infection of the bird's reproductive tissue, primarily ovaries and oviduct tissue. This type of infection is considered to be the major route of SE contamination. Experimental oral inoculation of laying hens with SE resulted in the invasion of a variety of internal organs, including the ovary and oviduct, and the production of contaminated eggs for a few weeks after infection. However, even when very large oral doses of SE are administered, the reported incidence of resulting egg contamination is usually relatively low and involves small numbers of bacterial cells. The overall incidence of SE-contaminated eggs in the United States has been estimated to occur in 1 of 20 000 eggs produced. Although, at the time of writing in 2010, the emerging outbreak will be likely to have an impact on that number. Eggs that are naturally contaminated may contain no more than a few hundred *Salmonella* cells.

Most strains of *S*E that typically infect egg-laying flocks do not generally cause clinical disease in these birds. In the United States, SE has been detected in the laying house environment of approximately 7% of egg-laying flocks. Sources of SE infection may include infected day-old chicks, people that come in contact with poultry, residual environmental contamination that resisted the clean-out efforts, rodents and insects present on the premises, feed, water, bird mortality and manure.

Receiving chicks in pullet buildings

Newly hatched baby poultry (chicks) are delivered to the rearing facilities loaded in boxes in a truck with environmentally controlled conditions where they are kept for approximately 18 weeks prior to move to the egg-laying buildings.

SE can be vertically transmitted by eggs from asymptomatic breeding stock. Following vertical transmission, environmental contamination and hatchery cross-contamination are a major source of SE infection (WHO, 1993). Chicks are highly susceptible to *Salmonella* infection, even at very low exposure doses, for several days after hatching (Gast, 2007) due to an absence of protective gut flora (Cox *et al.*, 1996). Chickens exposed to SE shortly after hatching can apparently remain infected until maturity, at which time they might produce contaminated eggs or spread the infection to other susceptible, previously unexposed, hens (Gast and Holt, 1998). Birds infected at 1 day of age also experience reduced ability to respond to vaccination (Holt *et al.*, 1999). Hatcheries, where young birds are at their stage of maximum susceptibility to infection, are especially critical *Salmonella* control points (Gast, 2009).

Salmonella control in egg production starts with a breeder production flock that is free from SE (Muira *et al.*, 1964). Written assurances from breeding flocks and hatcheries of the SE status of the flock must be obtained from approved chick suppliers. In the United States, control strategies, monitoring programmes and laboratory methods used at the primary breeding level follow the provision of the USDA's National Poultry Improvement Program (NPIP). These programmes ensure that flocks and their progeny are pullorum-typhoid clean and SE clean. Chick deliveries must be accompanied by NPIP certification or a certificate of veterinary inspection declaring the SE-clean status of the breeder flocks and hatcheries.

Furthermore, because SE infection in poultry is asymptomatic, chick deliveries must be regularly monitored by testing chick box liner papers at delivery to verify the SE status of the flock.

Biosecurity practices must be followed while receiving chicks at the pullet farm to prevent the introduction of disease in the pullet house. Allow only essential personnel into your pullet facilities. If people must come in the building, they require the use of hairnets, coverall and foot cover and require hand-washing and/or hand disinfection prior to having contact with birds.

Pullet and layer flock management

The successful control of (SE) contamination in eggs is based on the use of best management practices (BMPs) aimed at reducing the risk of introduction, build-up or spread of SE in the flock. These *Salmonella* control practices include the following.

Traffic control

Limit the access to the farm to authorised personnel and visitors. Control entry to the farm with fences, gates and signs. Require visitors to use hairnets, coveralls and foot covers to protect poultry from contamination on clothing and footwear. Enforce hand-washing and disinfection before and after visiting poultry houses. Restrict access of company and contract personnel hauling feed, spent hens, manure, dead birds and trash to outside the poultry buildings. Do not permit contact of pets, stray poultry and wildlife with pullets or egg-laying birds. Restrict the movement of dirty equipment and vehicles in the farm. Require farm personnel not to raise poultry at home and to avoid contact with other poultry and wild birds.

The risk of spreading *Salmonella* in cage systems where the birds are kept above the floor is smaller than in floor systems where *Salmonella* can be spread by movement of people (shoes). Poultry housed outdoors present a problem, as it is difficult to maintaining adequate biosecurity if the hens are exposed to SE contamination from pests and wildlife.

Water quality

Drinking water plays an important role in the transmission of many pathogenic bacteria, including *Salmonella*. Water contamination can occur if surface water drains into the well. Well sites should be graded to drain surface water away from the well casing. Locate wells as far as practical from septic tanks and land application areas for manure. Evaluate the quality of the water at least once a year for municipal water and twice a year for well water. If the well water is contaminated, determine whether the source is the well or the distribution system. For a properly sealed and located well, treatment with a chlorine source may resolve the problem. If continued chlorination is necessary, maintain approximately 3–5 ppm of residual chlorine in the water. Water pH should be maintained from 6 to 7.5. Monitor the levels of chlorine regularly. If the lines are the source of the contamination, flush the line periodically to remove any build-up, including biofilm, in the water system. Use approved products to remove and kill bacteria present in biofilm.

Farm sanitation

Cleaning and disinfection of poultry buildings and equipment are essential to prevent SE infection and other poultry pathogens, in pullet and egg-laying environments. Intensified rodent control should be initiated immediately (prior to depopulation), so as to avoid rodent migration to other poultry buildings. Disinfection should not take place until poultry manure or litter and bird mortality have been removed, the entire poultry building is thoroughly clean, and all repairs have been completed. Verify and document the cleaning and disinfection of poultry buildings and equipment, feed system, water lines and bird-moving equipment. Bacteriological monitoring of the efficacy of disinfection procedures is recommended when SE has been detected in the flock. Routine pest control procedures should also be carried out at this time.

Pest control

Rodents, flies and wildlife infected with SE serve as a source of continual reinfection in the hen house. Pest control programmes must prevent and reduce pests from the farm premises.

Rodents

Mice and rats are important reservoirs and multipliers of SE, and can quickly recontaminate cleaned and sanitised poultry buildings. It has been found that mice naturally infected with SE can excrete approximately 230 000 SE bacteria per pellet. Effective decontamination of SE-positive houses is difficult to achieve if rodents are not controlled inside and between hen houses. BMPs for rodent control include the following:

- Elimination of harbourage areas inside and outside poultry buildings
- Rodent-proofing poultry houses
- Rodent monitoring (trap records)
- Rodent reduction through baiting (bait use records)
- Record-keeping

Insects

Flies, cockroaches, darkling beetles and their larvae are vectors for transmission of SE and other pathogens inside the hen house (Kopanic *et al.*, 1994; McAllister *et al.*, 1994). Insect population growth is enhanced by the presence of wet manure, warm temperatures and humid conditions in the poultry house. BMPs to reduce fly breeding include maintaining effective airflow rates, composting manure rows under the cages, periodic removal of manure from the poultry houses, removal of manure from houses during winter, preventing water leaks into the pits and seepage of water into the pits. Work with a pest control operator with experience in animal production to reduce insect populations.

Other animals

Wild birds, cats, skunks and lizards have been found to be reservoirs of SE and can directly and indirectly cause transmission to poultry flocks (Sharma and Pathak, 1976; Kinde *et al.*, 1996; Craven *et al.*, 2000). Prevent the entrance of pets and wildlife to the poultry houses. Conduct regular inspections at night to monitor wildlife activity in the poultry houses.

Vaccination

Vaccination is an important tool used by the egg industry to increase the resistance of birds against *Salmonella* exposure and decrease shedding. Vaccination used as part of an overall management, hygiene and biosecurity package can make a valuable contribution to the reduction of SE infection. The use of live attenuated and/or inactivated vaccines provide the birds with cellular immunity of the gut and circulating antibodies to reduce the following (Gast, 2009):

- Susceptibility of individual birds to SE infection
- Horizontal transmission of infection within flocks
- Vertical transmission of infection to progeny of breeding flocks
- *Salmonella* load in the poultry house environment (and the likelihood of transmission to subsequent flocks)
- Frequency of egg contamination

Poor vaccine performance has been associated with weak rodent control or sanitation problems in poultry buildings, feed or water withdrawal and environmental stresses such as heat. Vaccination must be used in conjunction with strong biosecurity practices and should be implemented under the supervision of a poultry veterinarian.

Testing

The environmental persistence of salmonellae in poultry houses creates continual opportunities for laying hens to become infected by oral ingestion (Gast *et al.*, 2007). Because most SE infections of poultry are asymptomatic, routine SE monitoring of the flock environment is essential to detect and control SE contamination when it occurs. Samples for monitoring procedures must be routinely collected from the birds and their housing environments during the rearing and production phases, and equipment and poultry buildings after cleaning and disinfection procedures have been completed. Laboratory results then should be circulated to appropriate managers to verify freedom from infection or take corrective actions as needed.

Feed manufacturing

The most important risk factors in the production of poultry feed are feed ingredients. Require written assurances from suppliers attesting to following good agricultural and manufacturing practices in the processing of feed ingredients. SE is not normally found as a contaminant in feed; it is nonetheless recommended to monitor the *Salmonella* status of feed used in poultry houses. Purchase animal protein ingredients from suppliers that follow a *Salmonella* prevention programme. Require letters or documentation that they have in place a *Salmonella* prevention programme.

Feed should be stored in clean closed bins to prevent access by birds and pests. Spilled feed should be cleaned immediately to remove attractants for wild birds and pests. Feed mills should follow the guidelines of the *Recommended Salmonella Control for Processors of Livestock and Poultry Feeds* published by the American Feed Industry Association. Require a letter or document from each mill indicating the mill meets these standards.

The use of feeds subjected to bactericidal treatments is recommended to control *Salmonella* in poultry feed.

Receiving and storage of feed

Maintain dry conditions inside the feed bins to prevent bacterial and mould growth. Cover feed bins with a lid to prevent exposure to wild birds and rain. Regularly, clean and sanitise feed systems to prevent microbial build-up.

Bird mortality storage and disposal for pullet and layer operations

All dead birds must be removed promptly from poultry houses and stored for disposal in a secured manner so that they do not spread contamination to other houses. Methods for disposal include burial, incineration, composting and rendering. Removals of poultry carcasses from the farm and subsequent transport to a rendering facility pose a great risk for spread of pathogens into a farm or farms. Emerging methods for disposal of poultry carcasses that provide long-term stabilisation of the carcasses by reducing the levels of pathogens and result in a transportable product that can be processed by a rendering facility have been developed for safe and reasonable on-farm storage (CAST, 2008).

Manure storage and disposal for pullet and layer operations

Bird waste must be removed as frequently as possible and disposed of in an environmentally and hygienic responsible manner to prevent contamination at the farm. Bacterial pathogens may persist for long periods in animal manure under typical environmental conditions. This may be exacerbated when the temperatures are low, moisture remains optimal, and aeration is not used. For instance, *Salmonella* and *Escherichia coli* O157:H7 have been noted to survive for 4–6 months in animal manures and manure slurries kept at 1–9°C, up to 49 times longer than at 40–60°C (Rogers and Haines, 2005). The survival of *Salmonella* in the poultry house environment is dependent on both physical and chemical factors such as temperature, water activity (A_w) or equilibrium relative humidity, moisture content and pH. When combined, these factors have been found to be more effective at reducing *Salmonella* than when applied individually (Payne *et al.*, 2007). Maintaining chicken manure at a high pH and low A_w has been found to reduce *Salmonella* levels. In addition, appropriate ventilation in the poultry house may result in a drop of A_w in litter/manure to levels where there is a significant reduction in numbers of *Salmonella* organisms, thus reducing the risk of spread of infection among birds, particularly in cage-free facilities (Himathongkham *et al.*, 1999).

Receiving pullets in the egg-laying building

Pullets arrive at the egg-laying farm at the age of 18 weeks. Peak production is reached around 28 weeks of age. In principle, there are three categories of systems for housing laying hens: conventional cage systems, furnished or enriched cage systems (there is little or no contact with faecal materials, resulting in a reduced risk of SE contamination) and non-cage systems (also called cage-free systems). Cage-free systems with access to the outdoors are used in organic egg production. Keeping birds outdoor presents a risk of exposure to SE compared to birds kept only indoors due, for example, to exposure to wildlife including insect vectors. Chemical treatments of poultry manure and litter have shown to be successful at reducing microbial populations, including salmonellae.

Induced molting

As the egg-laying hen ages, its ability to produce eggs and quality of the eggs produced diminish such that it is no longer economically feasible to keep the flock in lay. Induced molting methods are used by egg producers to rejuvenate the reproductive system of laying hens and thus improve eggshell quality and egg production (Landers *et al.*, 2005). There are several types of induced molting methods used by the egg industry. A molt may be accomplished either by fasting the flock or by limiting critical nutrients such as protein, Ca and Na until or beyond the time egg production ceases. Non-feed withdrawal methods for molting are mostly used in the United States and include feeding the hen wheat middlings, corn–wheat middlings combination, gluten feed diet, wheat bran and alfalfa diets.

The feed withdrawal molt method has been shown to increase the bird susceptibility to SE infection (Holt, 1993; Corrier *et al.*, 1997; Durant *et al.*, 1999; Ricke, 2003). The use of non-feed withdrawal methods using alfalfa products (McReynolds *et al.*, 2005, 2006) and wheat bran (Murase *et al.*, 2006) has been found to reduce the incidence of SE contamination in eggs. In addition, *SE* vaccination with a live attenuated *Salmonella typhimurium* just prior to molt has shown remarkable reduction in horizontal transmission of SE to hens exposed during molt.

Egg collection

In caged facilities, hens lay eggs onto an angled wire floor, which rolls the egg towards the front of the cage onto a belt. The belt transports eggs out of the houses to either the egg processing facility or packing egg room. Because the processing facility and packing egg room remove eggs from the house, based on hourly demand, eggs may reside on the belt for as long as 12–14 hours, but most are collected within a few hours post-lay (Meunier and Latour, 2000). There are two primary methods of collection in layer facilities: in-line and off-line. The in-line facility characterises all largest producers. The egg-laying buildings are placed side by side and linked by a conveyor belt. Eggs move directly from the layer houses to the egg processing facility. Once the eggs enter the egg processing centre, usually within minutes or up to several hours post-lay, they are washed, visually inspected (checked for eggshell problems, cracks and blood spots) and then graded for packaging based on a system of grading AA, A, B and Loss Quality standards. Off-line facilities are used in smaller operations, differing from the in-line process in that eggs are transported out of the house directly to a simple packing head, which places them on 30-egg flats, and are taken to an egg cooling room. In this method, the eggs remain in the cool room for up to 2–3 days, after which they are transported to an egg processing facility by the refrigerated truck.

Egg collection in caged and cage-free facilities is typically automatic. It is essential to train farm personnel to use appropriate hygiene procedures when hand gathering eggs. In cage-free facilities, maintenance of the nest lining is important to reduce the potential of soiling and microbial penetration into the egg.

Packing nest-run eggs

Packing nest materials needs to be in a cleaned and sanitary condition. Soiled egg flats are a potential source for *Salmonella*. Plastic egg flats should be washed in hot water and a detergent to remove any organic material on the egg flat surface. Residual moisture on egg flats must be prevented to reduce the potential for SE penetration inside the egg.

Refrigeration of nest-run eggs (during storage and transport)

Refrigeration is an essential tool to restrict bacterial multiplication in eggs that are externally and internally contaminated with SE.

Refrigeration of nest-run eggs may be applied at the farm, throughout transportation to the processing plant and during storage of off-line eggs. Refrigeration of nest-run eggs increases

Salmonella survival on the shell surface. Condensation on eggs should be prevented to reduce the potential for SE penetration.

Freshly laid eggs are typically reported to contain no more than a few hundred SE cells (Humphrey *et al.*, 1991; Gast and Holt, 2000), so prompt refrigeration of eggs to temperatures that inhibit SE multiplication (7.2°C or lower), (Radkowski, 2002), should reduce the likelihood for a small population of SE inside the egg to grow to disease-causing levels (Gast *et al.*, 2006).

Additionally, refrigeration decreases yolk membrane degradation, thus preventing access to the egg yolk nutrient-rich environment by SE.

Receiving eggs at the processing plant

Receiving eggs must be done in a manner that prevents condensation on the shell surface. Nest-run eggs stored at 45°C should be tempered prior to egg washing to prevent thermal cracks, thus facilitating pathogen entry (Patterson *et al.*, 2008).

PROCESS FLOW DIAGRAM

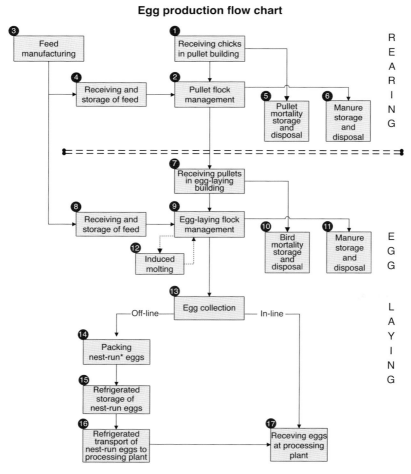

Egg production flow chart

*NEST RUN Unprocessed shell eggs. They may be sold to shell egg grading or packing plants or to official egg products plants.

Hazard analysis and control measure identification

Process step	Hazard	Control measure	Q1	Q1a	Q2	Q3	Q4	CCP	Justification for the decision

1. Receiving chicks in the pullet building	**Biological**
	1. Presence of *Salmonella* Enteritidis in baby chicks
		Agreed specifications
		Use approved suppliers (National Poultry Improvement Plan source)
		Certificate of Analysis
			Yes		No	Yes	No	Yes	• SE-infected chicks at day-old are likely to produce SE-positive eggs
									  Chickens exposed to SE shortly after hatching can apparently remain infected until maturity, at which time they might produce contaminated eggs or spread the infection to other susceptible, previously unexposed hens[2]
									  Birds infected at 1 day of age also experience reduced ability to respond to vaccination[3]

	2. Introduction of SE by people (e.g. hatchery driver and service personnel)	Best management practices: Visitors policy Biosecurity policy	Yes		No	No	–	No	• Policy discourages visits to poultry units and reduces the opportunity for introduction of SE • Requirement for hairnets, coveralls and foot covers protects poultry from contamination on clothing and footwear • Hand-washing and disinfection prevent introduction of contamination from hands

[2] See Gast and Holt (1998).
[3] See Holt *et al.* (1999).

2. Pullet flock management	**Biological** 1. Cross-contamination with SE from dirty equipment and building	Cleaning and disinfection SE vaccination	Yes	No	Yes	No	Yes	Yes	No

- Controlled by cleaning and disinfection of pullet buildings and equipment
- The prevalence of SE in pullet flocks is relatively low compared to layer flocks; data from PEQAP and CEQAP databases show that the environmental prevalence of SE in pullet houses ranges from 0 to 1.5%[4]. FDA estimates that 0.75% of pullet houses test environmentally SE positive
- Vaccination reduces the risk of production of infected eggs by infected flocks[5]

  Vaccination with either inactivated (killed) or live, attenuated preparations has a long history of application for reducing the susceptibility of poultry to Salmonella infections and for, thereby, protecting human consumers against food-borne disease transmission[6]

[4] See US Department of Health and Human Services, Food and Drug Administration (2009).
[5] See Davies and Breslin (2004).
[6] See Gast (2007).

Process step	Hazard	Control measure	Q1	Q1a	Q2	Q3	Q4	CCP	Justification for the decision
	2. Introduction of SE by people (e.g. farm or service personnel and visitors)	BMP – employee biosecurity Visitor policy Biosecurity policy	Yes	No	No	No	–	No	• Company policies and practices discourage visits to poultry units and reduce the opportunity for introduction of SE • Requirement for hairnets, coveralls and foot covers protects poultry from exposure to SE contamination on clothing and footwear • Hand-washing and disinfection prevent exposure to SE contamination from hands
	3. Introduction of SE by pets, stray poultry, wildlife and pests	BMPs for pets, stray poultry, wild birds/wildlife pest control programme	Yes		No	No	–	No	• Restricting access to animals that may carry SE in or/and on their bodies in poultry buildings protects pullets from SE contamination
3. Feed manufacturing	**Biological** 1. Introduction of SE in poultry feed	Agreed specification Use approved suppliers Letter of guarantee Dust control Pest control Moisture control Use approved *Salmonella* inhibitors (organic acid, formaldehyde-based products)	Yes		No	No	–	No	• SE is not likely to contaminate finished layer feed *Testing for SE in finished layer feed at mills has almost never yielded positive results[7]* • Monitor feed ingredients for the presence of SE • Use of *Salmonella* inhibitors in the feed ration will reduce the levels of *Salmonella* in feed
	2. Presence of mycotoxins	Agreed specification Use approved suppliers Surveillance Moisture control	Yes		No	No	–	No	• Incoming loads are tested for mycotoxin levels • Loads with high mycotoxin levels are rejected

[7] See Gast (2007).
[8] See US Department of Health and Human Services, Food and Drug Administration (2009).

Process step	Hazard	Control measure					Justification
	Chemical						
	1. Presence of dioxins	Agreed specification Use approved suppliers	Yes	No	—	No	• Dioxins are not reasonably likely to occur[9]
		Letter of guarantee	Yes	No	—	No	• GMP – equipment cleanout procedures at the feed mill
	2. Presence of veterinary drugs	Agreed specification Use approved suppliers GMPs – equipment cleanout	Yes	No	—	No	• No eggs produced at the rearing facilities are destined for human consumption
	3. Presence of pesticides	Agreed specification Use approved suppliers Letter of guarantee Chemical control Pest control	Yes	No	—	No	• FDA pesticide monitoring reports show a low incidence of pesticides above the regulatory guidance in feed ingredients (3.6% of the samples tested showed residue levels above the regulatory guidance[10])
4. Receiving and storage of feed	**Biological**						
	1. Cross-contamination of feed with SE	Sanitation standard operating procedure (SSOP)	Yes	No	—	No	• Controlled by cleaning and disinfection programme; feed storage is cleaned and disinfected regularly
	2. Cross-contamination of feed with mycotoxins	SSOP Controlled dry storage	Yes	No	—	No	• Feed is protected from rain and moisture
5. Pullet mortality storage and disposal	**Biological**						
	1. Growth of pathogens, including SE	BMP for storage and disposal of bird mortality	Yes	No	—	No	• Pullet mortality is removed from poultry house daily and stored for disposal in an enclosed container
	2. Spread of pathogens, including SE by pests	Pest control programme	Yes	No	—	No	• Controlled by pest control programme
6. Manure storage and disposal	**Biological**						
	1. Growth of enteric pathogens, including SE in manure	BMP for manure removal	Yes	No	—	No	• Manure is removed frequently or treated to reduce microbial loads
	2. Spread of pathogens, including SE by pests	Pest control programme	Yes	No	—	No	• Controlled by pest control programme

[9] See Herman (2006).
[10] See FDA (2007).

Process step	Hazard	Control measure	Q1	Q1a	Q2	Q3	Q4	CCP	Justification for the decision
7. Receiving pullets in the egg-laying building	**Biological** 1. Presence of SE in pullets	Agreed specifications Use approved suppliers Surveillance Certificate of Analysis Pest control Cleaning and disinfection	Yes	No	No	No	–	No	• All birds are obtained from sources with a working Salmonella prevention and control programme • Flocks are vaccinated either with inactivated (killed) or live, attenuated preparations to reduce the susceptibility of poultry to Salmonella infections • Test results will confirm SE-negative status of the flock
	2. Introduction of SE by people (service crew)	Best management practices Visitors policy Biosecurity policy	Yes		No	No	–	No	• Policy discourages visits to poultry units and reduces the opportunity for introduction of SE • Requirement for hairnets, coveralls and foot covers protects poultry from contamination on clothing and footwear • Hand-washing and disinfection prevent introduction of contamination from hands
8. Receiving and storage of feed	**Biological** 1. Cross-contamination of feed with SE	SSOP SSOP	Yes		No	No	–	No	• Controlled by cleaning and disinfection programme • Feed storage is cleaned and disinfected regularly
	2. Cross-contamination of feed with mycotoxins	Controlled dry storage	Yes		No	No	–	No	• Feed is protected from rain and moisture
9. Egg-laying flock management	**Biological** 1. Cross-contamination with SE from dirty equipment and poultry building	Cleaning and disinfection SE vaccination	Yes		No	Yes	Yes	No	• Dirty surfaces or application of low levels of disinfectant will allow survival of SE organisms • Vaccination reduces the risk of production of infected eggs by infected flocks[11]

---

[11] See Davies and Breslin (2004).

Process step	Hazard	Control measure					Justification
	2. Introduction of SE by people (e.g. farm or service personnel and visitors)	BMP – employee biosecurity Visitor's policy Biosecurity policy	Yes	No	–	No	• Company policies and practices discourage visits to poultry units and reduce the opportunity for introduction of SE • Requirement for hairnets, coveralls and foot covers protects poultry from exposure to SE contamination on clothing and footwear • Hand-washing and disinfection prevent exposure to SE contamination from hands
	3. Introduction of SE by pets, stray poultry, wildlife and pests	BMPs for pets, stray poultry, wild birds/wildlife pest control programme	Yes	No	–	No	• Pests carry SE in and on their bodies; restricting access to animals that may carry Se in or/and on their bodies in poultry buildings protects egg-laying hens from SE contamination
	4. Introduction of SE in the water	BMP – water sanitation Surveillance	Yes	No	–	No	• Water chlorination or acidification reduces SE contamination
10. Bird mortality storage and disposal	**Biological** 1. Growth of pathogens, including SE	BMP for storage and disposal of bird mortality	Yes	No	– / –	No	• Hen mortality is removed from poultry building daily and stored for disposal in an enclosed container
	2. Spread of SE by pests	Pest control programme	Yes	No	–	No	• Controlled by the pest control programme
11. Manure storage and disposal	**Biological** 1. Growth of enteric pathogens, including SE in manure	BMP for manure removal	Yes	No	– / –	No	• Manure is removed frequently or treated to reduce microbial loads
	2. Spread of pathogens, including SE by pests	Pest control programme	Yes	No	–	No	• Controlled by the pest control programme

Process step	Hazard	Control measure	Q1	Q1a	Q2	Q3	Q4	CCP	Justification for the decision
12. Induced molting	**Biological** 1. Shedding of SE in the environment	Non-feed withdrawal molt SE vaccination	Yes		No	Yes	Yes	No	• Use of non-feed withdrawal methods does not increase the incidence of SE in eggs[12] • Risk is reduced if the flock is vaccinated prior to molt
13. Egg collection	*None identified*								
14. Packing nest-run eggs	**Biological** 1. Presence of *Salmonella* organisms on dirty packaging materials	Agreed specification SSOP	Yes		No	No	—	No	• Surface egg contamination will be controlled at the egg-washing step at the processing plant • Egg flats and other egg-handling materials to be cleaned and sanitised
15. Storage of refrigerated nest-run eggs	**Biological** 1. Growth of SE in eggs	Temperature control	Yes		No	No	—	No	• Shell eggs are refrigerated at an ambient temperature at or below 45°F to prevent microbial growth[13] and yolk membrane degradation • Prompt refrigeration of eggs reduces the likelihood for SE to grow to dangerous levels[14] • Bacterial growth is prevented if the eggs are processed under the 36-hour limit for unrefrigerated holding[15]

---

12 See Woodward *et al.* (2005).

13 See Fehlhaber and Kruger (1998).

14 See Gast *et al.* (2006).

15 See US Department of Health and Human Services, Food and Drug Administration (2009). The 36-hour limit for unrefrigerated holding is supported by a model, contained in the 1998 joint SE risk assessment (Ref. 21), which was developed to examine the relationship among holding time, holding temperature and yolk membrane breakdown as an indicator of SE risk. (*The yolk membrane separates the nutrient-rich yolk and any SE bacteria that might be present in the albumen; breakdown or loss of the yolk membrane results in rapid growth of SE present in the albumen.*) The model showed that, at 70–90°F (i.e. temperatures that might be observed in unrefrigerated egg holding areas in farms or warehouses or in transport vehicles), there was much less breakdown of yolk membrane in eggs held no longer than 36 hours than in eggs held no longer than 72 hours. According to the model, eggs held at 70°F will experience at least a 16% breakdown of yolk membrane after 36 hours and a 25% breakdown after 72 hours. Eggs held at 80°F will suffer at least a 22% breakdown after 36 hours and a 39% breakdown in the yolk membrane at 72 hours. At 90°F, there is at least a 33% breakdown after 36 hours and at least a 62% breakdown of the yolk membrane after 72 hours.

16. Transport of refrigerated nest-run eggs to the processing plant	**Biological** 1. GROWTH of SE in and on eggs	Temperature control	Yes	No	No	–	No	• Shell eggs are refrigerated at an ambient temperature at or below 45°F to prevent microbial growth and yolk membrane degradation • Bacterial growth is prevented if the eggs are processed under the 36-hour limit for unrefrigerated holding
17. Receiving eggs at the processing plant	**Biological** 1. Presence of SE inside the egg (off-line egg production)	Temperature control	Yes	No	No	–	No	• Low risk of illness because of low occurrence and concentration in shell eggs • Contract-grown flocks are vaccinated to reduce the risk of *Salmonella* contamination and follow biosecurity practices • Outside egg purchases are received at an ambient temperature at or below 45°F
	Biological 2. Presence of SE inside the egg (in-line egg production)	Temperature control	Yes	No	No	–	No	• Low risk of illness because of low occurrence and concentration in shell eggs • Shell eggs are refrigerated at an ambient temperature at or below 45°F to prevent microbial growth and yolk membrane degradation
	3. Surface eggshell contamination with *Salmonella* sp.	Egg washing	Yes	No	No	–	No	• *Salmonella* present on shell is controlled by the alkaline egg-washing step in the processing plant

CCP IDENTIFICATION

The standard HACCP definition of critical control point (CCP) was used for identification as follows: CCP is defined as a point, procedure, practice, operation or stage in egg production at which control can be applied and the production of internally contaminated eggs can as a result be prevented, eliminated or reduced to acceptable levels.

The identification of the CCP was done following the decision tree adapted from Codex. The following decision tree process was used to identify CCPs in egg production:

Q1: Do control measures exist?

Q1a: Is control at this step necessary for safety?

Q2: Is the step specifically designed to eliminate or reduce the likely occurrence or a hazard to an acceptable level?

Q3: Could contamination with identified hazard(s) occur in excess of acceptable level(s) or could this increase to unacceptable levels?

Q4: Will a subsequent step eliminate identified hazards or reduce the likely occurrence of a hazard to acceptable level(s)?

- The goal is to reduce the threat to public health by reducing the number of SE-contaminated eggs for the table egg market.
- Newly hatched chicks are highly susceptible to SE infection, even at very low exposure doses, for several days after hatching. Chickens exposed to SE shortly after hatching can remain infected until maturity, at which time they might produce contaminated eggs or spread the infection to other susceptible, previously unexposed, hens. Birds infected at 1 day of age also experience a reduced ability to respond to vaccination.
- Receiving SE-infected chicks increases the risk for the production of SE-contaminated eggs and transmission of the infection to other susceptible birds.

HACCP control chart

Process step	CCP No.	Hazard to be controlled	Control measure	Critical limits	Monitoring			Corrective action	
					Procedure	Frequency	Responsibility	Procedure	Responsibility
Receiving chicks in the pullet building	1	*Salmonella* Enteritidis	Approved suppliers Agreed specifications Certificate of Analysis Surveillance	No SE-positive chick box paper tests.	Check NPIP form or letter documenting NPIP source Check Certificate of Analysis for evidence of compliance Sample[16] 1 in 10 chick box papers and submit to laboratory for SE testing. 9 CFR 147.12 (a)(4)	Every delivery	Ownership or farm manager	Reject delivery if chicks are from an unknown source If test result is SE-positive, depopulate flock immediately Clean and disinfect building and equipment Test the environment prior to restocking	Ownership or farm manager

[16] See eCFR Bacteriological Examination Procedure.

IMPLEMENTATION AND MAINTENANCE

- Develop a timeline for implementation. Plan the implementation process in stages. Although this is a longer process, it lightens the load on farm employees and minimises confusion.
- Seek support from your extension specialist (veterinarian/poultry/HACCP), universities, state agencies and institutions for appropriate HACCP training.
- Develop a master verification schedule to review regularly each part of the HACCP system.

In summary, the risk of SE in eggs can be minimised by use of the HACCP technique (in particular hazard analysis) and by implementation and maintenance of best management practices on the farm as an essential prerequisite programme.

REFERENCES

Board, R.G. (1974) Non-specific antimicrobial defenses of the avian egg, embryo, and neonate. Biol. Rev. **49**, 15–49.

Corrier, D.E., D.J. Nisbet, B.M. Hargis, P.S. Holt and J.R. DeLoach (1997) Provision of lactose to molting hens enhances resistance to *Salmonella* Enteritidis colonization. J. Food Prot. **60**, 10–15.

Council for Agricultural Science and Technology (CAST) (2008) *Poultry Carcass Disposal Options for Routine and Catastrophic Mortality*, Issue Paper 40, CAST, Ames, Iowa.

Cox, N.A., J.S. Bailey and M.E. Berrang (1996) Alternative routes for *Salmonella* intestinal tract colonization of chicks. J. Appl. Poult. Res. **5**, 282–288.

Craven, S.E., N.J. Stern, E. Line, J.S. Bailey, N.A. Cox and P. Fedorka-Cray (2000) Determination of the incidence of *Salmonella* spp., *Campylobacter jejuni*, and *Clostridium perfringens* in wild birds near broiler houses by sampling intestinal droppings. Avian Dis. **44**, 715–720.

Davies, R. and M. Breslin (2004) Observations on *Salmonella* contamination of eggs from infected commercial laying flocks where vaccination for *Salmonella enterica* serovar Enteritidis had been used. Avian Pathol. **33**, 133–144.

Durant, J.A., D.E. Corrier, J.A. Byrd, L.H. Stanker and S.C. Ricke (1999) Feed deprivation affects crop environment and modulates *Salmonella* eneritidis colonization and invasion of Leghorn Hens. Appl. Environ. Microbiol. **64**, 1919–1923.

Electronic Code of Federal Regulations (eCFR), Title 9 Animals and Animal Products, Part 147 Auxiliary Provisions on National Poultry Improvement Plan, Subpart B – Bacteriological Examination Procedure. http://ecfr.gpoaccess.gov/cgi/t/text/text-idx?c=ecfr&sid=6f570e45e419671334d657ce1f40c45c&rgn=div8&view=text&node=9:1.0.1.7.64.2.82.3&idno=9 (accessed February 2010).

FDA (2007) Pesticide monitoring. http://www.fda.gov/Food/FoodSafety/FoodContaminantsAdulteration/Pesticides/ResidueMonitoringReports/ucm169577.htm (accessed February 2010).

Fehlhaber, K. and G. Kruger (1998) The study of *Salmonella* enteiritids growth kinetics using rapid automated bacterial impedance technique. J. Appl. Microbiol. **84**, 945–949.

Gantois, I., R. Ducatelle, F. Pasmans, F. Haesebrouck, R.K. Gast, T. Humphrey and F. Van Immerseel (2009) Mechanisms of egg contamination by *Salmonella* Enteritidis. FEMS Microbiol. Lett. **33**, 718–738.

Gast, R.K. (2007) Serotype-specific and serotype-independent strategies for pre-harvest control of foodborne *Salmonella* in poultry. Avian Dis. **51**, 817–828.

Gast, R.K. (2009) Iowa Industry Symposium, Wednesday, 11 November 2009, p. 5.

Gast, R.K. and P.S. Holt (1998) Persistence of *Salmonella* Enteritidis from one day of age until maturity in experimentally infected layer chickens. Poult. Sci. **77**, 1759–1762.

Gast, R.K. and P.S. Holt (2000) Deposition of phage type 4 and 13a *Salmonella* Enteritidis strains in the yolk and albumen of eggs laid by experimentally infected hens. Avian Dis. **44**, 706–710.

Gast, R.K., P.S. Holt and R. Guraya (2006) Effect of refrigeration on in vitro penetration of *Salmonella* Enteritidis through the egg yolk membrane. J. Food Prot. **69** (6), 1426–1429.

Gast, R.K., R. Guraya, J. Guard-Bouldin, P.S. Holt and R.W. Moore (2007) Colonization of specific regions of the reproductive tract and deposition at different locations inside eggs laid by hens infected with *Salmonella* Enteritidis or *Salmonella* Heidelberg. Avian Dis. **51**, 40–44.

Herrman, T. (July/Aug 2006) Commercial feed manufacturing: top HACCP priorities. Feed Manage. http://www.feedmanagement-digital.com/feedmanagement/200608/?pg=26#pg26 (accessed October 2009).

Himathongkham, S., S. Nuanualsuwan and H. Riemann (1999) Survival of *Salmonella* Enteritidis and *Salmonella typhimurium* in chicken manure at different levels of water activity. FEMS Microbiol. Lett. **172**, 159–163.

Holt, P.S. (1993) Effect of induced molting on the susceptibility of White Leghorn hens to a *Salmonella* Enteritidis infection. Avian Dis. **37**, 412–417.

Holt, P.S., R.K. Gast, R.E. Porter Jr and H.D. Stone (1999) Hypo responsiveness of the systemic and mucosal humoral immune systems in chickens infected with *Salmonella enterica* serovar Enteritidis at one day of age. Poult. Sci. **78**, 1510–1517.

Humphrey, T.J., A. Whitehead, A.H.L. Gawler, A. Henley and B. Rowe (1991) Numbers of *Salmonella* Enteritidis in the contents of naturally contaminated hens eggs. Epidemiol. Infect. **106**, 489–496.

Kinde, H., D.H. Read, R.P. Chin, A.A. Bickford, R.L. Walker, A. Ardans, R.E. Breitmeyer, D. Willoughby, H.E. Little, D. Kerr and I.A. Gardner (1996) *Salmonella* Enteritidis phage type 4 infection in a commercial layer flock in Southern California: bacteriologic and epidemiologic findings. Avian Dis. **40**, 665–671.

Kopanic, R.J., B.W. Sheldon Jr and C.G. Wright (1994) Cockroaches as vectors of *Salmonella*: laboratory and field trials. J. Food Prot. **57**, 125–132.

Kreager, K. (1998) *Egg Industry Initiatives to Control Salmonella*. Presented at the International Symposium on Food-Borne Salmonella in Poultry Sponsored by the American Association of Avian Pathologis, 25–26 July 1998, Baltimore, MD.

Landers, K.L., C.L. Woodward, X. Li, L.F. Kubena, D.J. Nisbet and S.R. Ricke (2005) Alfalfa as a single dietary source to induce molting in laying hens. Bioresour. Technol. **96**, 565–570.

Li, X., J.B. Payne, F.B. Santos, J.F. Levine, K.E. Anderson and B.W. Sheldon (2007) *Salmonella* populations and prevalence in layer feces from commercial high-rise houses and characterization of the *Salmonella* isolates by serotyping, antibiotic resistance analysis, and pulsed field gel electrophoresis. Poult. Sci. **86**, 591–597.

McAllister, J.C., C.D. Steelman and J.K. Skeeles (1994) Reservoir competence of the lesser worm (*Coleoptera: Tenebrionidae*) for *Salmonella typhimurium* (*Eubacteriales: Enterobacteriaceae*). J. Med. Entomol. **31**, 369–372.

McReynolds, J., L. Kubena, J. Byrd, R. Anderson, S. Ricke and D. Nisbet (2005) Evaluation of *Salmonella* Enteritidis in molting hens after administration of an experimental chlorate product (for nine days) in the drinking water and feeding an alfalfa molt diet. Poult. Sci. **84**, 1186–1190.

McReynolds, J.L., R.W. Moore, L.F. Kubena, J. Byrd, C.L. AWoodward, D.J. Nisbet and S.C. Ricke (2006) Effect of various combinations of alfalfa and standard layer diet on susceptibility of laying hens to *Salmonella* Enteritidis during forced molt. Poult. Sci. **85**, 1123–1128.

Meunier, R.A. and M.A. Latour (2000) Commercial egg production and processing. http://ag.ansc.purdue. edu/poultry/publication/commegg/ (accessed November 2009).

Muira, S., G. Sato and T. Miyamae (1964) Occurrence and survival of *Salmonella* organisms in hatcher chick fluff in commercial hatcheries. Avian Dis. **8**, 546–554.

Murase, T., K. Chiba, T. Sato, K. Otsuki and P.S. Holt (2006) *Salmonella* infection in flocks of naturally contaminated laying hens in a commercial egg-producing farm by detection of yolk antibodies to *Salmonella* in eggs. J. Food Prot. **69**(12), 2883–2888.

Patterson, P.H., K.W. Koelkebeck, K.E. Anderson, M.J. Darre, J.B. Carey, D.U. Ahn, R.A. Ernst, D.R. Kuney and D.R. Jones (2008) Temperature sequence of eggs from oviposition through distribution, production – Part 1. Poult. Sci. **87**, 1182–1186.

Payne, J.B., J.A. Osborne, P.K. Jenkins and B.W. Sheldon (2007) Modeling the growth and death kinetics of *Salmonella* in poultry litter as a function of pH and water activity. Poult. Sci. **86**, 191–201.

Radkowski, M. (2002) Effect of moisture and temperature on survival of *Salmonella* Enteritidis on shell eggs. Arch. für Geflügelkd. **66**, 119–123.

Ricke, S.C. (2003) The gastrointestinal tract ecology of *Salmonella* Enteritidis colonization in molting hens. Poult. Sci. **82**, 1003–1007.

Rogers, S. and J. Haines (2005) *Detecting and Mitigating the Environmental Impact of Fecal Pathogens Originating from Confined Animal Feeding Operations*: EPA/600/R-06/021, Review by Land Remediation and Pollution Control Division National Risk Management Research Laboratory Cincinnati, OH 45268, September 2005.

Sharma, S.K. and R.C. Pathak (1976) Note on wall-lizard (*Hemidactylus flaviviridis*) as source of *Salmonella* infection (veterinary aspect). Pantnagar J. Res. **1**, 152–153.

Sharp, P.F. and C.K. Powell (1931) Increase in the H of the white and yolk of hens' eggs. Ind. Eng. Chem. **53**, 196–199.

Stadelmann, W.J. and O.J. Cotterill (1995) *Egg Science and Technology.* The Harworth Press, Binghamton, p. 85.

US Department of Health and Human Services, Food and Drug Administration (2009) *Prevention of Salmonella Enteritidis in Shell Eggs During Production, Storage, and Federal Register Final Rule,* Government Printing Office, 9 July 2009, 74 FR 33030 http://edocket.access.gpo.gov/2009/pdf/E9–16119.pdf (accessed 18 November 2009).

WHO (1993) Consultation on control of *Salmonella* infections in animals. *Prevention of Foodborne Salmonella Infections in Humans,* 21–26 November 1993, Jena, Germany.

Woodward, C.L., Y.M. Kwon, L.F. Kubena, J.A. Byrd, R.W. Moore, D.J. Nisbet and S.C. Ricke (2005) Reduction of *Salmonella enterica* serovar enteritidis colonization and invasion by an alfalfa diet during molt in leghorn hens. Poult. Sci. **84**, 185–193.

Woteki, C.E, M. O'K Glavin and B.D. Kineman (2004) HACCP as a model for improving food safety. *Perspectives in World Food and Agriculture 2004.* The World Food Prize. Iowa State Press, Ames, Iowa, pp. 101–118.

GLOSSARY

Cooperative Extension System

The Cooperative Extension System is a US nationwide, non-credit educational network. Each US state and territory has a state office at its land-grant university and a network of local or regional offices. These offices are staffed by one or more experts who provide useful, practical and research-based information to agricultural producers, small business owners, youth, consumers and others in rural areas and communities of all sizes.

Case study 2
Manufacturing – prepared meals

Carol A. Wallace

This case study is an example to illustrate application of food safety management and is provided without any liability in its application and use.

HACCP PLAN

HACCP team

R. Arborio – technical manager (HACCP team leader)
L. Grain – production manager
C. Basmati – engineering supervisor
M. Wild – production supervisor
T. Jasmine – technical consultant

SCOPE

The manufacture of ready-to-eat hot and cold prepared meals including products relating to special dietary need requirements for the following sectors:

- Retailer private label products
- Riviera Risottos branded products

TERMS OF REFERENCE

- The HACCP plan will cover all relevant microbiological, physical and chemical hazards to include allergens and compounds that cause intolerant reaction.
- This HACCP plan covers all processes from raw material intake to chilled storage of finished products prior to dispatch.

Food Safety for the 21st Century, First Edition By Carol A. Wallace, William H. Sperber and Sara E. Mortimore
© 2011 Carol A. Wallace, William H. Sperber and Sara E. Mortimore

DESCRIPTION OF PRODUCT

Ready-to-eat hot and cold prepared meals are manufactured from fresh, frozen and dried raw materials. Raw materials contained in the recipes include dairy products, fish and prawns, chicken, turkey, beef, lamb, bacon and pork. Allergens are used on site but are strictly controlled. Ingredients are sourced through approved suppliers globally.

All cooked prepared meals are heated to pasteurisation temperatures, then blast chilled, stored and distributed chilled. The shelf life of the products is determined by prescribed storage and usage conditions and is verified during production trials and confirmed microbiologically.

Fit for purpose food-grade packaging is used including foil and CPET (crystalline polyethylene terepthalate) microwavable food trays. All packaging carries full ingredient breakdowns, nutritional information, allergen information, heating and storage instructions and shelf-life information.

INTENDED CONSUMER USE

- The products are intended for the general population which may include high-risk groups.
- Some products may contain allergens, so are not suitable for the whole population.
- All allergens are stated on the pack and all packs carry the relevant warnings.
- Products may be consumed cold or reheated as per instructions.
- All products should be held under refrigerated storage prior to use.

ENVISAGED CONSUMER MISUSE

- Temperature abuse
- Consumed after the manufacturer's recommended shelf life has expired

PREREQUISITES

This HACCP study operates in conjunction with the following site prerequisite programmes under the Codex principles for food hygiene:

- *Establishment – design and facilities.* The site is located on a 7-acre site on the edge of an industrial food park. The production facilities are housed in a purpose-built factory unit, opened in October 2007.
- *Control of operation.* Control of food hazards, key aspects of hygiene control systems, incoming material requirements, packaging, water, management and supervision, documentation and records, recall procedures.
- *Establishment – maintenance and sanitation.* Maintenance and cleaning, cleaning programmes, pest control systems, waste management, monitoring effectiveness.
- *Establishment – personal hygiene.* Health status, illness and injuries, personal cleanliness, captive uniforms, personal behaviour and visitors.
- *Transportation.* General requirements, use and maintenance.

- *Product information and consumer awareness.* Lot identification, product information, labelling and consumer education.
- *Training, awareness and responsibilities.* Training programmes, instruction and supervision, refresher training.

HAZARD ANALYSIS PROCEDURE

A two-step high/low significance assessment procedure was used to identify the significant hazards from the list of potential hazards at each process step. The likelihood of occurrence and severity of effect were considered, and because a significant hazard is defined as one that is both likely to occur and cause an adverse health effect (Mortimore and Wallace, 1998), those hazards considered 'high' both for likelihood and severity were deemed significant hazards. All significant hazards were passed through the Codex decision tree (Codex, 2009b).

HACCP REVIEW

The HACCP plan will be reviewed annually and updates made to the plan as required.

A HACCP study will be carried out before a new product is launched, or a plant trial is done for proposed new products, if there is a new process involved or if a new raw material is to be introduced to the factory. All new products go through separate product safety assessment process and authorisation sign-off.

REFERENCES

Codex Committee on Food Hygiene (2009b) HACCP system and guidelines for its application. *Food Hygiene Basic Texts.* Fourth Ed., Food and Agriculture Organization of the United Nations, World Health Organization, Rome. http://www.fao.org/docrep/012/a1552e/a1552e00.htm (accessed July 2010).

Mortimore, S.E. and C.A. Wallace (1998) *HACCP – A Practical Approach*, Second Edition. Aspen Publishers Inc (now Springer), Gaithersburg, USA.

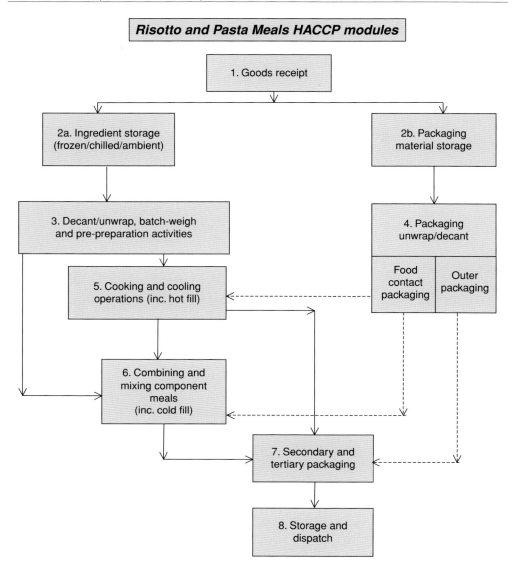

Risotto and Pasta Meals HACCP modules

1. Goods receipt

2a. Ingredient storage (frozen/chilled/ambient)

2b. Packaging material storage

3. Decant/unwrap, batch-weigh and pre-preparation activities

4. Packaging unwrap/decant

Food contact packaging | Outer packaging

5. Cooking and cooling operations (inc. hot fill)

6. Combining and mixing component meals (inc. cold fill)

7. Secondary and tertiary packaging

8. Storage and dispatch

Module 1 Goods Receipt

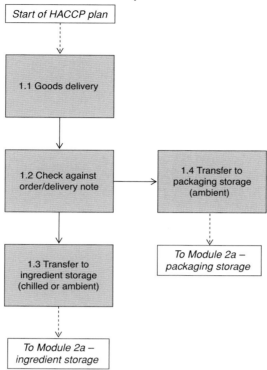

Module 2 Materials Storage

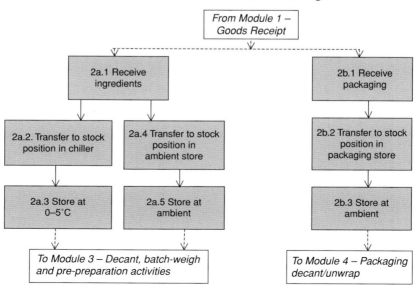

Module 3 Decant, Batch-weigh and Pre-Preparation Activities

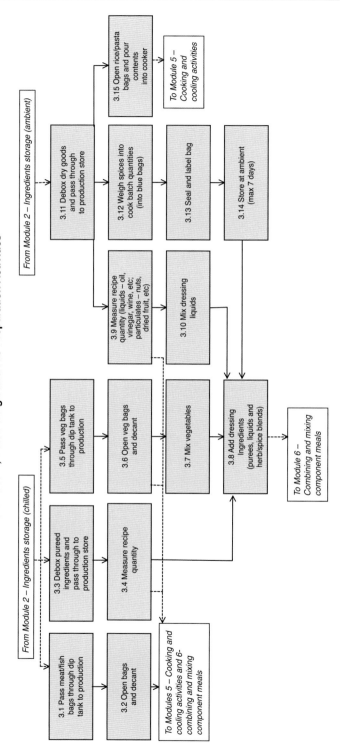

Module 4 Packaging Decant/Unwarp

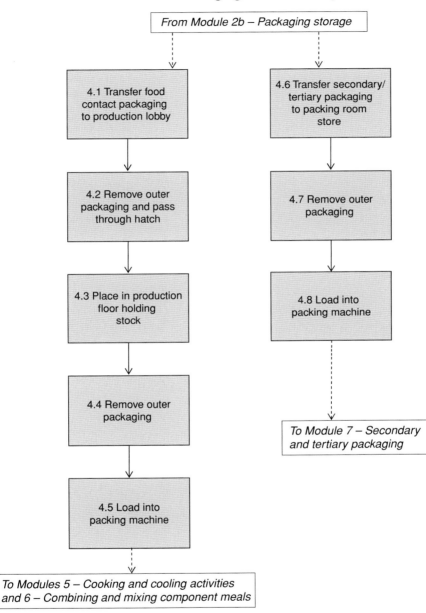

Module 5 Cooking and Cooling Activities (inc. Hot Fill)

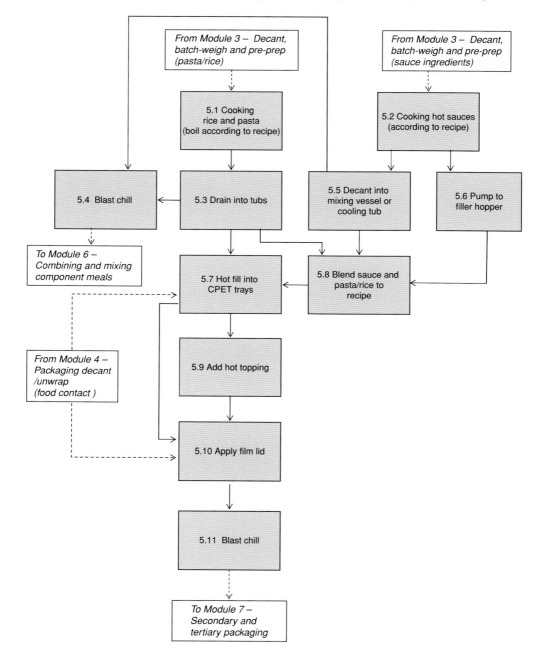

Module 6 Combining and Mixing Component Meals (Salads)

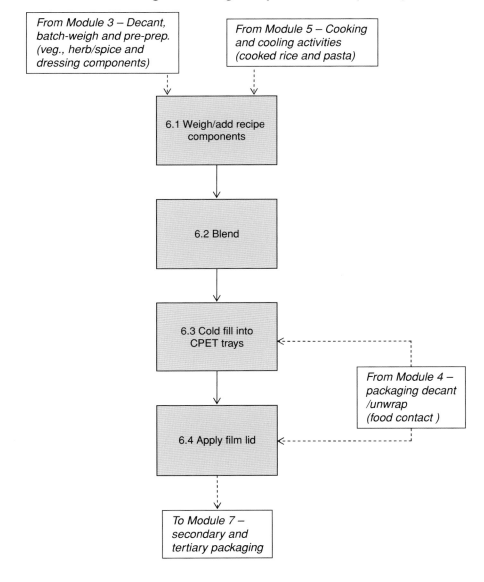

From Module 3 – Decant, batch-weigh and pre-prep. (veg., herb/spice and dressing components)

From Module 5 – Cooking and cooling activities (cooked rice and pasta)

6.1 Weigh/add recipe components

6.2 Blend

6.3 Cold fill into CPET trays

From Module 4 – packaging decant /unwrap (food contact)

6.4 Apply film lid

To Module 7 – secondary and tertiary packaging

Module 7 Secondary and Tertiary Packaging

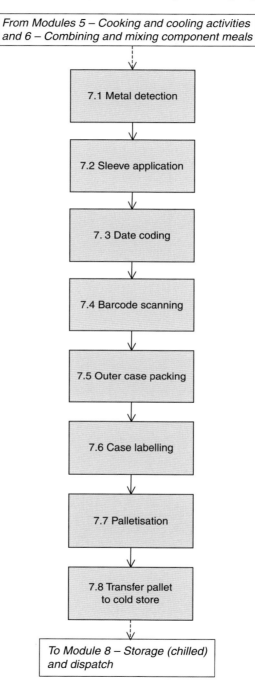

Module 8 Storage and Dispatch

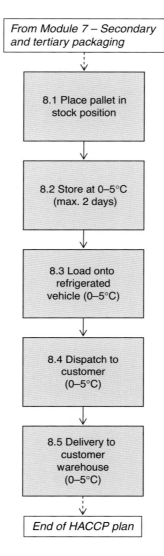

From Module 7 – Secondary
and tertiary packaging

8.1 Place pallet in
stock position

8.2 Store at 0–5°C
(max. 2 days)

8.3 Load onto
refrigerated
vehicle (0–5°C)

8.4 Dispatch to
customer
(0–5°C)

8.5 Delivery to
customer
warehouse
(0–5°C)

End of HACCP plan

		Hazard analysis				
Process step	**Hazard**	**Likelihood**	**Severity**	**Significant hazard?**	**Control measure**	**Justification**
Module 1: Goods receipt						
1.1 Goods delivery	Foreign material from damaged packaging	High	Low	No	Goods intake prerequisite; rejection of damaged goods	Unlikely to cause harm to the consumer
	Presence of pathogenic micro-organisms in raw products	High	High	**Yes**	Cooking at later step	Separate full ingredient hazard analysis already performed
	Growth of micro-organisms due to temperature abuse of refrigerated goods in transit	Low	Low	No	Rejection of material outside of specified limits – target <5°C; <7°C maximum	Some history of minor out of specification temperatures recorded at receipt on rare occasions but the risk of growth deemed to be low because no more than 2°C over target limit
1.2 Check against order/delivery note	Introduction of unknown allergens due to wrong product supplied	Low	High	No	Approved product specifications and goods intake procedures	Allergens managed by strict prerequisite programmes and labelling, which rely on knowledge of all allergens in materials supplied as per specifications and allergenic ingredient list. Any substitution of ingredients could endanger existing control measures; however, supplier quality assurance relationships are closely managed under prerequisites
1.3 Transfer to ingredient storage (chilled or ambient)	Possible growth of pathogenic micro-organisms	Low	Low	No	Rapid transfer; managed by prerequisite programmes	

Process step	Hazard					Control measures
1.4 Transfer to packaging storage (ambient)	No hazard identified	n/a	n/a	n/a	n/a	
Module 2: Materials storage						
2.a.1 Receive ingredients	No hazard identified	n/a	n/a	n/a	n/a	
2.a.2 Transfer to stock position in chiller	No hazard identified	n/a	n/a	n/a	n/a	
2.a.3 Store at 0–5°C	No hazard identified	n/a	n/a	n/a	n/a	Temperature control and stock rotation are part of prerequisite programmes
2.a.4 Transfer to stock position in ambient store	No hazard identified	n/a	n/a	n/a	n/a	
2.a.5 Store at ambient	No hazard identified	n/a	n/a	n/a	n/a	
2.b.1 Receive packaging	No hazard identified	n/a	n/a	n/a	n/a	
2.b.2 Transfer to stock position in packaging store	No hazard identified	n/a	n/a	n/a	n/a	
2.b.3 Store at ambient	No hazard identified	n/a	n/a	n/a	n/a	
Module 3: Decant (unwrap), batch-weigh and pre-preparation activities						
3.1 Pass meat/fish bags through dip tank to production	No hazard identified	n/a	n/a	n/a	n/a	
3.2 Open bags and decant/unwrap	No hazard identified	n/a	n/a	n/a	n/a	Allergens managed by strict prerequisite programmes, including dedicated containers and segregated storage area, plus labelling
3.3 Debox pureed ingredients and pass through to production store	Contamination with packaging	Low	Low	No	Prerequisite programmes and work instructions	Unlikely to harm consumers
3.4 Measure recipe quantity	No hazard identified	n/a	n/a	n/a	n/a	Allergens managed by strict prerequisite programmes, including dedicated containers and segregated storage area, plus labelling
3.5 Pass veg bags through dip tank to production	No hazard identified	n/a	n/a	n/a	n/a	

Process step	Hazard	Hazard analysis			Control measure	Justification
		Likelihood	Severity	Significant hazard?		
3.6 Open veg bags and decant/unwrap	No hazard identified	n/a	n/a	n/a	n/a	Allergens managed by strict prerequisite programmes, including dedicated containers and segregated storage area, plus labelling
3.7 Mix vegetables	No hazard identified	n/a	n/a	n/a	n/a	
3.8 Add dressing ingredients (purees, liquids and herb/spice blends)	No hazard identified	n/a	n/a	n/a	n/a	Allergens managed by strict prerequisite programmes, including dedicated containers and segregated storage area, plus labelling
3.9 Measure recipe quantity (liquids – oil, vinegar, wine, etc.; particulates – nuts, dried fruits, etc.)	No hazard identified	n/a	n/a	n/a	n/a	Allergens managed by strict prerequisite programmes, including dedicated containers and segregated storage area, plus labelling
3.10 Mix dressing liquids	No hazard identified	n/a	n/a	n/a	n/a	
3.11 Debox dry goods and pass through to production store	Contamination with packaging	Low	Low	No	Prerequisite programmes and work instructions	Unlikely to harm consumers
3.12 Weigh spices into cook batch quantities	No hazard identified	n/a	n/a	n/a	n/a	
3.13 Seal and label bags	No hazard identified	n/a	n/a	n/a	n/a	
3.14 Store at ambient (max. 7 days)	No hazard identified	n/a	n/a	n/a	n/a	
3.15 Open rice/pasta bags and pour contents into cooker	No hazard identified	n/a	n/a	n/a	n/a	Allergens managed by strict prerequisite programmes, including dedicated containers and segregated storage area, plus labelling

Module 4: Packaging decant/unwrap

Process step	Hazard				Control measures	Comments
4.1 Transfer food contact packaging to production lobby	No hazard identified	n/a	n/a	n/a	n/a	
4.2 Remove outer packaging and transfer through hatch	No hazard identified	n/a	n/a	n/a	n/a	
4.3 Place in production floor holding stock	No hazard identified	n/a	n/a	n/a	n/a	
4.4 Remove outer packaging	No hazard identified	n/a	n/a	n/a	n/a	
4.5 Load into packing machine	No hazard identified	n/a	n/a	n/a	n/a	
4.6 Transfer secondary/tertiary packaging to packing room store	No hazard identified	n/a	n/a	n/a	n/a	
4.7 Remove outer packaging	No hazard identified	n/a	n/a	n/a	n/a	
4.8 Load into packing machine	No hazard identified	n/a	n/a	n/a	n/a	

Module 5: Cooking and cooling activities (including hot fill)

Process step	Hazard				Control measures	Comments
5.1 Cook rice and pasta (boil according to recipe)	Survival of pathogenic micro-organisms due to inadequate heat processing	High	High	Yes	Approved cooking methods – times and temperatures	
5.2 Cook hot sauces (according to recipe)	Survival of pathogenic micro-organisms due to inadequate heat processing	High	High	Yes	Approved cooking methods – times and temperatures	
5.3 Drain into tubs	No hazard identified	n/a	n/a	n/a	n/a	
5.4 Blast chill	Germination and outgrowth of spore-forming pathogens	High	High	Yes	Approved procedures for handling and cooling; time and temperature parameters set	Size of vessels being chilled increases likelihood of growth
5.5 Decant into mixing vessel	No hazard identified	n/a	n/a	n/a	n/a	
5.6 Pump to filler hopper	No hazard identified	n/a	n/a	n/a	n/a	

Process step	Hazard	Hazard analysis			Control measure	Justification
		Likelihood	Severity	Significant hazard?		
5.7 Hot fill into CPET trays	Germination and outgrowth of spore-forming pathogens	Low	High	No	Immediate through process	Normal process is rapid cook-blend-fill; therefore, there is no time to allow temperature drop to danger zone for growth. If any process delay, e.g. blending and hot-fill operation is unavailable, work procedure is to blast chill
5.8 Blend sauce and pasta/rice to recipe	Germination and outgrowth of spore-forming pathogens	Low	High	No	Immediate through process	Normal process is rapid cook-blend-fill; therefore, there is no time to allow temperature drop to danger zone for growth. If any process delay, e.g. blending and hot-fill operation is unavailable, work procedure is to blast chill
5.9 Add hot topping	Germination and outgrowth of spore-forming pathogens	Low	High	No	Immediate through process	Normal process is rapid cook-blend-fill; therefore, there is no time to allow temperature drop to danger zone for growth. If any process delay, e.g. blending and hot-fill operation is unavailable, work procedure is to blast chill
5.10 Apply film lid	No hazard identified	n/a	n/a	n/a	n/a	
5.11 Blast chill	Germination and outgrowth of spore-forming pathogens	Low	High	No	Rapid chilling of individual units	Likelihood is low in this case due to the size of the individual units – time to cool is <20 minutes compared with the larger vessels being chilled at step 5.4

Module 6: Combining and mixing component meals (salads)

6.1 Weigh/add recipe components					
Cross-contamination with allergens into wrong products	Low	High	No	Prerequisite programmes and labelling	Allergens managed by strict prerequisite programmes; at this stage, the programmes include production scheduling for products containing specific allergens and deep cleaning after these products
Cross-contamination with pathogenic micro-organisms – vegetative or spore formers – from environment or utensils	Low	High	No	Prerequisite programmes	High standards of hygiene for production environment and utensils
Germination and outgrowth of spore-forming pathogens	Low	High	No	Immediate through process	Process area for Module 6 operations is held at <10°C; as all components are chilled then limited opportunity for temperature rise into danger zone
6.2 Blend					
Cross-contamination with allergens into wrong products	Low	High	No	Prerequisite programmes and labelling	Allergens managed by strict prerequisite programmes; at this stage, the programmes include production scheduling for products containing specific allergens and deep cleaning after these products
Cross-contamination with pathogenic micro-organisms – vegetative or spore formers – from environment or utensils	Low	High	No	Prerequisite programmes	High standards of hygiene for production environment and utensils

Process step	Hazard	Hazard analysis			Control measure	Justification
		Likelihood	Severity	Significant hazard?		
	Germination and outgrowth of spore-forming pathogens	Low	High	No	Immediate through process	Process area for Module 6 operations is held at <10°C; as all components are chilled, then limited opportunity for temperature rise into danger zone
6.3 Cold fill into CPET trays	Germination and outgrowth of spore-forming pathogens	Low	High	No	Immediate through process	Process area for Module 6 operations is held at <10°C; as all components are chilled, then limited opportunity for temperature rise into danger zone
6.4 Apply film lid	No hazard identified	n/a	n/a	n/a	n/a	
Module 7: Secondary and tertiary packaging						
7.1 Metal detection	Presence of metal (from previous steps or ingredients) not identified, leading to hazardous metal inclusion in product	High	High	**Yes**	All product passes through a functioning metal detector	No subsequent step to remove this hazard
7.2 Sleeve application	Presence of unlabelled allergens if wrong sleeve applied	High	High	**Yes**	Barcode scanning of all products to ensure correct sleeve on product	Historical evidence of product going into wrong sleeve. Although additional prerequisite programme controls in place for receipt of printed sleeves and machine changeovers, the HACCP team felt this is still an area of concern. All products also include 'may contain' statements

Process step	Hazard				Control / documented procedure	Justification
7.3 Date coding	Incorrect shelf life could lead to microbiological growth (*Listeria monocytogenes*) during shelf life	Low	High	No	Documented procedure for labelling of products as part of prerequisite programmes and legal control	Unlikely that incorrect shelf life could be applied
7.4 Scanning all products – barcode scanner	Presence of unlabelled allergens if scanner fails to pick up wrong sleeve applied	High	High	**Yes**	All product passes through functioning scanner device	Scanner will pick up wrong cartons that may be received in mid-stack from printer or that may have become stuck in machine at changeover
7.4 Outer case packing	No hazard identified	n/a	n/a	n/a	n/a	
7.5 Case labelling	No hazard identified	n/a	n/a	n/a	n/a	
7.6 Palletisation	No hazard identified	n/a	n/a	n/a	n/a	
7.7 Transfer pallets to cold store	No hazard identified	n/a	n/a	n/a	n/a	
Module 8: Storage and dispatch						
8.1 Place pallets in stock position	No hazard identified	n/a	n/a	n/a	n/a	
8.2 Store at 0–5°C (max. 2 days)	No hazard identified	n/a	n/a	n/a	n/a	Prerequisite programmes manage temperature of all chillers, cold stores and vehicle refrigeration
8.3 Load onto refrigerated vehicle (0–5°C)	No hazard identified	n/a	n/a	n/a	n/a	Prerequisite programmes manage temperature of all chillers, cold stores and vehicle refrigeration
8.4 Dispatch to the customer (0–5°C)	No hazard identified	n/a	n/a	n/a	n/a	Prerequisite programmes manage temperature of all chillers, cold stores and vehicle refrigeration
8.5 Delivery to the customer warehouse (0–5°C)	No hazard identified	n/a	n/a	n/a	n/a	Prerequisite programmes manage temperature of all chillers, cold stores and vehicle refrigeration

HACCP control chart

Process step	Hazard	Control measure	Critical limits	Monitoring	Monitoring responsibility	Corrective action	Corrective action responsibility	Record
5.1 Cooking rice and pasta	Survival of pathogenic micro-organisms due to inadequate heat processing	All cooked components cooked to minimum time and temperature.	All product achieves core temperature 72°C minimum	Temperature checks with calibrated probes	Cooker operator	Continue to heat until required temperature (72°C minimum) is reached	Production manager	Cooking records
5.2 Cooking hot sauces	Survival of pathogenic micro-organisms due to inadequate heat processing	All cooked components cooked to minimum time and temperature.	All product achieves core temperature 72°C minimum	Temperature checks with calibrated probes	Cooker operator	Continue to heat until required temperature (72°C minimum) is reached	Production manager	Cooking records
5.4 Blast chilling	Germination and outgrowth of spore-forming pathogens	Effective blast chill process reduces temperature within safe time limit (normally achieves <5°C within 90 minutes)	All product to be cooled below 5°C within 120 minutes	Centre temperature checks with calibrated probes at entry and exit from chiller. Residence time checked and recorded	Cooker operator	Discard batch. Investigate and repair any fault with blast chiller	Production/Technical/Engineering managers	Production records

	Hazard	Control	Critical limit	Monitoring		Corrective action		Records
7.1 Metal detection	Presence of metal (from previous steps or ingredients) not identified leading to hazardous metal inclusion in product	All product passes through a functioning metal detector	Absence of all metal above 7 mm – ferrous, non-ferrous and stainless. Correctlys functioning metal detector and rejection mechanism in place and working continuously.	Check detection and rejection mechanism with test strips – 2.5 mm all types – placed in centre of the product	Line operator	Recheck product since previous satisfactory check	Line manager	Production records
7.4 Barcode scanning	Presence of unlabelled allergens if scanner fails to pick up wrong sleeve applied	All product passes through a functioning scanner device	Scanner functioning at all times	Check with packaging samples – start-up and half hourly	Line operator	Recheck product since previous satisfactory check	Line manager	Production records

Case study 3
Food service – Lapland UK food service operation

Erica Sheward

This case study is an example to illustrate application of food safety management and is provided without any liability in its application and use.

INTRODUCTION

Lapland UK is a seasonal (Christmas) event company offering an award-winning recreation of Father Christmas' arctic homeland, where families enjoy a 4–5 hours magical experience that celebrates a child's belief in Father Christmas. The site includes outside snow trails and ice-skating facilities (not normally found in the UK under normal pre-Christmas weather conditions) plus opportunities to meet Father Christmas and help the Elves in the toy factory.

Part of the experience involves a festive family feast, with a range of choices for all ages. Because of the temporary nature of the Lapland site, it has been decided that meals will be purchased pre-prepared from a local prepared meal manufacturer that operates to BRC Global Standard Certification. These are delivered chilled as complete packaged meals in catering size portions or as packaged meal components. Meals are then served in the Lapland site dining room which, although this is a temporary site, includes a full professional standard kitchen and all necessary preparation and service equipment. An external chilled storage unit is also on site, immediately outside and linked to the kitchen.

SCOPE AND TERMS OF REFERENCE

This HACCP study relates to the preparation and service of meals at the Lapland UK site. It does not cover manufacturing of prepared meals and components, which are detailed in the supplier's HACCP system. The site HACCP study is supported by prerequisite programmes. Because of the nature of the operations at this temporary site, the HACCP system focuses on microbiological hazards. Allergen control is included in prerequisite programmes with specific reference to gluten-free dishes.

Food Safety for the 21st Century, First Edition By Carol A. Wallace, William H. Sperber and Sara E. Mortimore
© 2011 Carol A. Wallace, William H. Sperber and Sara E. Mortimore

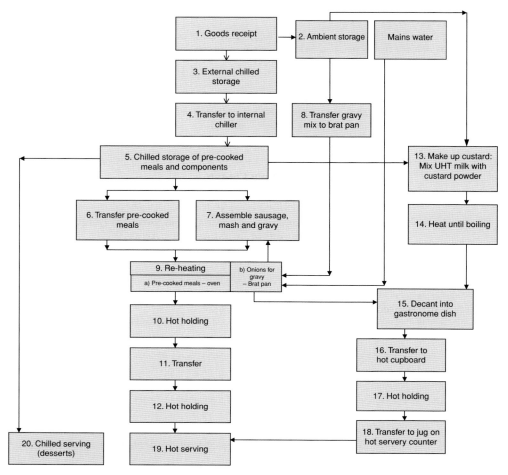

Fig. A.1 Prepared meals and desserts – Lapland site process.

HACCP team

A HACCP team was set up including the technical manager from the prepared meal supplier, the chef and assistant chef at Lapland site and a technical consultant. The team established the process flow for meal preparation and service at the site's dining room (Figure A.1), developed site prerequisite programmes and established and implemented the HACCP plan.

Menu

Table A.1 lists the menu items on offer every day at the Lapland site. All hot-serve meals, with the exception of sausages and mash with onion gravy, are received chilled in sealed catering size trays from the prepared meal facility. Sausages, mashed potatoes and cooked, sliced onions are received chilled individually packed in sealed catering trays for meal assembly on site. Cold desserts are received in containers of individual portions (disposable dessert cups). The hot

Table A.1   Lapland UK menu.

Norfolk Gold Turkey, Leek and Ham Pie
Turkey pieces with dry cured ham, and leeks, in a creamy parsley sauce, topped with all butter short crust
pastry
Cottage Pie*
Tender organic steak mince, with onion, carrot and leeks, topped with creamy mashed potato
Scottish Smoked Haddock, Salmon and Hake Pie
Naturally smoked haddock, with salmon, and hake in a creamy fish sauce, topped with mash
Sausages with Mash and Onion Gravy
Premium pork sausages with a sweet red onion gravy on creamy mashed potato
Red Lentil and Mixed Bean Casserole*
With sundried tomato and fresh basil sauce (vegan friendly)
Macaroni Cheese
Macaroni in a creamy mature cheddar sauce, baked with extra cheese topping

Chocolate Pot*
Strawberries, Raspberries and Blackcurrants Set in Strawberry Jelly*
Spiced Apple and Raisin Crumble with Custard
*Coeliac friendly

---

dessert option is received in sealed catering trays as per the hot meal options. Gravy mix and custard mix are received as dry goods in sealed packages.

The majority of the meals go through a simple heat – hot-hold – hot serve operation; however, sausage, mash and onion gravy require an assembly operation and both gravy and custard are made up for addition to meals and desserts, respectively. Cold desserts are received ready-to-serve and are simply unpackaged and served.

PREREQUISITE PROGRAMMES

Prerequisite programmes are crucial to the serving of safe food on the Lapland site. Key prerequisite programmes include:

- Temperature control – chilled storage, heating and hot-holding processes
- Personal hygiene – kitchen and service staff plus visitors (who may have been petting animals, e.g. reindeer).
- Cleaning, sanitation and waste management – effective cleaning of the kitchen, dining room and all equipment and utensils; pest management and waste control.
- Transportation – particularly effective temperature control and hygienic transfer of meals and components from the prepared meal-manufacturing site.
- Allergen control – several meal options are advertised as coeliac friendly since the manufacturing facility has special meal manufacturing capability and these products are specifically designed not to include gluten. Although all are received in sealed containers, this necessitates handling practices with special disciplines and utensils at the Lapland site to prevent cross-contamination. No other allergen claims are made, and customers are advised that meals may contain traces of other allergens, so are not suitable for allergy sufferers.

As examples of prerequisite programme documentation, an example of hygiene instructions is shown in Figure A.2 and the allergen risk assessment for each product, i.e. which allergens are expected to be present by recipe, is reproduced in Table A.2.

Table A.2 Allergens present as intrinsic ingredients by the product.

Product	Milk	Wheat and Gluten	Celery and Celeriac	Mustard	Sulphites	Nuts	Crustacean	Eggs	Fish	Peanut	Soy(a) bean	Sesame	Suitable for vegetarians	Suitable for vegans	Suitable for Coeliacs
Cottage pie	✓	✓	✓	☒	☒	☒	☒	☒	☒	☒	☒	☒	☒	☒	☒
Fiss pie	✓	✓	✓	✓	☒	☒	☒	☒	✓	☒	☒	☒	☒	☒	☒
Macaroni cheese	✓	✓	☒	✓	☒	☒	☒	✓	☒	☒	☒	☒	✓	☒	☒
Bean and lentil casserole	✓	☒	☒	✓	☒	☒	☒	☒	☒	☒	✓	☒	✓	✓	✓
Turkey pie	✓	✓	✓	✓	☒	☒	☒	✓	☒	☒	☒	☒	☒	☒	☒
Sausage and mash	✓	✓	☒	☒	☒	☒	☒	☒	☒	☒	☒	☒	☒	☒	☒
Apple crumble and custard	✓	✓	☒	☒	☒	✓	☒	✓	☒	☒	☒	☒	✓	☒	☒
Chcolate mousse	✓	✓	☒	☒	☒	☒	☒	✓	☒	☒	✓	☒	✓	☒	✓
Fruit jelly	☒	☒	☒	☒	☒	☒	☒	☒	☒	☒	☒	☒	✓	☒	✓
Apple crumble no custard	☒	✓	☒	☒	☒	☒	☒	☒	☒	☒	☒	☒	✓	✓	☒

Key

✓	ALLERGENS = CONTAINS
☒	ALLERGENS = DOES NOT CONTAIN

☒	NOT SUITABLE FOR LACTO VEGETARIANS
✓	SUITABLE FOR LACTO VEGETARIANS
☒	NOT SUITABLE FOR VEGANS
✓	SUITABLE FOR VEGANS
☒	NOT SUITABLE FOR COELIACS
✓	SUITABLE FOR COELIACS

FOOD HYGIENE NOTICE !!

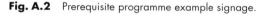

FOR REASONS OF 'ELF' AND SAFETY
PLEASE WASH YOUR HANDS AND
USE SANITISING HAND GEL REGULARLY

Fig. A.2   Prerequisite programme example signage.

HACCP study

Tables detailing hazard analysis (Table A.3) and CCP management (Table A.4) are reproduced. CCPs were identified using judgement and experience, and the consultant cross-checked the process using the Codex decision tree (Codex, 2003).

HACCP verification and review

The consultant performs a fortnightly audit of the HACCP system and prerequisite programmes during the operating season (3 months). Due to the seasonal nature of this operation, HACCP review will be performed the following year before start-up and will include assessment of any new menu options and related process activity changes.

Table A.3 Lapland hazard analysis: Lapland UK – meal preparation and service at the Lapland site.

Process step	Hazard	Hazard analysis			Control measure	Justification
		Likelihood	Severity	Significant hazard?		
1. Goods receipt	Growth of micro-organisms due to temperature abuse of refrigerated goods in transit	Low	Low	No	Rejection of material outside of specified limits – target <5°C; <7°C maximum	The risk of growth is low because the manufacturing site is 20 minutes away by road and temperature is checked before goods leave the site by the refrigerated vehicle
2. Ambient storage	No hazard identified	n/a	n/a	n/a	n/a	
3. External chilled storage	Growth of micro-organisms due to temperature abuse of refrigerated goods in transit	Low	Low	No	All chillers are controlled via prerequisite programmes, including detailed monitoring of operating temperatures; external storage is self-contained chiller unit	The risk of growth is low and unlikely that pathogens other than spores will be present
4. Transfer to internal chiller	No hazard identified	n/a	n/a	n/a	n/a	
5. Chilled storage of pre-cooked chilled meals and components	Growth of micro-organisms due to temperature abuse of refrigerated goods in transit	Low	Low	No	All chillers are controlled via prerequisite programmes, including detailed monitoring of operating temperatures; external storage is self-contained chiller unit	The risk of growth is low and unlikely that pathogens other than spores will be present
6. Transfer pre-cooked meals	No hazard identified	n/a	n/a	n/a	n/a	
7. Assemble sausage, mash and gravy	Cross-contamination with pathogens due to manual handling	Low	High	No	Strict management of prerequisite programmes on hygienic handling	This product has more manual handling but hygiene is strictly controlled
8. Transfer gravy mix to brat pan	No hazard identified	n/a	n/a	n/a	n/a	

No.	Process step	Hazard				Control measures	Comments
9.	Reheating (a) Oven (b) Brat pan	Survival of vegetative pathogens that may have contaminated meals and components since initial manufacture (environmental or handling contaminants)	High	High	Yes	Cook to centre temperature of 72°C minimum – both oven and brat pan processes	Due to the amount of handling, it was considered that there was a possibility of contamination between initial manufacture and reheating
10.	Hot holding	Germination and outgrowth of spore-forming pathogens	Low	High	No	Hot-holding procedures managed under prerequisite programmes	Prerequisite programmes require monitoring and recording of hot-hold temperatures every 30 minutes through the service period; in reality, hot-hold temperatures are normally over 80°C
11.	Transfer to dining room hot cupboard	No hazard identified	n/a	n/a	n/a	n/a	
12.	Hot holding	Germination and outgrowth of spore-forming pathogens	Low	High	No	Hot-holding procedures managed under prerequisite programmes	Prerequisite programmes require monitoring and recording of hot-hold temperatures every 30 minutes through the service period; in reality, hot-hold temperatures are normally over 80°C
13.	Make up custard: mix ultra-high temperature milk with custard powder	No hazard identified	n/a	n/a	n/a	n/a	

(Continued)

Table A.3 (Continued)

| Process step | Hazard | Hazard analysis | | | Control measure | Justification |
		Likelihood	Severity	Significant hazard?		
14. Heat until boiling	Survival of vegetative pathogens – possible contaminants in custard powder	High	High	Yes	Cook until boiling	The HACCP team discussed the possibility of *Salmonella* spp. contamination in custard powder ingredients and decided to consider this a significant hazard
15. Decant into gastronome dish	No hazard identified	n/a	n/a	n/a	n/a	
16. Transfer to hot cupboard in the dining room	No hazard identified	n/a	n/a	n/a	n/a	
17. Hot holding	Germination and outgrowth of spore-forming pathogens	Low	High	No	Hot-holding procedures managed under prerequisite programmes	Prerequisite programmes require monitoring and recording of hot-hold temperatures every 30 minutes through the service period; in reality, hot-hold temperatures are normally over 80°C
18. Transfer to jug on hot servery counter	No hazard identified	n/a	n/a	n/a	n/a	
19. Hot serving	No hazard identified	n/a	n/a	n/a	n/a	
20. Chilled serving (desserts)	No hazard identified	n/a	n/a	n/a	n/a	Chilled desserts taken straight from chiller for service

Table A.4 Lapland HACCP control chart – CCP management.

Process step	Hazard	Control measure	Critical limits	Monitoring	Monitoring responsibility	Corrective action	Corrective action responsibility	Record
21. Reheating (a) Oven (b) Brat pan	Survival of vegetative pathogens that may have contaminated meals and components since initial manufacture (environmental or handling contaminants)	Cook to centre temperature of 72°C minimum – both oven and brat pan processes	All product achieves core temperature 72°C minimum	Temperature checks with calibrated probes	Assistant chef	Continue to heat until required temperature (72°C minimum) is reached	Chef	Cooking daily check sheet
22. Heat until boiling	Survival of vegetative pathogens – possible contaminants in custard powder	Cook until boiling	Custard to be visibly boiling and meet minimum temperature 80°C	Temperature checks with calibrated probes	Assistant chef	Continue to heat until required temperature is reached	Chef	Cooking daily check sheet

REFERENCE

Codex Committee on Food Hygiene (2003). HACCP system and guidelines for its application. *Food Hygiene Basic Texts*. Food and Agriculture Organization of the United Nations, World Health Organization, Rome. http://www.fao.org/docrep/006/y5307e/y5307e00.htm (accessed February 2010).

Case study 4
Food safety in the home: a review and case study

William H. Sperber

This case study is an example to illustrate application of food safety management and is provided without any liability in its application and use.

INTRODUCTION

While all links in the farm-to-table food supply chain are important, the last link – the point of consumption – seems to be the most important because it is the last opportunity to assure the safety of a food before it is eaten. There are many points of consumption in the last link, including restaurants, institutional settings such as schools and hospitals, and the home. Restaurants and institutions can usually provide significant assurance of food safety because the food is usually prepared and served by trained personnel. In contrast, the home environment is much more vulnerable to food safety mistakes and the occurrence of foodborne illnesses because those handling and serving the food are typically untrained and often unaware of the potential hazards. Often, they are children or ignorant adults and work in a confined space that may include pets and infant children. Furthermore, the principal food safety controls in the home – refrigeration, cooking and baking – are not always used properly. It is most important to recognise, particularly in the home, that the consumer has a significant role in food safety, a role that is shared with the many participants in the food supply chain.

Recognising that a lack of knowledge contributed to the spread of many infectious diseases and foodborne illnesses in the home, an online resource, the International Scientific Forum on Home Hygiene, was established in 1997 to promote health by means of improved hygiene in the home. A detailed report on progress during its first 10 years is a most helpful resource for consumers (Bloomfield *et al.*, 2009). Additional efforts to increase consumer understanding have been initiated by the UK Food Standards Agency, whose goal is to promote kitchen hygiene in order to reduce foodborne illnesses in the home by 20% over the next 5 years (FSA, 2009).

This review and case study will provide information for educators and consumers to make them more aware of the potential foodborne hazards in the home, practical means of hazard control, and specific control measures that could be established. Ideally, some potential hazards could be 'managed' in the home environment, similar to the management of hazards in food-processing facilities by means of HACCP and prerequisite programmes.

Food Safety for the 21st Century, First Edition By Carol A. Wallace, William H. Sperber and Sara E. Mortimore
© 2011 Carol A. Wallace, William H. Sperber and Sara E. Mortimore

POTENTIAL HAZARDS

The wide range of microbiological, chemical and physical hazards that can be found through the food supply chain can often be encountered in the home kitchen.

Potential microbiological hazards include the following:

- *Salmonella* and *Campylobacter* in raw meat and poultry
- *Listeria monocytogenes* in cooked, ready-to-eat (RTE) meats and soft cheeses
- *E. coli* O157:H7 in raw ground beef, raw milk and juices, and fresh produce
- *Clostridium botulinum* in improperly cooled foods, e.g. soups and baked potatoes
- *Clostridium perfringens* in dressed, roasted poultry and gravies
- *Bacillus cereus* in improperly cooled, cooked rice and potatoes
- *Staphylococcus aureus* in custard or crème-filled cakes
- *Salmonella* in pet food

Both *B. cereus* and *S. aureus* produce heat-stable toxins that survive reheating of previously cooked foods.

Potential chemical hazards include the following:

- Allergens
- Cleaning chemicals
- Pesticides and rodenticides

Potential physical hazards include the following:

- Broken glass
- Other foreign material

The potential introduction of hazards into a prepared food can be heightened by several environmental factors, including the presence of pets and infants in the household, combined with inadequate hand-washing by the food preparer after handling the pet or changing diapers, etc. Accumulated dust on the floor is a common cause of infant botulism (Nevas *et al.*, 2005). While not directly a food safety issue, this important fact emphasises the need to maintain a clean kitchen. Cross-contamination from raw to cooked foods can result from inadequate hand-washing or using, without adequate washing or disinfection, the same utensils to handle raw and cooked or RTE foods. For example, a cutting board used to prepare raw poultry can contaminate fresh salad ingredients with *Salmonella* or *Campylobacter* if it is not properly washed after being exposed to the raw poultry.

Of course, such potential sources of cross-contamination are not limited to the home kitchen. The public health issues related to handling of raw meat and poultry and fresh produce in the same food preparation area are a major concern in retail and food service establishments that prepare food for consumption.

POTENTIAL CONTROL MEASURES

Several common-sense practices will minimise the possibilities of a foodborne illness originating in the home, but they are not always easy to implement:

- Use clean (potable) water for preparing foods, especially when rehydrating foods such as dried milk for consumption without heating. In many regions, limited access to potable water is a major public health issue.
- Clean and disinfect bottles used for infant feeding before filling with properly heated milk or infant formulas.
- Maintain allergen controls if a family member has a food allergy. Be aware of food allergies that visitors may have.
- Do not store toxic chemicals in the kitchen or in other areas where foods are stored.
- No pets are allowed on tables or countertops.
- Cross-contamination control, e.g. hand-washing and separate utensils for raw and cooked food handling.

The principal control measures available to ensure food safety in the kitchen are refrigeration, heating by cooking, baking or frying, sanitation and personal hygiene (Marchiony, 2004).

Refrigeration

Providing proper refrigeration of perishable foods begins when foods are purchased. Perishable foods should be refrigerated at 4°C or below as soon as possible, or within 2 hours of purchase. Attention should be given to the product's recommended shelf-life date so that it would be consumed before spoilage could occur. Care must be taken to promptly refrigerate leftover foods in order to prevent the growth of spoilage or pathogenic micro-organisms. A very good, and widely taught, guideline for holding foods is that cold foods should be stored at or below 4°C and hot foods should be stored at or above 60°C to prevent the growth of pathogens. This is an important consideration during holidays when family meals are served to large groups of people. Leftovers should be placed directly into refrigeration at 4°C within 2 hours of serving; it is not recommended to cool foods at room temperature before refrigeration. Large quantities of food that would require many hours to reach refrigeration temperature should be divided into smaller portions so that they will be properly chilled within several hours. Leftovers should be reheated to 74°C, if necessary, and consumed within 2 days of refrigerated storage. Family members may disagree on whether a refrigerated food is on the verge of spoilage or not. A wise saying applies in this case –'when in doubt, throw it out'. Some people, in the interest of saving money, have died of botulism after eating leftover food that was either questionable or obviously spoiled.

Refrigeration temperatures should be verified periodically with a reliable thermometer. It would be a public health service if refrigerator manufactures build reliable thermometers into the refrigeration unit such that the interior and door temperatures could be monitored. It is known that the door temperatures are substantially higher than the interior temperatures of household refrigerators. Therefore, items that do not spoil rapidly, such as condiments, acidic beverages or high-salt foods, should be stored in the door, rather than more perishable foods (Godwin *et al.*, 2007).

Many frozen foods need to be thawed before cooking. These should not be thawed at ambient temperatures, as pathogens could grow on the warming food surface while the interior of the food remains frozen. Preferably, frozen foods should be thawed in the refrigerator or under cold running water. They can also be thawed in a microwave oven, provided that they are cooked immediately after thawing.

Heating

In preparing processed foods for home serving, the manufacturers' label instructions should be followed for cooking, baking, microwaving or frying the product. It is the responsibility of food processors to validate that the food preparation instructions will have a sufficient margin of error to ensure the safety of the product. Usually, the heating process required to yield an organoleptically acceptable food is substantially higher than that needed to kill vegetative forms of pathogenic micro-organisms, thus providing the margin of safety.

Raw meat and poultry products must be cooked to a minimum centre temperature in order to ensure food safety. The recommended centre temperatures are (Marchiony, 2004):

- 71°C – raw ground beef, beef and pork
- 74°C – raw ground poultry, leftover foods
- 82°C – whole poultry or pieces

It is highly recommended that an accurate meat thermometer be used to measure the centre temperature before serving. This is especially important with ground meats; if not adequately cooked, the centre of the ground products can potentially contain pathogenic micro-organisms that had been on the meat surface before grinding.

Sanitation and personal hygiene

Many opportunities for contamination and cross-contamination exist in the kitchen (Walter *et al.* 2007). Elimination of the causes of contamination, when applied in millions of kitchens worldwide, will reduce the burden of foodborne illness. Examples of causes of contamination include the following:

- Contact of raw and cooked foods
- Unclean kitchen counters and utensils
- Inadequate hand-washing, e.g. after handling raw foods, changing diapers, taking out garbage and visiting the bathroom
- Preparing food when ill
- Improper use of dish towels
- Playing with pets while preparing foods
- Smoking, sneezing or coughing while preparing foods

Many of the potential contamination problems in the kitchen can be minimised or eliminated by using prerequisite programmes, to extend the use of this term from the rest of the food supply chain, such as cleaning and sanitising, and by altering food consumption patterns. Moreover, some control measures could be established and monitored as critical control points (CCPs) in the home kitchen.

POTENTIAL CCPs IN THE HOME

Simple but effective CCPs could be established in each home kitchen to create awareness of potential hazards and their means of control. Examples of home CCPs and control measures include the following:

- Controlled refrigeration temperatures.
- Controlled cooking temperatures.

- Removal of target allergens when susceptible individuals are known or expected to be present.
- Preventing consumption of raw milk and raw purchased juices, raw cake batter and unbaked cookie dough.
- Restricted or prohibited consumption of certain types of food by immunocompromised individuals. For example, pregnant women should not consume soft or surface-ripened cheeses, pre-cooked RTE meat and poultry products, unless the latter have been reheated to 74°C before consumption.

EDUCATION

Creation of a home 'HACCP plan' or 'food safety plan' would be a good educational device for the entire family, with an added benefit for society as a whole – many of the children in families will spend part of their early life in part-time jobs in the food service industry, preparing and serving a vast number of meals outside the home. If these children learned proper food safety procedures in the home, they would be better prepared to use safe food-handling practices when working outside the home. Education would include basic information about foodborne hazards, means of control and susceptible consumers, as described above.

Education for food safety in the home should begin in the primary school so that good practices are learned at a young age (this happens in the UK – see Chapter 4). A number of techniques can be employed to continually reinforce the early learning: public service announcements by radio, television or print media, public health agency websites, academic extension services, etc. Family members who participate in food safety awareness training will be better prepared to use safe food-handling practices in the home, to the point of actually using CCPs in the home, or creating a simple 'home HACCP plan'. Friendly competitions could be arranged in schools to encourage students to create the best HACCP plan, CCP or other food safety practices.

TV cookery show hosts could be educated to eliminate the poor practices which many currently use, and instead help to educate their viewers on the good practices needed in the kitchen – and why. Education of home appliance manufacturers could also be beneficial. For example, refrigeration units should have built-in, reliable thermometers that would facilitate observing and recording interior temperatures without opening the door, as mentioned previously. Optical scanners by which consumers could retrieve food safety information either from in-store displays or from label encryptions could be developed (Mortimore and Wallace, 1998).

CASE STUDY

Background

A 'typical' suburban Hill City, Kansas (USA), family has been stricken by a number of illnesses during the past several years, most or all of which may have been foodborne illnesses resulting from foods prepared in their home. The Knight family – father, Winston, age 61; mother, Margaret, age 35; daughter, Penelope, age 12; and son, Charles, age 10 – can recall three recent episodes in particular that seem to have been food related:

1. Three of the family members began vomiting within 2 hours of eating a meal that included Himalayan Nut Pilaf. Penelope, the only member who did not become ill, had not eaten the Pilaf.

2. All family members experienced repeated diarrhoea within 14 hours of eating a winter holiday meal that included a dressed and roasted 4 kg goose. Because this meal was a family tradition, all members ate heartily.

3. Two family members and three of four visiting neighbours experienced simultaneous vomiting and diarrhoea within 1 day of feasting at an outdoor barbeque that included grilled chicken and Caesar salads.

ORIGIN OF THE KNIGHT FAMILY HOME FOOD SAFETY PROGRAMME

The Knight family's growing awareness that some of their memorable bouts of illness might have been associated with food-handling practices was gradually reinforced with information gained by each family member from different sources. Margaret's suspicions were raised while watching a public television programme about the causes and nature of foodborne illnesses. Penelope and Charles learned simple facts about safe food handling in their school's health classes. In particular, they learned about the importance of proper refrigeration temperatures. Winston learned more about foodborne illness symptoms than he imagined possible after Googling 'diarrhoea, eating chicken' (http:scholar.google.com).

RETROSPECTIVE ANALYSIS OF PREVIOUS ILLNESSES

The Knight family began to discuss their newfound information about food-handling practices and began to develop hypotheses about the unexpected and, at the time, mysterious illnesses that had affected them and their neighbours. Additional online searching and attempted reconstruction of events surrounding the potentially incriminated meals and suspect foods led them to the following conclusions:

1. It seemed rather clear that the first series of illnesses involved the Himalayan Rice Pilaf, as it had not been eaten by Penelope, the only family member who had not become ill. Margaret recalled cooking rice the evening before the Pilaf was prepared and served. She spooned the hot cooked rice into a rectangular plastic storage dish, which she covered and placed in the refrigerator door. It is likely that the episodes of vomiting were caused by the growth of *Bacillus cereus* in the rice, which required many hours to cool below ambient temperature. *B. cereus* spores are normally present in rice. The spores survive the cooking process and are able to grow rapidly if the rice is not consumed or adequately chilled within several hours. During growth, *B. cereus* produces a heat-stable emetic toxin, which induces vomiting within several hours of consumption.

2. The dressed goose served for the holiday meal had been purposely roasted at an oven temperature lower than the recommended 163°C to retain the succulence of the meat. Winston learned online that *C. perfringens*, also a spore-forming micro-organism, was often the cause of diarrhoeal illness in meat and poultry products, particularly those involving dressing or gravy. It can grow very rapidly at temperatures up to about 50°C in foods that are roasted too slowly or held too long during serving. Following growth, it produces spores in the food. After consumption, the spores germinate in the host's intestine and produce toxins,

which cause diarrhoea typically within 8–24 hours. In this episode, the family's illnesses could have been caused by the growth of *C. perfringens* in the dressing during slow roasting, or in the gravy, which had been made from the goose drippings and held for many hours at ambient temperature during the long holiday meal.

3. Reconstruction of the third illness episode led to two plausible causes; perhaps both were involved to differing extents in the five illnesses. The same tongs had been used to handle raw and grilled chicken pieces. It is possible that grilled chicken could have been recontaminated with *Salmonella* or *Campylobacter*, both common contaminants of raw poultry. It is perhaps more likely that the Caesar salad was the cause of the illnesses, as it was made with two potential sources of contamination. Whole chickens were cut on a cutting board that was not washed and disinfected before being used to cut salad ingredients. Furthermore, the salad dressing was prepared with fresh, raw egg yolks, which have frequently been responsible for illnesses caused by *Salmonella* Enteritidis. *Salmonella* infections are typically characterised by vomiting and diarrhoea, while *Campylobacter* infections do not always involve vomiting. Therefore, it is more likely that the illnesses were caused by *Salmonella*, though it could not be determined whether the raw chicken or raw egg yolks were responsible for the contamination. In any case, both are serious food-handling mistakes which need to be prevented.

KNIGHT FAMILY FOOD SAFETY TEAM AND ACTION PLAN

Equipped with this knowledge about foodborne illnesses and their likely mistakes that caused the illnesses, the Knight family agreed to work together to avoid future occurrences. Each member assumed responsibility for specific aspects of the resulting family action plan:

- Penelope became the team leader. She wanted to be the keeper of the collected data, which she intended to use in a school project. She was also responsible to monitor refrigerator temperatures at least weekly and make adjustments when necessary.
- Charles agreed to monitor cooking and roasting temperatures as necessary, and to supervise prompt and proper refrigeration of foods.
- Margaret agreed to monitor food-handling practices and regularly clean and sanitise kitchen counters to minimise opportunities for food contamination.
- Winston agreed to continue online monitoring of safe food-handling information and to inform the entire family about useful practices.

At this writing, the Knight family has experienced no additional known cases of foodborne illness.

CONCLUSION

While the above case study is a mostly fictional example of a family and its home food safety plan, the authors believe that it can be effectively used to promote the possibilities of improving safe food-handling practices in the home, the ultimate link in the food supply chain.

REFERENCES

Bloomfield, S.F., M. Exner, G.M. Fara, K.J. Nath, E.A. Scott and C. Van Der Voorden (2009) The global burden of hygiene-related diseases in relation to the home and community. http://www.ifh-homehygiene.org (accessed 19 Nov 2009).

Food Standards Agency (FSA) (2009) Domestic sector hygiene research program (B20). http://www.food.gov.uk/science/research (accessed 23 October 2009).

Godwin, S.L., F.-C. Chen, E. Chambers IV, R. Coppings and D. Chambers (2007) A comprehensive evaluation of temperatures within home refrigerators. Food Protect. Trends **27**, 168–173.

Marchiony, A. (2004) *Food-Safe Kitchens*. Pearson/Prentice Hall, Upper Saddle River, New Jersey.

Mortimore, S. and C.A. Wallace (1998) *HACCP: A Practical Approach*, Second Edition. Aspen Publishers, Inc., Gaithersburg, Maryland, p. 287.

Nevas, M., M. Lindström, A. Virtanen, S. Hielm, M. Kuusi, S.S. Arnon, E. Vuori and H. Korkeala (2005) Infant botulism acquired from household dust presenting as sudden infant death syndrome. J. Clin. Microbiol. **43**, 511–513.

Walter, C. M., R.H. Schmidt, K.R. Schneider and J. Cornell (2007). Home food safety practices of government employees in Osceola County, Florida. Food Protect. Trends **27**, 389–399.

Appendix 2
Global food safety resources

All websites were viewed in December 2009.

INTERGOVERNMENTAL ORGANISATIONS

The United Nations – http://www.un.org
The World Health Organization (WHO) of the United Nations – http://www.who.int
The Food and Agriculture Organization (FAO) of the United Nations – http://www.fao.org
The Codex Alimentarius Commission (CAC) – http://www.codexalimentarius.net
The CAC Committee on Food Hygiene (CACCFH) – http://www.fsis.usda.gov/Codex_Alimentarius/Codex_Committee_Food_Hygiene/index.asp
The World Organization for Animal Health (OIE) – http://www.oie.int
The World Trade Organization – http://www.wto.org
The European Food Safety Authority (EFSA) – http://www.efsa.europa.eu
The Pan American Health Organization (PAHO) – http://www.paho.org

GOVERNMENTAL ORGANISATIONS

US Department of Health and Human Services (HHS) – http://www.hhs.gov
US Centers for Disease Control and Prevention (CDC) – http://www.cdc.gov
US Food and Drug Administration (FDA) – http://www.fda.gov
US FDA Center for Food Safety and Applied Nutrition (CFSAN) – http://www.fda.gov/aboutfda/centersoffices/cfsan/default.htm
USDA Food Safety and Inspection Service (FSIS) – http://www.fsis.usda.gov
USDA Agricultural Research Service Pathogen Modeling Program – http://www.ars.usda.gov/Services/software/software.htm
US National Institute of Food and Agriculture (formerly CSREES) – http://www.csrees.usda.gov
US National Advisory Committee on Microbiological Criteria for Food (NACMCF) – http://www.fsis.usda.gov/About_FSIS/NACMCF/index.asp
U.S. Department of Homeland Security (DHS) – http://www.dhs.gov
Food Standards of Australia and New Zealand – http://www.foodstandards.gov.au
New Zealand Food Safety Authority – http://www.nzfsa.govt.nz
UK Food Standards Agency – http://www.food.gov.uk

Food Safety for the 21st Century, First Edition By Carol A. Wallace, William H. Sperber and Sara E. Mortimore
© 2011 Carol A. Wallace, William H. Sperber and Sara E. Mortimore

U.K. Department for Environment, Food and Rural Affairs (DEFRA) – http://www.defra.gov.uk
UK Department of Health – http://www.dh.gov.uk

NON-GOVERNMENTAL ORGANISATIONS

Food Allergy and Anaphylaxis Network (FAAN) – http://www.foodallergy.org
Anaphylaxis Campaign – http://www.anaphylaxis.org.uk
Global Harmonization Initiative – http://www.globalharmonization.net
Center for Science in the Public Interest – http://www.cspinet.org
Wildlife Conservation Society (WCS) – http://www.wcs.org
Global Avian Influenza Network for Surveillance – http://www.gains.org
World Population Balance – http://www.worldpopulationbalance.org

PUBLIC–PRIVATE PARTNERSHIPS

Safe Supply of Affordable Food Everywhere, Inc. (SSAFE) – http://www.ssafe-food.org
Global Initiative for Food Systems Leadership (GIFSL) – http://foodsystemsleadership.org/

TRADE ASSOCIATIONS

Home Food Safety (American Dietetic Association and ConAgra Foundation) – http://www.
homefoodsafety.org/index.jsp
UK Food and Drink Federation – http://www.foodlink.org.uk
Food Safety Magazine (free subscription) – http://www.foodsafetymagazine.com
American Meat Institute Foundation (AMIF) Washington, DC – http://www.amif.org
Grocery Manufacturers Association. Washington, DC – http://www.gmaonline.org
International Life Sciences Institute (ILSI) – http://www.ilsi.org
North American Millers' Association (NAMA) – http://www.namamillers.org
International Food Information Council (IFIC) – http://ific.org
The Consumer Goods Forum (formerly CIES) Annual Global Food Safety Conference – http://
www.tcgffoodsafety.com
British Retail Consortium – http://www.brc.org.uk

PROFESSIONAL ASSOCIATIONS

Commonwealth Scientific and Industrial Research Organization (CSIRO), division of Food
Science and Technology – http://www.csiro.au/org/FNS.html
International Commission on Microbiological Criteria for Foods (ICMSF) – http://www.
icmsf.org
European Federation of Food Science and Technology – http://www.effost.org
The Society for Food Hygiene and Technology – http://www.sofht.co.uk
European Hygienic Engineering and Design Group – http://www.ehedg.org
International Union of Food Science and Technology (IUFOST) – http://www.iufost.org

Institute of Food Science and Technology (IFST) – http://www.ifst.org
Institute of Food Technologists (IFT) – http://www.ift.org
International Association for Food Protection (IAFP) – http://www.foodprotection.org
Royal Society for Public Health – http://www.rsph.org

ACADEMIC INSTITUTIONS

Food Research Institute, University of Wisconsin – http://fri.wisc.edu
Food Allergy Research and Resource Program, University of Nebraska – http://www.farrp.org
Center for Food Safety, University of Georgia – http://www.ugacfs.org
National Center for Food Science and Technology, Illinois Institute of Technology – http://www.ncfst.iit.edu
National Center for Food Protection and Defense, University of Minnesota – http://www.ncfpd.umn.edu
Center for Animal Health and Food Safety, University of Minnesota – http://www.cahfs.umn.edu
Netherlands Organization for Applied Science and Research (TNO) – http://www.voeding.tno.nl
Food Safety Consortium, University of Arkansas – http://fsconsortium.net
International HACCP Alliance, Texas A & M University – http://www.haccpalliance.org
Center for Infectious Disease Research and Policy, University of Minnesota – http://www.cidrap.umn.edu

CONSULTING ORGANISATIONS AND LABORATORIES

Leatherhead Food Research Association – http://www.leatherheadfood.com
Campden BRI – http://www.campden.co.uk
Silliker Group Corporation – http://www.silliker.com
Deibel Laboratories – http://www.deibellabs.com

AUDIT ORGANISATIONS

Global Food Safety Initiative (GFSI) – http://www.mygfsi.com

Index

Note: Italicized *b*, *f* and *t* refer to boxes, figures and tables, respectively.